Undergraduate Matrix Theory

and

Linear Algebra

$$\begin{bmatrix} a_{11} & a_{12} & \cdots & a_{1n} \\ a_{21} & a_{22} & \cdots & a_{2n} \\ \vdots & \vdots & \ddots & \vdots \\ a_{m1} & a_{m2} & \cdots & a_{mn} \end{bmatrix}$$

John S. Alin
Linfield College

Colin L. Starr
Willamette University

December 18, 2018

Contents

Chapter 1

SYSTEMS OF LINEAR EQUATIONS

1.1   Introduction

We are embarking on a tour of matrix theory and linear algebra. These topics have become extremely important not only in mathematics but also in a wide variety of other disciplines such as science, engineering, economics, and management. The study of matrices began in the mid-nineteenth century and came to its present form during the period 1920-1940. We will not dwell on the history of the subject[1].

To some extent, the power of matrix theory comes from the notation that is used. The ability to treat an array of numbers as a single quantity is a powerful tool and it yields significant results. While the details require hard work, the methods that follow are worthwhile. We will take an abstract approach to these topics in order to make our accumulated methods and knowledge applicable in a broader scope. This will enable us to fill in some holes left in previous courses. In several undergraduate courses material is omitted because of the unavailability of certain aspects of matrix theory. In calculus, it is customary to skip a discussion of optimizing functions of several variables - this is covered in Section 5.7. Systems of differential equations are often omitted from a first course in differential equations because of difficulties in dealing with the Jordan canonical form - this topic is introduced in Section 6.3. Physics instructors in courses on classical mechanics usually cannot cover the material on principal axes because of the lack of techniques on diagonalizing symmetric matrices - we outline this in Section 5.8.

In order to fill in these holes we will need to develop our theory in sufficient generality so that these problems can be solved. This means that our development must be "general" or "abstract." Students sometimes have difficulty with abstraction at first exposure. Working with axioms, definitions, theorems, and proofs can be confusing. The mathematician uses abstraction to simplify (and generalize) an investigation, not to complicate it. An appreciation of the power of the method of abstraction will be a secondary benefit of this course of study.

---

[1]For an interesting account of the development of this area of mathematics, the reader is referred to *A History of Mathematics* by Carl B. Boyer, New York, 1968.

The study of matrix theory and linear algebra begins with investigations into the solutions of systems of simultaneous linear equations. Students encounter systems of equations at an early point in their mathematical careers. For instance, in high school algebra, one encounters "story" problems such as: The sum of Tom's age and his older brother Dick's age is 17 and the difference is 7. How old is each?

If x represents Dick's age and y represents Tom's age, then from the statement of the problem, we obtain the following system of linear equations:

$$\begin{aligned} x + y &= 17 \\ x - y &= 7 \end{aligned} \qquad (1.1.1)$$

Adding the equations, we obtain the equation $2x = 24$ and we see that $x = 12$. Substitution into the first equation gives $12 + y = 17$, and we obtain $y = 5$. So, we see that Tom's age is 5 and Dick's age is 12, and we say that the solution of the system of equations 1.1.1 is $x = 12$ and $y = 5$, or we might say that the solution is the pair of numbers $(12, 5)$ and that the solution set is $\{(12, 5)\}$.

It seems that we have found a rich source of problems. Systems of equations like those in 1.1.1 can be constructed with more variables and more equations. How do we find the solutions? Are there any? Are there many? A lot of questions arise, and we will be able to answer most of them!

GEOMETRIC CONSIDERATIONS

To begin our investigations, it is worthwhile to consider the geometrical aspects of systems of equations of the type 1.1.1. Each equation describes a straight line in the xy-plane, and since the line given by $x + y = 17$ has slope -1 and the one given by $x - y = 7$ has slope 1, the lines are not parallel and so they must intersect (see Figure 1.1.). The solution is the unique pair of numbers that describe the point of intersection of the lines.

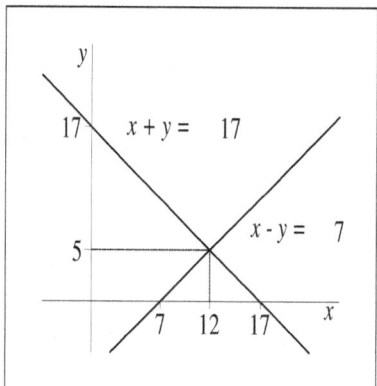

Figure 1.1: Ages of Dick and Tom

By considering other systems of equations we can see the possibilities for solution sets of similar systems of linear equations. In the system

$$x + y = 3$$
$$x + y = 5,$$
(1.1.2)

we see that the lines are parallel (both have slope -1) and distinct (see Figure 1.2), and so the two lines have no points in common. This is another way of saying that the system 1.1.2 has no simultaneous solution or that the solution set is the empty set. Such a system is said to be **inconsistent**.

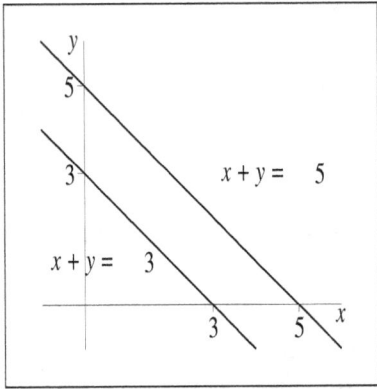

Figure 1.2: Parallel Lines

The third possibility is that the lines described by the two equations are exactly the same. For example, consider the system:

$$x - y = 1$$
$$-2x + 2y = -2.$$
(1.1.3)

Figure 1.3 shows the graph. Since the two lines are the same, any point on the line is a solution. It follows that the solution set can be written as $\{(x,y) | x - y = 1\}$ or alternately $\{(x, x-1) | x$ is any real number$\}$. Such a system is said to be **redundant**.

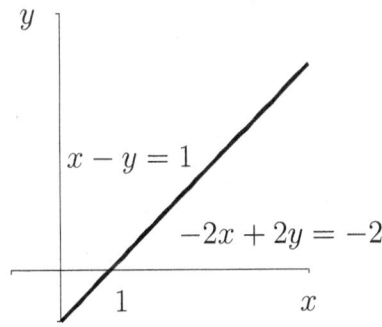

Figure 1.3: Identical Lines

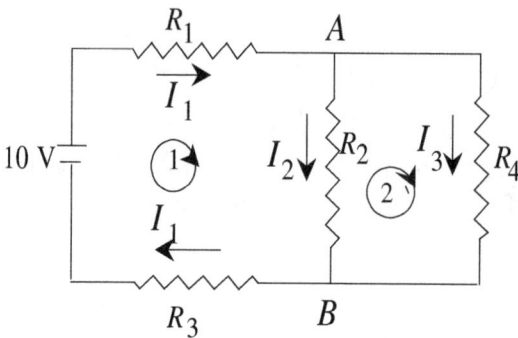

Figure 1.4: Circuit Diagram

Notice that system 1.1.1 has exactly one solution, that 1.1.2 has no solutions, and that 1.1.3 has infinitely many solutions.

Systems of linear equations can be found in many disciplines - physics, chemistry, engineering, economics, etc. In courses that involve electricity and magnetism one encounters systems of simultaneous linear equations that arise from certain electrical networks obeying Kirchhoff's laws. Those networks having only voltage sources (batteries, for example) and resistors will produce a system of linear equations with the currents in each branch of the networks as the unknowns. Previous experience with Kirchhoff's laws will not be needed in the following discussion. We will present an example and this will give the general idea of how these systems arise.

KIRCHHOFF'S LAWS

Briefly, Kirchhoff's laws state that around any closed loop in an electrical network, the sum of the voltages produced by the energy sources in the loop equals the sum of the voltage drops across the resistors in the loop, and that at any branch point in the network, the sum of the currents is zero. Using Σ-notation, we can express these laws symbolically by:

$$\sum_{\text{loop}} V = \sum_{\text{loop}} IR$$

$$\sum_{\text{point}} I = 0,$$

where V represents the voltage, I the current, and R the resistance in a given resistor.

Kirchhoff's laws will be investigated more fully in Section 1.8; here we will consider one example. In reality, the orientation of the assigned currents must be considered, as well as the orientation of the power sources in the circuit. The sum of the voltage drops must be taken "algebraically;" that is, with allowance for the algebraic sign. Consider the circuit in Figure 1.4.

By Kirchhoff's laws, after some rearrangement, the following system of linear equations is obtained:

$$
\begin{array}{llrcrcrcl}
\text{Loop 1:} & R_1I_1 & + & R_2I_2 & + & R_3I_1 & = & 10 \\
\text{Loop 2:} & & - & R_2I_2 & + & R_4I_3 & = & 0 \\
\text{Point } A: & I_1 & - & I_2 & - & I_3 & = & 0
\end{array} \quad (1.1.4)
$$

Point B produces essentially the same equation as Point A and so it need not be considered. Let us assume certain specific values for the resistance; say $R_1 = 2, R_2 = 5, R_3 = 3$, and $R_4 = 2$. Combining the two terms involving I_1, the system of equations [1.1.4] becomes:

$$
\begin{array}{rcrcrcl}
5I_1 & + & 5I_2 & & & = & 10 \\
& - & 5I_2 & + & 2I_3 & = & 0 \\
I_1 & - & I_2 & + & I_3 & = & 0
\end{array} \quad (1.1.5)
$$

Dividing the first equation by 5 and subtracting it from the last equation, we get

$$
\begin{array}{rcrcrcl}
I_1 & + & I_2 & & & = & 2 \\
& & 5I_2 & - & 2I_3 & = & 0 \\
& - & 2I_2 & + & I_3 & = & -2
\end{array} \quad (1.1.6)
$$

(Note that we also multiplied the second equation by -1.) Now dividing the second equation by 5, adding twice the second equation to the last, and dividing by -2, we obtain

$$
\begin{array}{rcrcl}
I_1 & + & I_2 & & & = & 2 \\
& & I_2 & - & \frac{2}{5}I_3 & = & 0 \\
& & & & \frac{-1}{10}I_3 & = & 1
\end{array} \quad (1.1.7)
$$

Solving for the unknown currents, we obtain $I_3 = -10, I_2 = -4$, and $I_1 = 6$.

Section 1.1 Exercises

In Exercises 1-7 solve the systems of equations:

1. (a) $x + y = 3$ (b) $2x + 4y = 6$
    $ x - y = 2$ $-x + 2y = 3$

2. (a) $2x + 3y = 4$ (b) $2x - 4y = 8$
    $ x - 2y = 8$ $x + 3y = -1$

3. $\begin{array}{rcrcrcl}
    x & + & y & + & z & = & 4 \\
    & & y & - & z & = & 6 \\
    x & & & + & z & = & 2
    \end{array}$

4. $\begin{array}{rcrcrcl}
    2x & + & y & + & 2z & = & 3 \\
    x & + & & & z & = & -2 \\
    -x & + & y & - & z & = & 6
    \end{array}$

5. $\begin{aligned} w & - & x & + & y & - & z & = & 7 \\ & & x & + & 2y & + & z & = & 3 \\ & & & & y & - & z & = & 6 \\ & & & & y & + & z & = & -1 \end{aligned}$

6. In the electrical circuit in Figure 1.4, assume that the resistances are given by $R_1 = 2, R_2 = 3, R_3 = 1$, and $R_4 = 1$. Solve for the currents I_1, I_2, and I_3.

7. In the electrical circuit in Figure 1.4, assume that the resistances are given by $R_1 = 5, R_2 = 10, R_3 = 5$, and $R_4 = 6$. Solve for the currents I_1, I_2, and I_3.

8. Find all solutions of: $\begin{aligned} 2x & + & 3y & - & z & = & 12 \\ x & - & 2y & + & 2z & = & 7. \end{aligned}$

9. Consider the system of linear equations:

$$\begin{aligned} ax & + & y & = & 2 \\ x & + & 2y & = & -3 \end{aligned}$$

where x and y are the unknowns and a is some unknown constant. Find values of a (if possible) for which the system of equations has: a) exactly one solution, b) no solutions, c) infinitely many solutions.

10. Consider the system of linear equations:

$$\begin{aligned} x & + & y & = & a \\ ax & + & y & = & 2. \end{aligned}$$

Find the values of a for which the system of equations has: a) no solution, b) a unique solution. Could the system have infinitely many solutions?

11. The sum of three numbers is 32. The quotient of two of them is 3 and the third number is 3/5 of the sum of the other two. Find the three numbers.

12. A couple has two children. The youngest was born when the oldest was four, and two years from now the sum of the children's ages will be 20. How old is each child?

1.2 Systems of Linear Equations and Matrices

The discussion and examples in Section 1.1 should suggest that systems of linear equations arise naturally in problems and that it is important to be able to solve such systems. While the problems in Section 1.1 involved only two or three unknowns, it is not hard to imagine problems involving many unknowns. For example, a Kirchhoff's laws problem with 16 loops could have 36 or more variables. In this section we will discuss general systems of linear equations and show how a system of linear equations may be regarded as a single "matrix" equation.

SCALARS AND TERMINOLOGY

The systems of equations considered in Section 1.1 all involved coefficients taken from the field of real numbers $\mathbb{R}$, and in a solution the variables took on real number values. In other situations it may be desirable to consider systems of equations with coefficients in the field of complex numbers or in some other field, such as the rational numbers. In each problem or theorem, the field in question will be specified or understood and numbers from this field will be called **scalars**. In general, the reader may assume that the field in question is the field of real numbers and that the scalars are real numbers, since any field has the arithmetic properties of the field of real numbers; that is, the operations of addition, subtraction, multiplication, and division behave as they do in the system of reals. Notice, however, that the order $(a < b)$ properties of the reals do not hold in general.

The possibility of having scalars that are not real numbers is raised not to complicate the discussion, but to generalize the discussion and methods. If we wish to refer to the collection of all scalars, we will often denote this set by the letter F. For a discussion of fields and their properties, the reader is referred to Appendix C. Throughout, we will use $\mathbb{Z}$ to refer to the set (or system) of integers, $\mathbb{Q}$ for the rational numbers, $\mathbb{R}$ for the set of real numbers, and $\mathbb{C}$ for the complex numbers. Notice that we are using a special font for these special sets. Also note that the system of integers $\mathbb{Z}$ does not satisfy the axioms for a field - not all elements have multiplicative inverses. The properties of a number system have a direct relationship to the solvability of equations over the number system. Notice that an equation of the form $ax = b$, $a \neq 0$, does not, in general, have a solution in the integers, but a solution is always obtainable over a field. By the usual conventions, we may assume that $\mathbb{Z} \subseteq \mathbb{Q} \subseteq \mathbb{R} \subseteq \mathbb{C}$.

By a **linear equation in the variables** (or **unknowns**) $x_1, x_2, \ldots, x_n$ we mean an equation of the form

$$a_1 x_1 + a_2 x_2 + \ldots + a_n x_n = c,$$

where $a_1, a_2, \ldots, a_n, c$ are scalars. The scalars $a_1, \ldots, a_n$ are called the **coefficients**; more specifically, a_i is called the **coefficient of the variable** x_i. The term "linear" is used because no powers, products, or quotients of the variable occur in the equation. In addition, if one considers the geometrical aspects of two- and three-dimensional space one sees that a line in the xy-plane is given by a linear equation in x and y, and a plane in three-dimensional space is given by a linear equation in x, y, and z. If the coefficients of the linear equation come from a certain field of scalars, we will say that the equation is a linear equation "over" that field.

By a **solution of the linear equation** $a_1 x_1 + a_2 x_2 + \ldots + a_n x_n = c$ we mean a collection of scalars $b_1, b_2, \ldots, b_n$ with the property that $a_1 b_1 + a_2 b_2 + \ldots + a_n b_n = c$; that is, when each b_i is substituted for x_i, the equation is satisfied. Solutions are often listed as n-**tuples** $(b_1, b_2, \ldots, b_n)$, an ordered sequence of scalars. Normally, the scalars making up the solution come from the same field that the coefficients come from.

Example 1.2.1. The equation $2x + 3y - z = 2$ is linear over the real numbers and the equation $(1 + i)x + 3y = 2 + i$ (where $i^2 = -1$) is linear over the complex field. The triple $(2, -1, -1)$ is a solution of the first equation, as is $(-1, 2, 2)$; and $(1, 1/3)$ and $(-3/2 - i/2, 1 + i)$ are solutions of the second equation.

$\square$

SYSTEMS OF EQUATIONS

In order to solve problems involving many unknowns – such as problems arising from electrical networks with many loops – it is necessary to consider a **general system of** m **linear equations in** n **unknowns**:

$$
\begin{array}{ccccc}
a_{11}x_1 & + & \ldots & + & a_{1n}x_n & = & h_1 \\
a_{21}x_1 & + & \ldots & + & a_{2n}x_n & = & h_2 \\
\vdots & & \vdots & & \vdots & & \vdots \\
a_{m1}x_1 & + & \ldots & + & a_{mn}x_n & = & h_m,
\end{array}
\tag{1.2.1}
$$

where $h_1, \ldots, h_m$ are scalars called the **constants**, the **coefficients** a_{ij} are all scalars and $x_1, \ldots, x_n$ are **variables** or **unknowns**. A **solution** of this system of equations (or a **simultaneous solution**) is an n-tuple $(b_1, \ldots, b_n)$ that satisfies each of the linear equations in the system; that is, $a_{i1}b_1 + \ldots + a_{in}b_n = h_i$ for $i = 1, 2, \ldots, m$. The **solution set** for a system of equations is the collection of all solutions of the system.

Recall the system of equations that we solved in Section 1.1:

$$
\begin{array}{ccccc}
x & + & y & = & 17 \\
x & - & y & = & 7
\end{array}
\tag{1.1.1}
$$

If one looks at our solution of this system of equations, it becomes apparent that nothing really happens to the variables - they are merely "place holders" - and it is the changes in the coefficients that are important. We could just as easily write the coefficients and constants in an array of numbers and perform the operations on the array. The solution of the system 1.1.1 would then become:

$$
\begin{bmatrix} 1 & 1 & 17 \\ 1 & -1 & 7 \end{bmatrix} \underset{(1)}{\longrightarrow} \begin{bmatrix} 1 & 1 & 17 \\ 2 & 0 & 24 \end{bmatrix} \underset{(2)}{\longrightarrow} \begin{bmatrix} 1 & 1 & 17 \\ 1 & 0 & 12 \end{bmatrix} \underset{(3)}{\longrightarrow} \begin{bmatrix} 0 & 1 & 5 \\ 1 & 0 & 12 \end{bmatrix}.
$$

In the first transition (1), Row 1 was added to Row 2 - this corresponds to the first step in the solution, which consisted in adding the first equation to the second equation. In Step (2), Row 2 was divided by 2, and in Step (3), Row 2 was subtracted from Row 1. The last array shows us that $x = 12$ and $y = 5$.

MATRICES

Arrays of numbers of the above sort are important, and we call them **matrices**: An $m \times n$ **matrix** is an array of m rows and n columns of scalars a_{ij} of the form

$$
\begin{bmatrix}
a_{11} & a_{12} & \cdots & a_{1n} \\
a_{21} & a_{22} & \cdots & a_{2n} \\
\vdots & \vdots & \ddots & \vdots \\
a_{m1} & a_{m2} & \cdots & a_{mn}
\end{bmatrix}.
\tag{1.2.2}
$$

If the scalars a_{ij} are from the field F, we say the matrix is "over F". We refer to the number of rows (m) of the matrix and the number of columns (n) of the matrix as the **size of the matrix**. If $m = n$, we say that the matrix is **square** and that it is of **order** n. Notice that matrix is singular and matrices is plural. Also, the symbol a_{ij} should be interpreted as $a_{i,j}$; that is, i and j are separate indices.

We will need some terminology in dealing with matrices: Capital letters $A, B, C, \ldots$ are most often used to denote matrices. Let A denote the matrix in 1.2.2 above. The i-th row of A contains the scalars $a_{i1}, a_{i2}, \ldots, a_{in}$ and the j-**th column** contains $a_{1j}, a_{2j}, \ldots, a_{mj}$. Columns go up and down, rows are horizontal. The **entry** in row i and column j is a_{ij}, and A is often written in the very abbreviated form $A = [a_{ij}]$. It is typical to denote the entries of a matrix named with a specific capital letter with the the lower case version of the same letter. The letter i is most often used for the **row index** (or number) and j for the **column index**. Note that square brackets – [and] – are used to denote matrices. If one uses parentheses – (and) – as grouping symbols in expressions, it will avoid confusion.

In the system of equations

$$
\begin{aligned}
a_{11}x_1 &+ \ldots + a_{1n}x_n = h_1 \\
a_{21}x_1 &+ \ldots + a_{2n}x_n = h_2 \\
&\vdots \quad \ddots \quad \vdots \quad \vdots \quad , \\
a_{m1}x_1 &+ \ldots + a_{mn}x_n = h_m
\end{aligned}
\tag{1.2.3}
$$

the matrix

$$
\begin{bmatrix}
a_{11} & a_{12} & \cdots & a_{1n} \\
a_{21} & a_{22} & \cdots & a_{2n} \\
\vdots & \vdots & \ddots & \vdots \\
a_{m1} & a_{m2} & \cdots & a_{mn}
\end{bmatrix}
\tag{1.2.4}
$$

is called the **coefficient matrix**, and the **augmented matrix** of the system is the matrix

$$
\begin{bmatrix}
a_{11} & a_{12} & \cdots & a_{1n} & h_1 \\
a_{21} & a_{22} & \cdots & a_{2n} & h_2 \\
\vdots & \vdots & \ddots & \vdots & \vdots \\
a_{m1} & a_{m2} & \cdots & a_{mn} & h_m
\end{bmatrix}.
\tag{1.2.5}
$$

While matrices allow one to express a system of linear equations in a concise form, there is more: it is possible to write a single matrix equation that is equivalent to the system of equations. In order to show how a system of linear equations may be regarded as a single matrix equation, it is necessary to define "operations" on matrices. We will develop an "arithmetic" for matrices, including the operations of addition, matrix multiplication, and scalar multiplication. These definitions are made for general matrices without regard to whether they are the coefficient or augmented matrices for some system of equations. The operations give extraordinary power to the matrix notation. While an analogy can be made with the arithmetic of the real numbers, we must proceed cautiously. Before defining these operations, it must be made clear what it means for two matrices to be equal.

EQUALITY

Definition 1.2.1. The $m \times n$ matrices $[a_{ij}]$ and $[b_{ij}]$ are **equal**, and we write $[a_{ij}] = [b_{ij}]$, if and only if $a_{ij} = b_{ij}$ for $i = 1, \ldots, m$ and $j = 1, \ldots, n$. Thus, matrices are equal when they are the same size and when all corresponding entries are equal. This definition of equality is just an extension of the definition of equality of ordered pairs (x, y) that the reader has no doubt encountered in previous courses in algebra or calculus.

Example 1.2.2. For example, $\begin{bmatrix} 2 & 1 \\ -1 & 1 \end{bmatrix} = \begin{bmatrix} 2 & 1 \\ -1 & 1 \end{bmatrix}$ and $\begin{bmatrix} 4/2 & 1 \\ 1 & 3/3 \end{bmatrix} = \begin{bmatrix} 2 & 1 \\ 1 & 1 \end{bmatrix}$, but $\begin{bmatrix} 2 & 1 \\ -1 & 1 \end{bmatrix}$ and $\begin{bmatrix} 2 & -1 \\ 1 & 1 \end{bmatrix}$ are not equal and we write $\begin{bmatrix} 2 & 1 \\ -1 & 1 \end{bmatrix} \neq \begin{bmatrix} 2 & -1 \\ 1 & 1 \end{bmatrix}$.

$\square$

MATRIX OPERATIONS

The algebraic operations of addition, scalar multiplication, and multiplication of matrices are defined as follows.

Definition 1.2.2. If $A = [a_{ij}]$ and $B = [b_{ij}]$ are two $m \times n$ matrices, the **sum** $A + B$ of A and B is defined by $A + B = [a_{ij} + b_{ij}]$.

Example 1.2.3.

$$\begin{bmatrix} 1 & 2 & 1 \\ 1 & -1 & 0 \end{bmatrix} + \begin{bmatrix} 2 & 1 & 6 \\ 0 & 1 & 2 \end{bmatrix} = \begin{bmatrix} 1+2 & 2+1 & 1+6 \\ 1+0 & -1+1 & 0+2 \end{bmatrix}$$
$$= \begin{bmatrix} 3 & 3 & 7 \\ 1 & 0 & 2 \end{bmatrix}.$$

$\square$

Notice that for the sum of two matrices to be defined, the matrices must be of the same size, or **order**, and the result of the operation is a matrix of their common size.

Definition 1.2.3. If a is a scalar and A is a matrix, then aA, the scalar product of a and A, is defined by $aA = [a \cdot a_{ij}]$.

Example 1.2.4. Using $a = -2$ and A the first matrix from Example 1.2.3, we have

$$aA = -2 \begin{bmatrix} 1 & 2 & 1 \\ 1 & -1 & 0 \end{bmatrix}$$
$$= \begin{bmatrix} -2(1) & -2(2) & -2(1) \\ -2(1) & -2(-1) & -2(0) \end{bmatrix}$$
$$= \begin{bmatrix} -2 & -4 & -2 \\ -2 & 2 & 0 \end{bmatrix}.$$

□

These definitions are straightforward, but the definition of the product is less so. Historically, it seems that the product of matrices first arose in the consideration of the composite of two transformations of vectors. Suppose that we consider two changes in coordinate systems: $(x, y) \to (u, v) \to (s, t)$. Let us suppose that

$$\begin{aligned} u &= ax + by \\ v &= cx + dy \end{aligned} \qquad \begin{aligned} s &= eu + fv \\ t &= gu + hv \end{aligned}.$$

How is the transformation $(x, y) \to (s, t)$ represented? Substituting and simplifying we get:

$$s = e(ax + by) + f(cx + dy)$$
$$= (ea + fc)x + (eb + fd)y$$

$$t = g(ax + by) + h(cx + dy)$$
$$= (ga + hc)x + (gb + hd)y.$$

The coefficients in the composite, considered as a matrix, are obtained as the product of corresponding matrices of the individual transformations:

$$\begin{bmatrix} ea + fc & eb + fd \\ ga + hc & gb + hd \end{bmatrix} = \begin{bmatrix} e & f \\ g & h \end{bmatrix} \begin{bmatrix} a & b \\ c & d \end{bmatrix}.$$

As a consequence, the product of two matrices is defined as a "sum of products" as illustrated above. Let $A = [a_{ij}]$ be an $m \times n$ matrix and $B = [b_{ij}]$ be an $n \times r$ matrix (m, n, and r are, of course, positive integers). The product AB of A and B is the $m \times r$ matrix $AB = [c_{ij}]$, where

$$c_{ij} = a_{i1}b_{1j} + a_{i2}b_{2j} + \ldots + a_{in}b_{nj} = \sum_{k=1}^{n} a_{ik}b_{kj}.$$

Note that the ij-th entry c_{ij} of the product AB is the sum of the products of the entries in row i of A by the corresponding entries in column j of B. We will often talk about "row i times column j" and by it we mean the above sum of products. Notice that the product AB of two matrices A and B is only defined when the number of columns in A equals the number of rows in B.

Examples

Example 1.2.5. 1. In a product AB of two matrices, the number of rows in the product is the same as the number of rows in the matrix A and the number of columns is the same as the number of columns in B. So, a 3×2 times 2×3 produces a 3×3. Also, row 2 of A and column 3 of B give row 2 and column 3 in the product. For example:

$$\begin{bmatrix} \boxed{\begin{matrix} 2 & 3 \\ -1 & 2 \end{matrix}} \\ \hline 4 & -3 \end{bmatrix} \begin{bmatrix} -1 & 3 & \boxed{4} \\ 2 & -3 & \boxed{2} \end{bmatrix} = \begin{bmatrix} 4 & -3 & 14 \\ 5 & -9 & \boxed{0} \\ -10 & 21 & 10 \end{bmatrix}$$

2.

$$\begin{bmatrix} 1 & -1 \\ 2 & 0 \end{bmatrix} \begin{bmatrix} 1 & 2 & 1 \\ -1 & 3 & 4 \end{bmatrix} = \begin{bmatrix} (1)(1)+(-1)(-1) & 1(2)+(-1)(3) & (1)(1)+(-1)(4) \\ (2)(1)+(0)(-1) & (2)(2)+(0)(3) & (2)(1)+(0)(4) \end{bmatrix}$$
$$= \begin{bmatrix} 2 & -1 & -3 \\ 2 & 4 & 2 \end{bmatrix}.$$

3. The product $\begin{bmatrix} 1 & 2 & 1 \\ -1 & 3 & 4 \end{bmatrix} \begin{bmatrix} 1 & -1 \\ 2 & 0 \end{bmatrix}$ is not defined since the first matrix has three columns but the second matrix has only two rows. But notice that the product in the reverse order is defined and is the product in part 2.

4. $\begin{bmatrix} 1 & 2 & 1 \\ -1 & 3 & 4 \end{bmatrix} \begin{bmatrix} -1 & 2 \\ 3 & 6 \\ 1 & -2 \end{bmatrix} = \begin{bmatrix} 6 & 12 \\ 14 & 8 \end{bmatrix}.$

5. Finally, note that the product of a 2×1 matrix and a 1×3 matrix is defined and produces a 2×3 matrix $\begin{bmatrix} 2 \\ -1 \end{bmatrix} \begin{bmatrix} 3 & -1 & 2 \end{bmatrix} = \begin{bmatrix} 6 & -2 & 4 \\ -3 & 1 & -2 \end{bmatrix}.$

$\square$

There are a couple of useful perspectives on matrix multiplication that are not obvious from the definition. To see them, we will look at an example and highlight a few parts.

Example 1.2.6.

$$\begin{bmatrix} 3 & 1 & 2 \\ 4 & -1 & 5 \end{bmatrix} \begin{bmatrix} 2 \\ -3 \\ 0 \end{bmatrix} = \begin{bmatrix} 3(\mathbf{2})+1(-\mathbf{3})+2(\mathbf{0}) \\ 4(\mathbf{2})-1(-\mathbf{3})+5(\mathbf{0}) \end{bmatrix}$$
$$= \mathbf{2}\begin{bmatrix} 3 \\ 4 \end{bmatrix} - \mathbf{3}\begin{bmatrix} 1 \\ -1 \end{bmatrix} + \mathbf{0}\begin{bmatrix} 2 \\ 5 \end{bmatrix}.$$

That is, we can think of this product as a **linear combination** of the columns of the left matrix using coefficients from the right matrix. We will explore linear combinations in much more detail later; for now, just observe that we obtain the columns in the product by multiplying each column on the left by the entries in the columns on the right and then adding the results. Let's see what this looks like with more than one column.

$$\begin{bmatrix} 3 & 1 & 2 \\ 4 & -1 & 5 \end{bmatrix} \begin{bmatrix} 2 & 6 \\ -3 & 1 \\ 0 & 4 \end{bmatrix} = \begin{bmatrix} 3(\mathbf{2})+1(-\mathbf{3})+2(\mathbf{0}) & 3(\mathbf{6})+1(\mathbf{1})+2(\mathbf{4}) \\ 4(\mathbf{2})-1(-\mathbf{3})+5(\mathbf{0}) & 4(\mathbf{6})-1(\mathbf{1})+5(\mathbf{4}) \end{bmatrix}$$
$$= \begin{bmatrix} \left(\mathbf{2}\begin{bmatrix} 3 \\ 4 \end{bmatrix} - \mathbf{3}\begin{bmatrix} 1 \\ -1 \end{bmatrix} + \mathbf{0}\begin{bmatrix} 2 \\ 5 \end{bmatrix} \right) & \left(\mathbf{6}\begin{bmatrix} 3 \\ 4 \end{bmatrix} + \mathbf{1}\begin{bmatrix} 1 \\ -1 \end{bmatrix} + \mathbf{4}\begin{bmatrix} 2 \\ 5 \end{bmatrix} \right) \end{bmatrix}.$$

Thus, to obtain column j, we compute a linear combination of the columns from the left matrix using coefficients from column j of the right matrix. Similarly, to find row i of the product, we compute a linear combination of the rows of the right matrix using coefficients from row i of the left matrix. We just shift the emphasis a bit:

$$\begin{bmatrix} 3 & 1 & 2 \\ 4 & -1 & 5 \end{bmatrix} \begin{bmatrix} 2 & 6 \\ -3 & 1 \\ 0 & 4 \end{bmatrix} = \begin{bmatrix} 3(\mathbf{2}) + 1(-\mathbf{3}) + 2(\mathbf{0}) & 3(\mathbf{6}) + 1(\mathbf{1}) + 2(\mathbf{4}) \\ 4(\mathbf{2}) - 1(-\mathbf{3}) + 5(\mathbf{0}) & 4(\mathbf{6}) - 1(\mathbf{1}) + 5(\mathbf{4}) \end{bmatrix}$$

$$= \begin{bmatrix} \mathbf{3} \begin{bmatrix} 2 & 6 \end{bmatrix} + \mathbf{1} \begin{bmatrix} -3 & 1 \end{bmatrix} + \mathbf{2} \begin{bmatrix} 0 & 4 \end{bmatrix} \\ \mathbf{4} \begin{bmatrix} 2 & 6 \end{bmatrix} - \mathbf{1} \begin{bmatrix} -3 & 1 \end{bmatrix} + \mathbf{5} \begin{bmatrix} 0 & 4 \end{bmatrix} \end{bmatrix}.$$

$\square$

More generally, we have the following:

> **Theorem 1.2.1.** *Let $A = [a_{ij}]$ be an $m \times n$ matrix, and suppose that the columns of A (in order) are $A_1, A_2, \ldots, A_n$ (i.e., $A = \begin{bmatrix} A_1 & A_2 & \cdots & A_n \end{bmatrix}$, where each A_j is a column vector). Let $B = [b_{ij}]$ be an $n \times r$ matrix. Then column j of AB is given by*
>
> $$b_{1j}A_1 + b_{2j}A_2 + \ldots b_{nj}A_n.$$
>
> *If instead the rows of B (in order) are $B_1, B_2, \ldots, B_n$ (i.e., $B = \begin{bmatrix} B_1 \\ B_2 \\ \vdots \\ B_n \end{bmatrix}$, where each B_i is a row vector), then row i of AB is given by*
>
> $$a_{i1}B_1 + a_{i2}B_2 + \ldots a_{in}B_n.$$

Proof. We prove the first case. Consider j as fixed. The ij-entry in the product is $\sum_{k=1}^{n} a_{ik}b_{kj}$. But this is just the ith entry in $\sum_{k=1}^{n} b_{kj}A_k$ as it draws from the ith entry in each A_k. Since the ith entries match for each i, the result holds. $\square$

We will find these perspectives very useful.

MATRIX EQUATIONS

This odd-looking definition of the product is at least partially justified by the following observation: Consider the general system of equations:

$$\begin{array}{ccccccc}
a_{11}x_1 & + & \ldots & + & a_{1n}x_n & = & h_1 \\
a_{21}x_1 & + & \ldots & + & a_{2n}x_n & = & h_2 \\
& & \vdots & & \vdots & & \vdots \\
a_{m1}x_1 & + & \ldots & + & a_{mn}x_n & = & h_m.
\end{array} \tag{1.2.6}$$

Let $A = [a_{ij}]$ be the coefficient matrix of the system, let X be the $n \times 1$ matrix (or column vector)

$$X = \begin{bmatrix} x_1 \\ \vdots \\ x_n \end{bmatrix} \tag{1.2.7}$$

whose entries are the variables of the system and let

$$H = \begin{bmatrix} h_1 \\ \vdots \\ h_m \end{bmatrix} \tag{1.2.8}$$

be the $m \times 1$ matrix of constants. Now the product AX is an $m \times 1$ matrix, and the entry in row i is $a_{i1}x_1 + \ldots + a_{in}x_n$. This is the left-hand side of the i-th equation in the system 1.2.6 and, using the definition of equality of matrices, we see that the system of equations 1.2.6 is equivalent to the matrix equation

$$\begin{bmatrix}
a_{11} & a_{12} & \cdots & a_{1n} \\
a_{21} & a_{22} & \cdots & a_{2n} \\
\vdots & \vdots & \ddots & \vdots \\
a_{m1} & a_{m2} & \cdots & a_{mn}
\end{bmatrix}
\begin{bmatrix} x_1 \\ x_2 \\ \vdots \\ x_n \end{bmatrix}
=
\begin{bmatrix}
a_{11}x_1 + a_{12}x_2 + \ldots + a_{1n}x_n \\
\vdots \\
a_{i1}x_1 + a_{i2}x_2 + \ldots + a_{in}x_n \\
\vdots \\
a_{m1}x_1 + a_{m2}x_2 + \ldots + a_{mn}x_n
\end{bmatrix}
=
\begin{bmatrix} h_1 \\ \vdots \\ h_i \\ \vdots \\ h_m \end{bmatrix}$$

or

$$AX = H. \tag{1.2.9}$$

For example, consider the following system of linear equations:

$$\begin{array}{ccccccc}
2x_1 & - & 3x_2 & + & x_3 & = & 1 \\
3x_1 & - & x_2 & + & 2x_3 & = & 7.
\end{array} \tag{1.2.10}$$

The coefficient matrix of this system is

$$A = \begin{bmatrix} 2 & -3 & 1 \\ 3 & -1 & 2 \end{bmatrix}.$$

If we let

$$X = \begin{bmatrix} x_1 \\ x_2 \\ x_3 \end{bmatrix} \text{ and } H = \begin{bmatrix} 1 \\ 7 \end{bmatrix},$$

we see that the product of A and X is given by

$$AX = \left[\begin{array}{c} 2x_1 - 3x_2 + x_3 \\ 3x_1 - x_2 + 2x_3 \end{array} \right],$$

and so the matrix equation

$$AX = H$$

is equivalent to the system of linear equations 1.2.9; that is, $AX = H$ for some values of x_1, x_2, and x_3 if and only if the system 1.2.9 is satisfied for the same values of x_1, x_2, and x_3.

We initially introduced solutions of linear equations as "n-tuples" of the form $(b_1, \ldots, b_n)$. The above discussion suggests that we might profit from considering solutions to be column vectors of the form

$$\left[\begin{array}{c} x_1 \\ \vdots \\ x_n \end{array} \right].$$

We need to reconcile these two approaches and to be clearer about what we mean by the term "n-tuple." Since we have defined column vectors, it seems reasonable to define a **row vector** to be a $1 \times n$ matrix $\left[\begin{array}{ccc} x_1, \ldots, x_n \end{array} \right]$. We will make no distinction between n-tuples and $1 \times n$ row vectors; that is, between $\left[\begin{array}{ccc} x_1, \ldots, x_n \end{array} \right]$ and $(x_1, \ldots, x_n)$ will be regarded as the same. Notice that the identification of row vectors and n-tuples is consistent with the definitions of equality for these objects. While this identification is possible for row vectors and n-tuples, notice that a similar identification of sets with n elements and n-tuples would fail. For example, $\{1, 2\} = \{2, 1\}$, but $(1, 2) \neq (2, 1)$.

While it is traditional to express the solution of the system of equations 1.2.6 as a row vector $(x_1, \ldots, x_n)$, in the matrix equation $AX = H$ it seems natural to represent the solution as a column vector

$$\left[\begin{array}{c} x_1 \\ \vdots \\ x_n \end{array} \right].$$

(To some extent this comes from the practice of writing functions to the left of the variable - we would more naturally write AX rather than XA.) To reconcile these two notations we can use the transposition operator. The **transpose** of a matrix A is the matrix A^t obtained by interchanging the rows and columns of the original matrix. This operation is defined and discussed more thoroughly in Section 4.1. Here it will suffice to note that

$$(x_1, \ldots, x_n)^t = \left[\begin{array}{ccc} x_1, \ldots, x_n \end{array} \right]^t = \left[\begin{array}{c} x_1 \\ \vdots \\ x_n \end{array} \right]$$

and

$$\begin{bmatrix} x_1 \\ \vdots \\ x_n \end{bmatrix}^t = \begin{bmatrix} x_1, \ldots, x_n \end{bmatrix} = (x_1, \ldots, x_n).$$

Section 1.2 Exercises

In Exercises 1-4, write the coefficient matrix and the augmented matrix of each system of linear equations.

1. $2x + 3y = 1$
    $2x - 4y = 8$

2. $x + y - z = 3$
    $2x + y - 3z = 2$

3. $-2x + 3y = 2$
    $x + y + z = 3$
    $-x + 2y + 6z = 4$

    (Note: In the first equation, the variable z does not occur and 0 must be entered for the coefficient of z.)

4. $2x - 1 = 3y + 1$
    $x - z = y$

    (Note: Before we write the coefficient and augmented matrices, the system must be rewritten in the standard form as in 1.2.1.)

5. Express the system in Exercise 1 as a matrix equation.

6. Express the system in Exercise 2 as a matrix equation.

7. Express the system in Exercise 3 as a matrix equation.

8. Express the system in Exercise 4 as a matrix equation.

9. Compute the sum $\begin{bmatrix} -1 & 2 \\ 3 & 2 \end{bmatrix} + \begin{bmatrix} 0 & 1 \\ -3 & 2 \end{bmatrix}$.

10. Compute the sum $\begin{bmatrix} 2 & 1 & -3 \\ 4 & 0 & 2 \end{bmatrix} + \begin{bmatrix} -1 & 3 & 2 \\ 4 & 6 & -3 \end{bmatrix}$.

11. Compute the following: $-2\begin{bmatrix} 2 & 1 \\ 0 & 3 \end{bmatrix} + 2\begin{bmatrix} 1 & 4 \\ 1 & 1 \end{bmatrix}$.

12. Compute the product $\begin{bmatrix} 2 & 1 \\ -1 & 3 \end{bmatrix}\begin{bmatrix} 1 & 2 & -1 \\ 4 & 1 & 0 \end{bmatrix}$.

13. Compute the product $\begin{bmatrix} 2 & 1 \end{bmatrix} \begin{bmatrix} -1 & 6 \\ 3 & 2 \end{bmatrix}$.

In Exercises 14-20, let $A = \begin{bmatrix} 1 & 2 \\ -1 & 1 \end{bmatrix}$, $B = \begin{bmatrix} 0 & -1 & 3 \\ 1 & 2 & -4 \end{bmatrix}$, and $C = \begin{bmatrix} 2 & 1 \\ -1 & 3 \\ 4 & -2 \end{bmatrix}$.

14. Compute the product AB.

15. Compute the product BC.

16. Compute $2B - 3AB$.

17. Which of the following products are defined?

$$AB, BA, BC, CB, CA$$

18. Which of the following sums are defined?

$$A + B, B + AB, B + BC, A + C$$

19. Let $I_2 = \begin{bmatrix} 1 & 0 \\ 0 & 1 \end{bmatrix}$. Calculate $I_2 A$ and $A I_2$. What do you notice?

20. Let $I_3 = \begin{bmatrix} 1 & 0 & 0 \\ 0 & 1 & 0 \\ 0 & 0 & 1 \end{bmatrix}$. Calculate $I_3 C$.

21. Let A be an $m \times n$ matrix and assume that the product AA is defined. What can be said about m and n?

22. Let A be an $m \times n$ matrix and let B be an $m' \times n'$ matrix. Assume that both products AB and BA are defined. What can be said about m, n, m', and n'?

1.3 Properties of Matrix Operations

In order to make full use of matrices and matrix operations, it is necessary to investigate the algebraic properties of these operations. Which of the properties of ordinary arithmetic hold for the operations on matrices? Do the familiar properties of commutativity and associativity hold for matrix addition and multiplication? What properties are true for scalar multiplication? Is matrix multiplication distributive over matrix addition?

We don't study these properties because we find them so fascinating; the real importance comes from use of the properties in solving equations and manipulating identities. The distributive property does not make good conversation, but if one wishes to solve $2x + 5x = 14$, then knowing that $2x + 5x = (2 + 5)x$ is crucial. Let us consider the solution of the equation $2x + 3 - 5x = x - 4$, along with the steps that we go through and the reasons for these steps:

$$\begin{array}{lll}
1) & 2x + 3 - 5x = x - 4 & \text{the original equation} \\
2) & 2x - 5x + 3 = x - 4 & \text{the commutative property of addition} \\
3) & (2 - 5)x + 3 = x - 4 & \text{the distributive property} \\
4) & -3x + 3 = x - 4 & \text{evaluation} \\
5) & -3x + 3 - 3 = x - 4 - 3 & \text{existence of additive inverses} \\
6) & -3x + 0 = x - 7 & \text{definition of additive inverses and evaluation} \\
7) & -3x = x - 7 & \text{definition of zero} \\
8) & -3x - x = x - 7 - x & \text{existence of additive inverses} \\
9) & \text{etc.} &
\end{array}$$

Consider the identity $(a+b)(a-b) = a^2 - b^2$. What properties are used in its verification? Look at the steps:

$$\begin{array}{ll}
1) & (a + b)(a - b) = a(a - b) + b(a - b) \\
2) & = a^2 - ab + ba - b^2 \\
3) & = a^2 - ab + ab - b^2 \\
4) & = a^2 - b^2
\end{array}$$

It is fundamental in the above computations that the variables x, a, and b represent real numbers and so they may be manipulated using the properties of the real number system. What happens to the above computations if x, a, and b represent matrices rather than real numbers? Are they still valid? We need to know the rules for the algebra of matrices and scalars.

We will see that that the answer to these questions is "yes," for the most part, but there is need for caution. Consider the following:

Example 1.3.1. Let A and B be the 2×2 matrices given by

$$A = \begin{bmatrix} 1 & -1 \\ 2 & 0 \end{bmatrix} \text{ and } B = \begin{bmatrix} -1 & 0 \\ 2 & 1 \end{bmatrix}.$$

Both of the products AB and BA are defined, but notice that

$$AB = \begin{bmatrix} -3 & -1 \\ -2 & 0 \end{bmatrix} \text{ and } BA = \begin{bmatrix} -1 & 1 \\ 4 & -2 \end{bmatrix}.$$

We see that $AB \neq BA$, and so matrix multiplication is not in general commutative.

$\square$

The example above shows that it is wise to proceed with caution. Fortunately, many of the familiar properties of the number systems do hold. In fact, to some extent, the operations on matrices inherit their properties from corresponding properties on the field of scalars. Before beginning a discussion of the properties of the various operations on matrices, it is important to think about some basic properties of equality. Does the notion of equality of matrices obey the usual rules of equality? If two matrices are equal, will the equality be maintained if a matrix be added to each side of the equality? The answer is "yes" as we see in the following theorem.

PROPERTIES OF EQUALITY

Theorem 1.3.1. *Let A, B, C, and D be $m \times n$ matrices over some field of scalars. Then*

(a) $A = A$. *(Reflexive property)*

(b) *If $A = B$, then $B = A$.* *(Symmetric property)*

(c) *If $A = B$ and $B = C$, then $A = C$.* *(Transitive property)*

(d) *If $A = B$ and $C = D$, then $A + C = B + D$.* *(Addition is well-defined.)*

(e) *If $A = B$ and $E = F$, where E and F are $n \times r$ matrices, then $AE = BF$.* *(Multiplication is well-defined.)*

(f) *If r and s are scalars, $r = s$ and $A = B$, then $rA = sB$.* *(Scalar multiplication is well-defined.)*

(With each theorem and some definitions, we will try to give a name or short phrase which describes the result stated. This "key phrase" will be in parentheses and in smaller type, and it is hoped that it will assist the student in remembering the theorem.)

Proof. We will prove parts (c) and (e) and leave the remaining parts as an exercise. Notice that we have assumed that the matrices A, B, C, and D are all of the same size. We will use the shorthand notation introduced in Section 1.2: if a_{ij} represents the entry in row i and column j of A, then we denote A by $[a_{ij}]$.

(c) Let $A = [a_{ij}]$, $B = [b_{ij}]$, and $C = [c_{ij}]$. Then $A = B$ implies $a_{ij} = b_{ij}$ for all i, j, and $B = C$ implies $b_{ij} = c_{ij}$ for all i, j. It follows that $a_{ij} = c_{ij}$ for all i, j and so $A = C$.

(e) Let $A = [a_{ij}]$, $B = [b_{ij}]$, $E = [e_{ij}]$, and $F = [f_{ij}]$, and note since that A and B are $m \times n$ and E and F are $n \times r$, both products AE and BF are defined and $m \times r$. Also $A = B$ and $E = F$ imply $a_{ij} = b_{ij}$ and $e_{ij} = f_{ij}$ for all i, j. Now from these last equalities we have

$$\sum_{k=1}^{n} a_{ik}e_{kj} = \sum_{k=1}^{n} b_{ik}f_{kj}.$$

The first sum is the (i, j)-th entry of AE and the second sum is the (i, j)-th entry of BF. By the definition of equality of matrices, $AE = BF$.

$\square$

(The symbol $\square$ is used to indicate the ends of proofs.)

Many of the algebraic properties of the operations on matrices depend heavily on the corresponding properties of the field of scalars. It is helpful at this point to review Appendix

3 where properties of (or axioms for) fields are presented. In the next theorem one sees how the associative and commutative properties for addition of matrices follow from the corresponding properties for addition of scalars.

PROPERTIES OF ADDITION

Theorem 1.3.2. *Let* $A, B,$ *and* C *be* $m \times n$ *matrices. Then*

(a) $A + B = B + A$ *(Commutative property)*

(b) $A + (B + C) = (A + B) + C$ *(Associative property)*

Proof. We will prove part (b) part and leave part (a) as an exercise. (b) Let $A = [a_{ij}], B = [b_{ij}],$ and $C = [c_{ij}]$. Now using the definitions of addition and equality of matrices, we have

$$
\begin{aligned}
A + (B + C) &= [a_{ij}] + ([b_{ij}] + [c_{ij}]) \\
&= [a_{ij}] + [b_{ij} + c_{ij}] \\
&= [a_{ij} + (b_{ij} + c_{ij})] \\
&= [(a_{ij} + b_{ij}) + c_{ij}] \text{ using associativity of addition of scalars} \\
&= [a_{ij} + b_{ij}] + [c_{ij}] \\
&= ([a_{ij}] + [b_{ij}]) + [c_{ij}] \\
&= (A + B) + C.
\end{aligned}
$$

$\square$

PROPERTIES OF SCALAR MULTIPLICATION

Theorem 1.3.3. *Let* A *be an* $m \times n$ *matrix and let* a *and* b *be scalars. Then*

(a) $a(bA) = (ab)A$ *(Associative property)*

(b) $(a + b)A = aA + bA$ *(Distributive property)*

(c) $a(A + B) = aA + aB.$ *(Distributive property)*

Proof. We will prove part (b). Let $A = [a_{ij}]$. Then

$$
\begin{aligned}
(a+b)A &= (a+b)[a_{ij}] \\
&= [(a+b)a_{ij}] \\
&= [aa_{ij} + ba_{ij}] \text{ (using the distributive property for scalars)} \\
&= [aa_{ij}] + [ba_{ij}] \\
&= a[a_{ij}] + b[a_{ij}] \\
&= aA + bA.
\end{aligned}
$$

$\square$

THE Σ-NOTATION

Proofs involving matrix multiplication often make use of the Σ-notation. Recall that if $a_1, \ldots, a_n$ are scalars, then

$$
\sum_{i=1}^{n} a_i = a_1 + a_2 + \ldots + a_n.
$$

Consider a matrix

$$
\begin{bmatrix}
a_{11} & a_{12} & \cdots & a_{1n} \\
a_{21} & a_{22} & \cdots & a_{2n} \\
\vdots & \vdots & \ddots & \vdots \\
a_{m1} & a_{m2} & \cdots & a_{mn}
\end{bmatrix}.
\tag{1.3.1}
$$

The sum

$$
\sum_{j=1}^{n} a_{ij} = a_{i1} + a_{i2} + \ldots + a_{in}
$$

denotes the sum of the entries in row i of the matrix 1.3.1 and

$$
\sum_{i=1}^{m} a_{ij} = a_{1j} + a_{2j} + \ldots + a_{nj}
$$

denotes the sum of the entries in column j in 1.3.1. Now we see that the double sum

$$
\sum_{i=1}^{m} \sum_{j=1}^{n} a_{ij}
$$

denotes the sum of all the sums of the rows and

$$
\sum_{j=1}^{n} \sum_{i=1}^{m} a_{ij}
$$

denotes the sums of the sums of the columns, but each of these numbers equals the sum of all the entries in the matrix, and so they are equal. Consequently,

$$
\sum_{i=1}^{m} \sum_{j=1}^{n} a_{ij} = \sum_{j=1}^{n} \sum_{i=1}^{m} a_{ij}.
$$

In other words, the order of summation may be interchanged.

Two further properties are of use. Suppose that we have two lists of scalars, $a_1, \ldots, a_n$ and $b_1, \ldots, b_n$. Using the commutative, associative and distributive properties for scalars, we see that

$$a \sum_{i=1}^{m} a_i = \sum_{i=1}^{m}(aa_i) \text{ and } \sum_{i=1}^{m}(a_i + b_i) = \sum_{i=1}^{m} a_i + \sum_{i=1}^{m} b_i.$$

Proofs of properties such as the ones above require mathematical induction. For an introduction, see Appendix 5.

PROPERTIES OF MATRIX MULTIPLICATION

Theorem 1.3.4. *If A is an $m \times n$ matrix, B is an $n \times r$, and C is an $r \times s$, then $A(BC) = (AB)C$.*

(Associative property)

Proof. With $A, B,$ and C as in the statement of the theorem, let $A = [a_{ij}], B = [b_{ij}],$ and $C = [c_{ij}]$. Note that BC is an $n \times s$ matrix. Let us assume that the entry in row i and column j is d_{ij} so that $BC = [d_{ij}]$. By definition, we know that

$$d_{ij} = \sum_{k=1}^{r} b_{ik} c_{kj}.$$

Then the i, j-th entry of $A(BC)$ is

$$\sum_{h=1}^{n} a_{ih} d_{hj} = \sum_{h=1}^{n} \left(a_{ih} \sum_{k=1}^{r} b_{hk} c_{kj} \right).$$

Now let AB be the $m \times r$ matrix $[e_{ij}]$, where

$$e_{ij} = \sum_{h=1}^{n} a_{ih} b_{hj}.$$

Then the i, j-th entry of the matrix $(AB)C$ is given by

$$\sum_{k=1}^{r} e_{ik} c_{kj} = \sum_{k=1}^{r} \left(\sum_{h=1}^{n} a_{ih} b_{hk} \right) c_{kj}$$

$$= \sum_{k=1}^{r} \left(\sum_{h=1}^{n} a_{ih} b_{hk} c_{kj} \right)$$

$$= \sum_{h=1}^{n} \left(\sum_{k=1}^{r} a_{ih} b_{hk} c_{kj} \right)$$

$$= \sum_{h=1}^{n} a_{ih} \left(\sum_{k=1}^{r} b_{hk} c_{kj} \right).$$

Notice that we have interchanged the order of summation using the property of the Σ-notation mentioned above. Since the i,j-th entries of $(AB)C$ and $A(BC)$ are equal, the matrices are equal. It follows that $(AB)C = A(BC)$. $\qquad\qquad\square$

Some of the properties of matrix operations involve combinations of the operations of matrix addition, matrix multiplication, and scalar multiplication. These properties are given in the following theorem.

Theorem 1.3.5. *Let A, B, and C be matrices.*

(a) Let B and C be $m \times n$ matrices. If A is an $r \times m$ matrix then $A(B+C) = AB + AC$, and if A is an $n \times r$ matrix, then $(B+C)A = BA + CA$. (Distributive properties)

(b) If A is an $m \times n$ matrix, B is an $n \times r$ matrix, and a is a scalar, then $a(AB) = (aA)B = A(aB)$. (Associative property)

Proof. We'll prove part (a) and leave part (b) as an exercise.

(a) Let $A = [a_{ij}], B = [b_{ij}]$, and $C = [c_{ij}]$. Notice that A is an $r \times m$ matrix and B and C are $m \times n$ matrices. Consequently, the sum $B + C$ is defined and is an $m \times n$ matrix, and so all of the products $A(B + C), AB$, and AC are defined. Now $B + C = [b_{ij} + c_{ij}]$, and let us set $A(B + C) = [d_{ij}]$ and note $[d_{ij}]$ is $r \times n$ and $d_{ij} = \sum_{k=1}^{m} a_{ik}(b_{kj} + c_{kj})$. But $d_{ij} = \sum_{k=1}^{m} a_{ik}b_{kj} + \sum_{k=1}^{m} a_{ik}c_{kj}$. The first sum is the i,j-th entry in the product AB, and the second sum is the i,j-th entry in the product AC. It follows that $A(B + C) = [d_{ij}] = AB + AC$.

The proof that $(B + C)A = BA + CA$ is similar. $\qquad\qquad\square$

The $m \times n$ matrix $[z_{ij}]$ in which $z_{ij} = 0$ for all i, j is called the **zero matrix**. We will denote this matrix by 0, without any indication of the order $(m \times n)$, and assume that the context in which it is used will make clear what the order is. So,

$$\begin{bmatrix} 0 \\ 0 \end{bmatrix}, \begin{bmatrix} 0 & 0 \\ 0 & 0 \end{bmatrix}, \text{ and } \begin{bmatrix} 0 & 0 \end{bmatrix}$$

are all zero matrices and will be denoted by 0. Further, it will usually be clear from the context whether 0 denotes the zero matrix or the zero scalar.

It is not hard to see that the zero matrix has many of the properties of the zero scalar and that the scalar 1 is an identity with respect to scalar multiplication. We state the following theorem without proof.

PROPERTIES OF IDENTITIES AND INVERSES

Theorem 1.3.6. *Let A be an $m \times n$ matrix. Then*

(a) $0 + A = A + 0 = A$ *(here 0 denotes the $m \times n$ zero matrix)* *(0 is an additive identity.)*

(b) $0A = 0$ *(here the first 0 is the zero scalar and the second 0 denotes the $m \times n$ zero matrix).* *(The zero scalar times a matrix is the zero matrix.)*

(c) $1A = A$ *(1 is an identity for scalar multiplication.)*

(d) $A + (-1)A = (-1)A + A = 0.$ *((-1)A is an **additive inverse** of A.)*

Part (d) of the above theorem gives an important property of addition of matrices. The matrix $(-1)A$ is called the **additive inverse** of A and will be denoted by $-A$. The additive inverse of a matrix is analogous to the additive inverse or negative of a scalar. An important consequence of the existence of an additive inverse is the following:

Corollary 1.3.7. *Let $A, B,$ and C be $m \times n$ matrices. If $A + B = A + C$ or $B + A = C + A$ then $B = C$.* *(Cancellation property of addition)*

Proof. We make use of the above theorem and add the additive inverse of A to each side of the equality. Note also that the associative property is used and that properties of equality are used. Assume $A + B = A + C$. Then

$$-A + (A + B) = -A + (A + C)$$
$$(-A + A) + B = (-A + A) + C$$
$$0 + B = 0 + C$$
$$B = C.$$

$\square$

So far, except for the failure of the commutative property of multiplication, we have seen that there is a close agreement between the arithmetic properties of scalar fields and the properties of matrix operations. It remains to see whether nonzero matrices have "multiplicative inverses." That is, is there a matrix that behaves like the scalar 1 and is there a matrix corresponding to the inverse or reciprocal r^{-1} of a nonzero scalar? The number 1 is an identity for multiplication since $1x = x$ for any x. Is there a matrix which has a similar property? Let's try it for 2×2 matrices!

Assume that $\begin{bmatrix} a & b \\ c & d \end{bmatrix}$ is any 2×2 matrix, and suppose that

$$\begin{bmatrix} e & f \\ g & h \end{bmatrix} \begin{bmatrix} a & b \\ c & d \end{bmatrix} = \begin{bmatrix} ea + fc & eb + fd \\ ga + hc & gb + hd \end{bmatrix} = \begin{bmatrix} a & b \\ c & d \end{bmatrix}.$$

By trial and error, we see that $e = 1, f = 0, g = 0,$ and $h = 1$ is a solution, so the matrix $\begin{bmatrix} 1 & 0 \\ 0 & 1 \end{bmatrix}$ works.

Definition 1.3.1. The $n \times n$ **identity matrix** is the matrix

$$I_n = \begin{bmatrix} 1 & 0 & \cdots & 0 \\ 0 & 1 & \cdots & 0 \\ \vdots & \vdots & \ddots & \vdots \\ 0 & 0 & \cdots & 1 \end{bmatrix}.$$

When no confusion will result, we will use I (without a subscript) to denote the identity matrix.

It might be otherwise defined as the $n \times n$ matrix $I_n = [\delta_{ij}]$, where $\delta_{ij} = 0$ if $i \neq j$ and $\delta_{ij} = 1$ if $i = j$. The function δ_{ij} is called the Kronecker δ.

The following theorem states that the identity matrix behaves as an identity with respect to multiplication.

Theorem 1.3.8. *Let $A = [a_{ij}]$ be an $m \times n$ matrix. Then*

(a) $I_m A = A$

(b) $AI_n = A$.   *(I is a multiplicative identity)*

Proof. (a) Let $I_m A = [c_{ij}]$. Then

$$c_{ij} = \sum_{k=1}^{m} \delta_{ik} a_{kj}$$

and since $\delta_{ik} = 0$ for $k \neq i$, we see that $c_{ij} = \delta_{ii} a_{ij}$. But $\delta_{ii} = 1$, so $c_{ij} = a_{ij}$. By the definition of equality of matrices, we see that $I_m A = A$ since $I_m A$ and A are both $m \times n$ matrices and since we have shown that corresponding entries are equal.

The proof of part (b) is left as an exercise.

As an alternative approach, consider the "linear combinations" perspective on matrix multiplication. If the rows of A are $A_1, A_2, \ldots, A_m$, then row i of $I_m A$ is $0A_1 + 0A_2 + \ldots + 0A_{i-1} + 1A_i + 0A_{i+1} + \ldots + 0A_m = A_i$.

$\square$

By the above theorem, we see that the analogy between scalar operations and matrix operations holds with respect to the multiplicative identity. The situation is more complicated regarding multiplicative inverses. After a definition or two we will see that not every nonzero matrix has a multiplicative inverse.

An $n \times n$ matrix A is said to be **nonsingular** provided there is an $n \times n$ matrix B with $AB = BA = I$. We will see later that the matrix B is unique. B is called the **inverse** of A and is denoted by A^{-1}. A matrix is said to be **singular** if it fails to have an inverse. Notice that we are assuming that the matrix is "square;" that is, the matrix has the same number of rows as columns. The nonsquare case makes an interesting research project! Can one find a 1×2 matrix A and a 2×1 matrix B with $AB = I_1$ and $BA = I_2$?

Example 1.3.2.   (a) Let $A = \begin{bmatrix} 1 & 0 \\ -1 & 1 \end{bmatrix}$. Then if $B = \begin{bmatrix} 1 & 0 \\ 1 & 1 \end{bmatrix}$,

$$AB = \begin{bmatrix} 1 & 0 \\ -1 & 1 \end{bmatrix} \begin{bmatrix} 1 & 0 \\ 1 & 1 \end{bmatrix} = \begin{bmatrix} 1 & 0 \\ 0 & 1 \end{bmatrix} \text{ and } BA = \begin{bmatrix} 1 & 0 \\ 1 & 1 \end{bmatrix} \begin{bmatrix} 1 & 0 \\ -1 & 1 \end{bmatrix} = \begin{bmatrix} 1 & 0 \\ 0 & 1 \end{bmatrix}.$$

Thus A is nonsingular and B is the inverse of A; that is, $A^{-1} = B = \begin{bmatrix} 1 & 0 \\ 1 & 1 \end{bmatrix}$. Notice also that B is nonsingular and A is the inverse of B.

(b) Let $A = \begin{bmatrix} 1 & 1 \\ 1 & 1 \end{bmatrix}$. We can see that A has no inverse since

$$\begin{bmatrix} 1 & 1 \\ 1 & 1 \end{bmatrix} \begin{bmatrix} a & b \\ c & d \end{bmatrix} = \begin{bmatrix} a+c & b+d \\ a+c & b+d \end{bmatrix} = \begin{bmatrix} 1 & 0 \\ 0 & 1 \end{bmatrix}$$

implies both $a + c = 1$ and $a + c = 0$.

$\square$

The above examples show that some nonzero matrices have inverses and some nonzero matrices do not. The question of the existence of inverses for matrices is an important one that we will investigate further in later sections.

Section 1.3 Exercises

1. Let A be an $m \times n$ matrix. Show that $2A + 3A = 5A$. Which of the above properties were needed?

2. For $m \times n$ matrices A and B show that $A + (B + A) = 2A + B$. Which of the above theorems were needed?

3. If A, B, and C are $m \times n$ matrices and $A + C = B + A$, show that $B = C$. State which of the above theorems were used.

4. Let $A = \begin{bmatrix} -1 & 2 \\ -2 & 4 \end{bmatrix}, B = \begin{bmatrix} 2 & 1 \\ -3 & 2 \end{bmatrix}$, and assume that C is a 2×2 matrix. If $A + C = B$, find C. What properties of the matrix operations were used in the solution?

5. Complete the proof of Theorem 1.3.1.

6. Complete the proof of Theorem 1.3.2.

7. Show by example that if A and B are $n \times n$ matrices, then $(A - B)(A + B) = AA - BB$ is not in general true. (See Example 1.3.1.)

8. Complete the proof of Theorem 1.3.3.

9. Let $A = [a_{ij}]$ be the 3×3 matrix $\begin{bmatrix} 1 & -1 & 2 \\ 0 & 1 & 3 \\ -2 & 4 & -3 \end{bmatrix}$. Compute the following:

(a) $\displaystyle\sum_{i=1}^{3} a_{i1}$

(b) $\displaystyle\sum_{i=1}^{2}\sum_{j=1}^{3} a_{ij}.$

(c) $\displaystyle\sum_{i=1}^{3}\sum_{j=1}^{3} a_{ij}.$

10. Complete the proof of Theorem 1.3.5.

11. Let A be an $m \times n$ matrix and r some scalar. In each of the following expressions, state what the symbol 0 must represent in order for the expression to make sense. Note that in some cases there may be ambiguity, that is, there may be more than one correct answer.

   (a) $0 + A$

   (b) $(r + 0)A$

   (c) $A = 0$

   (d) $0A$

12. Prove Theorem 1.3.6.

13. In the proof of Corollary 1.3.7, give reasons for each step in the proof, that is, cite the appropriate theorem.

14. Prove part (b) of Theorem 1.3.8.

15. Show that the matrix $\begin{bmatrix} 1 & -2 \\ -2 & 4 \end{bmatrix}$ is singular.

16. Show that the matrix $\begin{bmatrix} 1 & 2 & -1 \\ 3 & 1 & 1 \\ 3 & -1 & 2 \end{bmatrix}$ is singular.

17. Show that the matrix $\begin{bmatrix} 1 & -1 \\ 2 & 1 \end{bmatrix}$ is nonsingular and find its inverse.

18. Let A and B be 2×2 matrices and assume $AB = I$. Prove that $BA = I$.

1.4 Equivalent Systems of Equations and Row Operations

In solving a system of linear equations, one proceeds from the original system of equations, through several systems, and finally to a system of equations in which the solution is obvious. An important underlying principle in this method is that each of the systems of equations in the sequence has the same solution set. Two systems of equations that have the same solution set are called **equivalent**. It follows that the systems

$$x - y = 1$$
$$x - 2y = 0$$

and

$$x + y = 3$$
$$x + 2y = 4$$

are equivalent, the solution set of each being $\{(2, 1)\}$. The (very small) system $x - 2y = 3$ and the system $-3x + 6y = -9$ are also equivalent; the solution set of each is

$$\{(3 + 2y, y) | y \text{ is any real number}\}.$$

(Sets and set notation are discussed in Appendix A.)

The following theorem gives basic principles related to equivalence. Beginning with a system of equations that we would like to solve, our goal is to find an equivalent system of equations in which the solution is clear. This theorem tells us what can be done.

Theorem 1.4.1. *For a given system of linear equations we have the following:*

(a) *If two equations in the system are interchanged, then the resulting system is equivalent to the original system.*   *(The order of equations can be switched.)*

(b) *If one equation in the system is multiplied (on both sides) by the nonzero scalar c, then the resulting system is equivalent to the original one.*
(An equation can be multiplied by a nonzero constant.)

(c) *If one equation is modified by adding a multiple of another equation to it, then the resulting system is equivalent to the original system.*
(A multiple of one equation can be added to another.)

Proof. Part (a) is straightforward, Part (b) is an exercise. We will prove Part (c). Assume that c times the equation $a_1 x_1 + \ldots + a_n x_n = h$ is added to the equation $b_1 x_1 + \ldots + b_n x_n = k$. The resulting equation is $(ca_1 + b_1)x_1 + \ldots + (ca_n + b_n)x = ch + k$, and so we see that any solution of the original system is a solution of the resulting system.

If $(c_1, \ldots, c_n)$ is a solution of the new system, then it satisfies the equations

$$a_1 x_1 + \ldots + a_n x_n = h$$

and

$$(ca_1 + b_1)x_1 + \ldots + (ca_n + b_n)x_n = ch + k$$

of the new system. (The first equation is still an equation in the system – only one equation has been changed.) So

$$a_1 c_1 + \ldots + a_n c_n = h$$

and

$$(ca_1 + b_1)c_1 + \ldots + (ca_n + b_n)c_n = ch + k.$$

Multiply both sides of the first equation by $-c$ and add it to the second equality. After cancelling, the result is $b_1 c_1 + \ldots + b_n c_n = k$. It follows that $(c_1, \ldots, c_n)$ satisfies the original system of equations, and so the two systems are equivalent. $\square$

ELEMENTARY ROW OPERATIONS

The operations that may be performed on a system of equations as described in the above theorem suggest corresponding operations on the coefficient and augmented matrices of the system. These operations are called **elementary row operations**. They are of interest for general matrices and so are defined without reference to a system of equations. We will see that these row operations are quite useful; in fact, we will see that they can be of help in understanding the theory that underlies much of matrix theory and linear algebra. These three types of operations are defined as follows:

1. Rows i and k are interchanged. (This operation is denoted by R_{ik}.)

2. Each entry in row i is multiplied by the scalar $c \neq 0$. (This operation is denoted by $R_i(c)$.)

3. For a scalar c, c times each entry in row i is added to each corresponding entry in row k. (This operation is denoted by $R_{ik}(c)$.)

To illustrate these row operations, we again refer to the system 1.1.1 and find its solution by using the augmented matrix and row operations. As shown in Section 1.2, the solution can be found as follows:

$$\begin{bmatrix} 1 & 1 & 17 \\ 1 & -1 & 7 \end{bmatrix} \xrightarrow[\text{Row 1 added to row 2}]{R_{12}(1)} \begin{bmatrix} 1 & 1 & 17 \\ 2 & 0 & 24 \end{bmatrix} \xrightarrow[\text{Row 2 multiplied by 1/2}]{R_2(1/2)} \begin{bmatrix} 1 & 1 & 17 \\ 1 & 0 & 12 \end{bmatrix}$$

$$\xrightarrow[\text{Row 2 subtracted from row 1}]{R_{21}(-1)} \begin{bmatrix} 0 & 1 & 5 \\ 1 & 0 & 12 \end{bmatrix} \xrightarrow[\text{Rows 1 and 2 are switched.}]{R_{12}} \begin{bmatrix} 1 & 0 & 12 \\ 0 & 1 & 5 \end{bmatrix}$$

The symbol and numbers over the arrow indicates the elementary row operation that was performed on the first matrix to obtain the second matrix.

Two matrices A and B are said to be **row equivalent** if B can be obtained from A by a sequence of row operations. In the above example we can see that $\begin{bmatrix} 1 & 1 & 17 \\ 1 & -1 & 7 \end{bmatrix}$

is row equivalent to $\begin{bmatrix} 1 & 0 & 12 \\ 0 & 1 & 5 \end{bmatrix}$. If no confusion arises, we will sometimes shorten "row equivalence" to "equivalence."

The question arises about the relationship between the equivalence of matrices and the equivalence of systems of equations. For a system of linear equations written in the matrix form $AX = H$, we will use $[A|H]$ to denote the augmented matrix.

Theorem 1.4.2. *Consider systems of equations* $AX = H$, *and* $BX = K$. *If the augmented matrices* $[A|H]$ *and* $[B|K]$ *are equivalent, then the systems of equations* $AX = H$ *and* $BX = K$ *are equivalent.*   (*Equivalent matrices give equivalent systems.*)

Proof. If $[A|H]$ is equivalent to $[B|K]$, then $[B|K]$ can be obtained from $[A|H]$ by a sequence of row operations. If the corresponding operations are performed on the system of equations $AX = H$ the system $BX = K$ will be obtained, and, by the first theorem, each system of equations in this sequence is equivalent to the previous system.   □

The theorem above, which relates the two different types of equivalence, is an implication, that is, a statement of the form "*statement* 1 implies *statement* 2". Mathematicians will wonder about the **converse** of the statement: "*statement* 2 implies *statement* 1". Is the converse also true? Can it be proved? Can a counterexample be found? We won't have occasion to need the converse of Theorem 1.4.2, and so, we will leave it as an "open question" for the reader. Try to settle the matter!

Suppose that A and B are equivalent matrices; let's say that we start with A, perform row operations, and arrive at the matrix B. How do the rows of B relate to those of A? It is not hard to see that each row of the matrix B is a sum of scalar multiples of the rows of A. In Chapter 2, these sums of scalar multiples will be called "linear combinations;" thus, the rows of B are linear combinations of the rows of A.

Notice that each elementary row operation is reversible, so that if B may be obtained from A by row operations, then A may be obtained from B by row operations. To see this, we need only consider the reversibility of each of the three types of elementary row operations.

If B is obtained from A by switching rows i and k, then A may be obtained from B by again switching rows i and k, this time in the matrix B. Similarly, to undo the operation of multiplying row i by a nonzero scalar c, we need only multiply row i by the reciprocal of $c, 1/c$. Finally, if B is obtained from A by adding c times row i to row k, then adding $-c$ times row i in the matrix B to row k produces the matrix A.

Try some examples to convince yourself. For example, if $A = \begin{bmatrix} 1 & 0 & -1 \\ 2 & 3 & 1 \\ 0 & 2 & 1 \end{bmatrix}$, then the row operation $R_{13}(2)$ produces the matrix $B = \begin{bmatrix} 1 & 0 & -1 \\ 2 & 3 & 1 \\ 2 & 2 & -1 \end{bmatrix}$. Performing the operation $R_{13}(-2)$ on the matrix B yields the matrix A.

ELEMENTARY MATRICES

There are some important matrices associated with the elementary row operations. Recall that I_n denotes the $n \times n$ identity matrix. We let R_{ik} denote I_n with rows i and k interchanged,

$R_i(c)$ denotes I_n with row i multiplied by c, and $R_{ik}(c)$ denotes I_n with c times row i added to row k. These matrices $R_{ik}, R_i(c)$, and $R_{ik}(c)$ are called **elementary (row) matrices**. Each of the three types of elementary matrices is obtained by performing the corresponding elementary row operation on the identity matrix. One might be confused by the fact that the same notation is used to refer to an elementary row operation as is used to refer to the corresponding elementary matrix. This situation will cause very little trouble, since the context will always make it clear whether one is referring to a row operation or to an elementary matrix.

The following are examples of 2×2 elementary matrices:

$$R_1(-2) = \begin{bmatrix} -2 & 0 \\ 0 & 1 \end{bmatrix} \qquad \text{(type } R_i(c)\text{)}$$

$$R_{12} = \begin{bmatrix} 0 & 1 \\ 1 & 0 \end{bmatrix} \qquad \text{(type } R_{ik}\text{)}$$

$$R_{12}(-2) = \begin{bmatrix} 1 & 0 \\ -2 & 1 \end{bmatrix} \qquad \text{(type } R_{ik}(c)\text{)}$$

$$R_{21}(3) = \begin{bmatrix} 1 & 3 \\ 0 & 1 \end{bmatrix}$$

The matrices

$$\begin{bmatrix} 1 & 1 \\ 1 & 1 \end{bmatrix}, \begin{bmatrix} 1 & 2 \\ 0 & 2 \end{bmatrix}, \begin{bmatrix} 2 & 0 \\ 0 & 2 \end{bmatrix}, \text{ and } \begin{bmatrix} 1 & 0 \\ 0 & 0 \end{bmatrix}$$

are not elementary matrices since they cannot be obtained from the 2×2 identity matrix by performing a single elementary row operation.

Notice what happens when a matrix is multiplied on the left by an elementary matrix:

$$R_{12} \begin{bmatrix} 1 & -1 \\ 3 & 2 \end{bmatrix} = \begin{bmatrix} 0 & 1 \\ 1 & 0 \end{bmatrix} \begin{bmatrix} 1 & -1 \\ 3 & 2 \end{bmatrix} = \begin{bmatrix} 3 & 2 \\ 1 & -1 \end{bmatrix}$$

$$R_2(-2) \begin{bmatrix} 1 & -1 \\ 3 & 2 \end{bmatrix} = \begin{bmatrix} 1 & 0 \\ 0 & -2 \end{bmatrix} \begin{bmatrix} 1 & -1 \\ 3 & 2 \end{bmatrix} = \begin{bmatrix} 1 & -1 \\ -6 & -4 \end{bmatrix}$$

$$R_{12}(1) \begin{bmatrix} 1 & -1 \\ 3 & 2 \end{bmatrix} = \begin{bmatrix} 1 & 0 \\ 1 & 1 \end{bmatrix} \begin{bmatrix} 1 & -1 \\ 3 & 2 \end{bmatrix} = \begin{bmatrix} 1 & -1 \\ 4 & 1 \end{bmatrix}.$$

From these examples we see that left multiplication by the above elementary matrices produces the same result as performing the corresponding row operation on the matrix. This observation is true in general, as stated in the following theorem.

> **Theorem 1.4.3.** *A row operation can be performed on an $m \times n$ matrix A by multiplying A on the left by the corresponding elementary matrix.*
>
> *(Left multiplication by an elementary matrix performs the row operation.)*

Proof. In effect, we already know this because of our alternative perspectives on matrix multiplication. Let the rows of A be $A_1, \ldots, A_m$. If we swap two rows of the identity matrix, the columns in which the 1s appear are also swapped, so row i of the product $R_{ik}A$ is A_k, row k of the product is A_i, and all other rows remain the same. In the case of $R_i(c)$, we have 1 times row k of A in every row except row i, where we have c times row i of A. (The linear combination is just $0A_1 + \ldots + cA_i + \ldots + 0A_m = cA_i$.)

Now consider $R_{ik}(c)A$. In every row j except row k, we just get 1 times A_j in the product. For row k of the product, we have $0A_1 + \ldots + cA_i + \ldots + 1A_k + \ldots + 0A_m = cA_i + A_k$, as desired. $\qquad\qquad\square$

Theorem 1.4.3 makes it very easy to compute the products of elementary matrices: just perform the row operations. For example, for 3×3 matrices:

$$R_{12}R_2(-2)R_{13}(2)R_{21}(-1)R_{23} = R_{12}R_2(-2)R_{13}(2)R_{21}(-1)\begin{bmatrix} 1 & 0 & 0 \\ 0 & 0 & 1 \\ 0 & 1 & 0 \end{bmatrix}$$

$$= R_{12}R_2(-2)R_{13}(2)\begin{bmatrix} 1 & 0 & -1 \\ 0 & 0 & 1 \\ 0 & 1 & 0 \end{bmatrix}$$

$$= R_{12}R_2(-2)\begin{bmatrix} 1 & 0 & -1 \\ 0 & 0 & 1 \\ 2 & 1 & -2 \end{bmatrix}$$

$$= R_{12}\begin{bmatrix} 1 & 0 & -1 \\ 0 & 0 & -2 \\ 2 & 1 & -2 \end{bmatrix}$$

$$= \begin{bmatrix} 0 & 0 & -2 \\ 1 & 0 & -1 \\ 2 & 1 & -1 \end{bmatrix}.$$

Section 1.4 Exercises

In Exercises 1-4 determine whether the given systems of equations are equivalent.

1. $\begin{aligned} x + 2y &= 3 \\ x - 3y &= -2 \end{aligned}$ $\qquad\qquad$ $\begin{aligned} x - y &= 0 \\ x - 2y &= -1 \end{aligned}$

2. $\begin{aligned} x - y &= 1 \\ x + y &= 0 \end{aligned}$ $\qquad\qquad$ $\begin{aligned} x &= 1 \\ y &= 2 \end{aligned}$

3. $\begin{aligned} x + y + z &= 1 \\ x - y - z &= 1 \end{aligned}$ $\qquad\qquad$ $\begin{aligned} 2x + y + z &= 2 \\ -x + y - z &= 3 \end{aligned}$

4. $\begin{aligned} x - y + z &= 2 \\ x + y - z &= 4 \\ x - z &= 2 \end{aligned}$ $\qquad\qquad$ $\begin{aligned} 2x - y + z &= 5 \\ x - z &= 2 \\ -2y + z &= -3 \end{aligned}$

5. Prove part (b) of Theorem 1.4.1.

6. Prove the remaining parts of Theorem 1.4.3.

   In exercises 7-12, compute the given product of 3×3 elementary matrices.

7. $R_{12}R_{23}(-1)R_3(3)$

8. $R_{13}R_{21}(-2)R_3(2)$

9. $R_{23}(-2)R_{23}(2)$

10. $R_2(2)R_2(1/2)$

11. $R_{21}(-2)R_{13}(4)R_{23}(-2)R_{23}(2)$

12. $R_{12}(2)R_{23}(-4)R_{13}(2)R_2(-2)R_{12}R_{31}(-1)$

13. For each of the following matrices, state whether the matrix is an elementary matrix and, if it is, identify it using the "R" notation.

   (a) $\begin{bmatrix} 1 & 0 \\ 1 & 1 \\ 2 & 0 \end{bmatrix}$
   (b) $\begin{bmatrix} 1 & 0 & 0 \\ 1 & 1 & 0 \\ 0 & 0 & 1 \end{bmatrix}$
   (c) $\begin{bmatrix} 0 & 1 & 0 \\ 1 & 0 & 0 \\ 0 & 0 & 1 \end{bmatrix}$
   (d) $\begin{bmatrix} 1 & 2 & 0 \\ 0 & 2 & 0 \\ 0 & 0 & 1 \end{bmatrix}$

14. Let A be the $n \times n$ elementary matrix R_{ik}. What is the product AA?

15. Let A be the $n \times n$ matrix $R_i(c)$. Find an $n \times n$ matrix B with $AB = I_n$.

16. Let A be the $n \times n$ matrix $R_{ik}(c)$. Find an $n \times n$ matrix B with $AB = I_n$.

1.5 Gaussian Elimination and the Reduced Echelon Form

Gaussian Elimination[2] is a process by which one can proceed from a system of linear equations to an equivalent system (one with the same solution set) in which the solution is clear.

   We will begin the discussion of Gaussian elimination with an example of a system of linear equations. The problem will be stated and solved "in equation form"; that is, the augmented matrix will not be used.

   Suppose we must find all solutions of

$$\begin{array}{rcrcrcrcl} x_1 & + & 2x_2 & + & x_3 & - & 3x_4 & = & 2 \\ x_1 & + & 3x_2 & + & 2x_3 & - & 2x_4 & = & 3 \\ x_1 & + & 3x_2 & + & 3x_3 & & & = & 5. \end{array} \qquad (1.5.1)$$

Subtracting the first equation from the second and third equations, we obtain

---

[2]The process is named after the great German mathematician Carl Friedrich Gauss (1777-1855).

$$
\begin{aligned}
x_1 + 2x_2 + x_3 - 3x_4 &= 2 \\
x_2 + x_3 + x_4 &= 1 \\
x_2 + 2x_3 + 3x_4 &= 3,
\end{aligned}
\tag{1.5.2}
$$

and if x_2 is eliminated from the last equation by subtracting the middle equation, the system becomes:

$$
\begin{aligned}
x_1 + 2x_2 + x_3 - 3x_4 &= 2 \\
x_2 + x_3 + x_4 &= 1 \\
x_3 + 2x_4 &= 2.
\end{aligned}
\tag{1.5.3}
$$

This first part of the process is called **forward elimination**; it is the first part of the Gaussian elimination process. The equations are now said to be in **triangular form**: x_1 is eliminated from all but the first, x_2 from all below the second, and so forth. It is now easy to find all solutions of the system of equations. This latter part of the process is often called **back substitution**. It will give us the **complete** or **general solution**. We will solve the last equation for x_3 and substitute the result into the second equation; then we solve the second for x_2 and substitute into the first. We get:

$$
\begin{aligned}
x_3 &= 2 - 2x_4 \\
x_2 &= 1 - (2 - 2x_4) - x_4 \\
&= -1 + x_4 \\
x_1 &= 2 - 2(-1 + x_4) - (2 - 2x_4) + 3x_4
\end{aligned}
\tag{1.5.4}
$$

or

$$
\begin{aligned}
x_3 &= 2 - 2x_4 \\
x_2 &= -1 + x_4 \\
x_1 &= 2 + 3x_4,
\end{aligned}
\tag{1.5.5}
$$

where x_4 is arbitrary; that is, it may take on any value.

The above method is called **Gaussian elimination**; it provides a method for finding all solutions of a system of linear equations. A later refinement of the method[3] eliminated the need for "back substitution." It is called the **Gauss-Jordan** method and is sometimes referred to as **Gauss-Jordan reduction**. The Gaussian elimination process involves elimination of a variable from all the remaining equations. The solution of the system 1.5.1 using the Gauss-Jordan method proceeds as follows.

Eliminate x_1 from the second and third equations as before, obtaining

$$
\begin{aligned}
x_1 + 2x_2 + x_3 - 3x_4 &= 2 \\
x_2 + x_3 + x_4 &= 1 \\
x_2 + 2x_3 + 3x_4 &= 3.
\end{aligned}
\tag{1.5.6}
$$

Now eliminate x_2 from the first and third equations by using the second equation, and eliminate x_3 from the first and second equations by using the resulting third equation:

$$
\begin{aligned}
x_1 \qquad\quad - 3x_4 &= 2 \\
2x_2 \qquad - x_4 &= -1 \\
x_3 + 2x_4 &= 2.
\end{aligned}
\tag{1.5.7}
$$

---

[3]Due to the German mathematician Wilhelm Jordan

The solution is now clear and there is no need for "back substitution"; one simply moves the terms involving x_4 to the other side of the equation.

PIVOTING

The process of eliminating a variable from the remaining equations is called a **pivot operation**. To pivot on x_1 in the first equation, for example, means to make the coefficient of x_1 a 1 and then to eliminate x_1 from the remaining equations.

As we observed before, it is somewhat easier to use the augmented matrix in solving a system of equations than to use the system itself. So, in discussing the Gauss-Jordan method we will want to refer to the augmented matrix. We must first define the pivot operations (or elimination operations) for a matrix, and then describe the final form for the augmented matrix in the solution of the system.

Let $A = [a_{ij}]$ be an $m \times n$ matrix. To **pivot** on a nonzero entry a_{ij} means to perform the following sequence of row operations:

$$R_i(1/a_{ij}), R_{i,1}(-a_{1j}), \ldots, R_{i,i-1}(-a_{i-1,j}), R_{i,i+1}(-a_{i+1}, j), \ldots, R_{i,m}(-a_{m,j}).$$

This looks complicated, but it is not. To pivot on a_{ij} means to divide row i by a_{ij} (so there is a 1 in row i and column j) and then perform the necessary row operations to make 0's in column j in all rows except row i.

Example 1.5.1. Consider the matrix

$$A = \begin{bmatrix} 1 & 2 & -1 \\ -3 & 2 & 4 \\ 1 & 1 & 2 \end{bmatrix}. \tag{1.5.8}$$

If one pivots on the entry in row 1, column 1 (performing the row operations $R_{12}(3)$, $R_{13}(-1)$), one obtains

$$\begin{bmatrix} 1 & 2 & -1 \\ 0 & 8 & 1 \\ 0 & -1 & 3 \end{bmatrix}. \tag{1.5.9}$$

If one then pivots on the 8 in row 2, column 2 (performing the row operations $R_2(1/8)$, $R_{21}(-2)$, and $R_{23}(1)$), one obtains

$$\begin{bmatrix} 1 & 0 & -5/4 \\ 0 & 1 & 1/8 \\ 0 & 0 & 25/8 \end{bmatrix}. \tag{1.5.10}$$

$\square$

Recall that by Theorem 1.4.2 of Section 1.4 the systems of equations represented by matrices before and after a pivot operation is performed are equivalent since the matrices are equivalent. It follows that a sequence of pivot operations produces a matrix equivalent to the original one.

THE REDUCED ECHELON FORM

Let us now describe the augmented matrix associated with the system of equations after the Gauss-Jordan reduction has been performed. This matrix is said to be in **reduced row echelon form** - the word "echelon" comes from the French word for "step." Some authors use the term "**row echelon normal form**." For later use, we will define both a row echelon form and a reduced row echelon form.

An $m \times n$ matrix $A = [a_{ij}]$ is said to be in **row echelon form** if and only if it satisfies the following three conditions:

1. For some integer r with $0 \leq r \leq m$, the first r rows contain nonzero entries and the remaining $m - r$ rows contain only zeros.   (The nonzero rows are at the top.)

2. For $i = 1, \ldots, r$, the first nonzero entry in row i is a 1 in column j_i and there are only zeroes below it in column j_i.   (The first nonzero entry is a 1 and there are 0's below it.)

3. $j_1 < j_2 < \ldots < j_r$.   (The 1s step down and to the right.)

   If the matrix A satisfies conditions 1) - 3) above and satisfies

4. For $i = 1, \ldots, r$, the 1 in row i and column j_i is the only nonzero entry in column j_i,
   (The first nonzero entry is a 1 and there are 0s above and below it.)

then the matrix is said to be in **reduced** row echelon form.

The numbers r and $j_1, \ldots, j_r$ are important; we will refer to them as the **constants associated with the echelon form**.

The above definitions involve the rows of a matrix. There is a corresponding version related to columns of a matrix, but we will not have occasion to use it. Since no "column echelon form" will ever be used, we will normally suppress the word "row" and refer simply to a matrix in **echelon form** or in **reduced echelon form**.

Example 1.5.2. The matrix

$$\begin{bmatrix} 0 & 1 & 2 & 0 & 1 \\ 0 & 0 & 0 & 1 & 1 \\ 0 & 0 & 0 & 0 & 0 \end{bmatrix} \tag{1.5.11}$$

is in reduced echelon form, with constants $r = 2, j_1 = 2$, and $j_2 = 4$. The following matrices are also in reduced echelon form:

$$\begin{bmatrix} 0 & 0 \\ 0 & 0 \end{bmatrix}, \begin{bmatrix} 1 & 0 \\ 0 & 0 \end{bmatrix}, \text{ and } \begin{bmatrix} 1 & 0 \\ 0 & 1 \end{bmatrix}.$$

The matrix

$$\begin{bmatrix} 0 & 0 \\ 0 & 1 \end{bmatrix}$$

is not in echelon form (since 1) fails), and so not in reduced echelon form. The matrix

$$\begin{bmatrix} 2 & 0 \\ 0 & 1 \end{bmatrix}$$

is not in echelon form (since 2) fails), and

$$\begin{bmatrix} 0 & 1 \\ 1 & 0 \end{bmatrix}$$

is not in echelon form (since 3) fails). The matrices

$$\begin{bmatrix} 1 & 1 \\ 0 & 1 \end{bmatrix}, \begin{bmatrix} 1 & 1 & 2 \\ 0 & 1 & 1 \\ 0 & 0 & 0 \end{bmatrix}, \text{ and } \begin{bmatrix} 1 & 1 & 3 \\ 0 & 1 & 2 \\ 0 & 0 & 1 \end{bmatrix}$$

are in echelon form, but not in reduced echelon form since 4) fails.

$\square$

THE GENERAL SOLUTION USING THE GAUSS-JORDAN METHOD

The process of solving a system of linear equations using the Gauss-Jordan method and the associated augmented matrix is as follows.

1. Write the augmented matrix.

2. Using pivot operations, reduce the augmented matrix to a matrix in reduced echelon form with constants $r, j_1, \ldots, j_r$ as in the definition above.

3. Write the system of equations associated with this matrix in reduced echelon form.

4. Solve for the variables $x_{j_1}, \ldots, x_{j_r}$ in terms of the remaining variables.

Not every system of equations has a solution, but if a solution exists, the above method will, in theory, find it. If there is a solution, the variables $x_{j_1}, \ldots, x_{j_r}$ that are associated with the pivot columns will be called **basic variables**. The variables associated with the other columns will be called **free variables** since they may be chosen arbitrarily.

Example 1.5.3. Consider the following system of linear equations:

$$\begin{array}{rcrcrcrcrcl} x_1 & + & & & x_3 & + & & & x_5 & = & 1 \\ x_1 & + & x_2 & + & 3x_3 & + & & & 3x_5 & = & 2 \\ 2x_1 & + & x_2 & + & 4x_3 & + & x_4 & + & 5x_5 & = & 4 \\ 4x_1 & + & 2x_2 & + & 8x_3 & + & x_4 & + & 9x_5 & = & 7 \end{array} \qquad (1.5.12)$$

We write the augmented matrix

$$\begin{bmatrix} 1 & 0 & 1 & 0 & 1 & 1 \\ 1 & 1 & 3 & 0 & 3 & 2 \\ 2 & 1 & 4 & 1 & 5 & 4 \\ 4 & 2 & 8 & 1 & 9 & 7 \end{bmatrix}$$

and perform pivot operations in order to reduce to reduced echelon form.

First we pivot on the 1 in row 1, column 1, and obtain

$$\begin{bmatrix} 1 & 0 & 1 & 0 & 1 & 1 \\ 0 & 1 & 2 & 0 & 2 & 1 \\ 0 & 1 & 2 & 1 & 3 & 2 \\ 0 & 2 & 4 & 1 & 5 & 3 \end{bmatrix} \qquad (1.5.13)$$

Next, we pivot on the 1 in row 2, column 2, and obtain

$$\begin{bmatrix} 1 & 0 & 1 & 0 & 1 & 1 \\ 0 & 1 & 2 & 0 & 2 & 1 \\ 0 & 0 & 0 & 1 & 1 & 1 \\ 0 & 0 & 0 & 1 & 1 & 1 \end{bmatrix} \qquad (1.5.14)$$

Finally, we pivot on the 1 in row 3, column 4, and obtain

$$\begin{bmatrix} 1 & 0 & 1 & 0 & 1 & 1 \\ 0 & 1 & 2 & 0 & 2 & 1 \\ 0 & 0 & 0 & 1 & 1 & 1 \\ 0 & 0 & 0 & 0 & 0 & 0 \end{bmatrix} \qquad (1.5.15)$$

The system of equations corresponding to this last matrix is:

$$\begin{aligned} x_1 + \quad\quad\ x_3 + \quad\quad\quad\ x_5 &= 1 \\ x_2 + 2x_3 + \quad\quad\ 2x_5 &= 1 \ . \\ x_4 + \quad x_5 &= 1 \end{aligned} \qquad (1.5.16)$$

The constants associated with the reduced echelon form of the augmented matrix (that is, the last matrix in the process) are $r = 3, j_1 = 1, j_2 = 2, j_3 = 4$. Solving the associated system of equations for x_1, x_2, x_4 we obtain:

$$x_1 = 1 - x_3 - x_5$$
$$x_2 = 1 - 2x_3 - 2x_5$$
$$x_4 = 1 - x_5,$$

where x_3 and x_5 are arbitrary.

This is the **complete solution** or **general solution** of the original system of equations. By the theory that has been developed, the last system of equations has exactly the same solution set as the original system of equations (in other words, the two systems are equivalent). In the last system, we can see that any choice of values for the variables x_3 and x_5

uniquely determines the values of the variables x_1, x_2, and x_4. Furthermore, any solution of the original system of equations must satisfy this last set of equations. The solution set, expressed as a set of 5-tuples, is given by:

$$\{(1 - x_3 - x_5, 1 - 2x_3 - 2x_5, x_3, 1 - x_5, x_5)|x_3, x_5 \in \mathbb{R}\}.$$

Notice that, while the solution set is unique, the expression of the general solution is not. In the above general solution we could solve the last equation for x_5 in terms of x_4, obtaining $x_5 = 1 - x_4$. This could be substituted into the other two equations to obtain an alternate general solution of the form:

$$x_1 = -x_3 + x_4$$
$$x_2 = -1 - 2x_3 + 2x_4$$
$$x_5 = 1 - x_4.$$

In this general solution, x_3 and x_4 may be chosen arbitrarily.

$\square$

The question now arises whether one can always find a complete solution for every consistent system of linear equations; that is, a system that actually has a solution. Our discussion above indicates that this question is equivalent to the question of whether every matrix is equivalent to a matrix in reduced echelon form. As the following theorem states, the answer to these questions is "yes" - at least in theory. In practice, if the number of equations or variables is very large, it may be difficult to deal with the mass of coefficients and computations involved. Further, if the computations are done using a computer or a calculator, roundoff errors may occur and they may compound in the process of repeated pivot operations and give an "approximate" solution that is not even close.

As we will see in the theorem, a matrix is equivalent to a unique matrix in reduced echelon form; that is, there is one and only one such matrix. A matrix may be equivalent to many matrices that are in echelon form, but only one of these matrices will be in reduced echelon form.

EXISTENCE OF THE REDUCED ECHELON FORM OF A MATRIX

> **Theorem 1.5.1.** *Any $m \times n$ matrix A is row equivalent to one and only one $m \times n$ matrix B that is in reduced echelon form. B is called the* **reduced echelon form** *of A.*
> *(Every matrix has a unique reduced echelon form.)*

Proof. We will first prove "existence"; that is, that an arbitrary $m \times n$ matrix is row equivalent to some $m \times n$ matrix in reduced echelon form. The second part of the theorem states that this matrix in reduced echelon form is "unique;" that is, there is only one such matrix. The "uniqueness" part of the proof is dealt with in Appendix 6. A formal proof of existence would require mathematical induction (see Appendix 5); here we take an informal approach.

The first step is as follows: Find a nonzero entry a_{ij} in the column farthest to the left (that is, j is the least integer with $a_{ij} \neq 0$ for some i). Interchange rows 1 and i (that is, perform operation R_{1i}), and then pivot on the entry in row 1, column j. Call this pivot column j_1. The first nonzero entry in row 1 is now a 1; it is the only nonzero entry in column j_1.

Now repeat this process as follows: Find a nonzero entry in a_{ij} in one of the rows $2, 3, \ldots, m$ that is the farthest to the left (that is, j is the least). Set $j_2 = j$, switch rows 2 and i, and then pivot on the entry in row 2, column j_2. Notice that the pivot operation does not change any of the columns to the left of column j_2, since there was a zero in these columns in row 2. If no nonzero entry can be found in rows $2, \ldots, m$, the process stops.

Continue this process until no nonzero entry can be found in the remaining rows or until no rows remain. Since the matrix has only m rows, the process must stop.

Assume that the above process required r steps, that is, r pivot operations were performed along with the necessary switching of rows. We claim that the resulting matrix is in reduced echelon form: The first r rows will contain nonzero entries and the first nonzero entry will be a 1 since pivot operations have been performed on each of the first r rows. Since j_i was chosen to be least, $j_1 < j_2 < \ldots < j_r$, and the 1 in row i and column j_i is the only nonzero entry in column j_i, since a pivot operation was performed on that entry. If we denote the resulting matrix by B, B is in reduced echelon form. $\qquad\square$

Example 1.5.4. As an application of the uniqueness part of the Theorem, we can easily see that

$$
\begin{bmatrix} 1 & 0 & 1 & 0 \\ 0 & 1 & 2 & 0 \\ 0 & 0 & 0 & 1 \end{bmatrix} \text{ and } \begin{bmatrix} 1 & 0 & 0 & 0 \\ 0 & 1 & 1 & 0 \\ 0 & 0 & 0 & 1 \end{bmatrix}
$$

are not row equivalent since both are matrices in row reduced echelon form and they are not identical. If the first matrix were row equivalent to the second, then it would be row equivalent to two matrices: itself and the second matrix.

As an illustration of the method of the proof, let us reduce the matrix

$$
A = \begin{bmatrix} 0 & 0 & 0 & 0 & 0 \\ 0 & 1 & 2 & 0 & 0 \\ 0 & 1 & 2 & 1 & 0 \\ 0 & 0 & 0 & 1 & 1 \end{bmatrix}
$$

to reduced echelon form.

$$A \quad = \quad \begin{bmatrix} 0 & 0 & 0 & 0 & 0 \\ 0 & 1 & 2 & 0 & 0 \\ 0 & 1 & 2 & 1 & 0 \\ 0 & 0 & 0 & 1 & 1 \end{bmatrix} \xrightarrow{R_{12}} \begin{bmatrix} 0 & 1 & 2 & 0 & 0 \\ 0 & 0 & 0 & 0 & 0 \\ 0 & 1 & 2 & 1 & 0 \\ 0 & 0 & 0 & 1 & 1 \end{bmatrix}$$

$$\xrightarrow{\mathbb{R}_{13}(-1)} \begin{bmatrix} 0 & 1 & 2 & 0 & 0 \\ 0 & 0 & 0 & 0 & 0 \\ 0 & 0 & 0 & 1 & 0 \\ 0 & 0 & 0 & 1 & 1 \end{bmatrix} \xrightarrow{R_{23}} \begin{bmatrix} 0 & 1 & 2 & 0 & 0 \\ 0 & 0 & 0 & 1 & 0 \\ 0 & 0 & 0 & 0 & 0 \\ 0 & 0 & 0 & 1 & 1 \end{bmatrix}.$$

$$\xrightarrow{R_{24}(-1)} \begin{bmatrix} 0 & 1 & 2 & 0 & 0 \\ 0 & 0 & 0 & 1 & 0 \\ 0 & 0 & 0 & 0 & 0 \\ 0 & 0 & 0 & 0 & 1 \end{bmatrix} \xrightarrow{R_{34}} \begin{bmatrix} 0 & 1 & 2 & 0 & 0 \\ 0 & 0 & 0 & 1 & 0 \\ 0 & 0 & 0 & 0 & 1 \\ 0 & 0 & 0 & 0 & 0 \end{bmatrix}$$

$$= \quad B$$

The constants associated with B are $r = 3, j_1 = 2, j_2 = 4, j_3 = 5$. This last matrix B is a matrix in reduced echelon form, and so it must be the reduced echelon form of A. Notice that

$$B = R_{34}R_{24}(-1)R_{23}R_{13}(-1)R_{12}A.$$

At this point confusion often arises, since the elementary matrices appear to come in reverse order. This results from the fact that it is left multiplication by the corresponding elementary matrix which performs the row operation. We see that in the above sequence of matrices, the first is A, the second $R_{12}A$, the third $R_{13}(-1)R_{12}A$, the fourth $R_{23}R_{13}(-1)R_{12}A$, etc.

$\square$

Notice that in the matrix B above that it is not possible to reduce further; that is, we cannot perform further row operations and make more rows of zeros. In the terminology to be introduced in Chapter 2, we say that these nonzero rows are "linearly independent." We cannot form a sum of nonzero scalar multiples of these rows and produce a row of zeros.

GAUSSIAN ELIMINATION VS. THE GAUSS-JORDAN METHOD

In Section 4.5 we will consider the efficiency of solving systems of equations by various methods. We will see that the method of reducing the augmented matrix to reduced echelon form as described above is not the most efficient course of action. As presented above, we pivoted at each step making zeros above and below the pivot element. In the method of Gaussian elimination, one follows the strategy of "forward elimination" followed by "back substitution." It turns out that this latter method requires fewer arithmetic operations, and for large systems the savings can be significant. In effect, the calculation of some of the entries above and to the right of the pivot element is wasteful since these entries may be made zero by later pivots. It is more efficient to eliminate forward and then back substitute by starting at the lower right and making zeros above.

Consider the following two sequences of row operations reducing a matrix to reduced echelon form. Using the Gaussian elimination method:

$$\begin{bmatrix} 1 & 1 & 2 & 1 \\ 0 & 1 & 1 & 2 \\ 0 & 3 & 1 & 2 \end{bmatrix} \longrightarrow \begin{bmatrix} 1 & 1 & 2 & 1 \\ 0 & 1 & 1 & 2 \\ 0 & 0 & -2 & -4 \end{bmatrix} \longrightarrow \begin{bmatrix} 1 & 1 & 2 & 1 \\ 0 & 1 & 1 & 2 \\ 0 & 0 & 1 & 2 \end{bmatrix}$$

$$\longrightarrow \begin{bmatrix} 1 & 1 & 0 & -3 \\ 0 & 1 & 0 & 0 \\ 0 & 0 & 1 & 2 \end{bmatrix} \longrightarrow \begin{bmatrix} 1 & 0 & 0 & -3 \\ 0 & 1 & 0 & 0 \\ 0 & 0 & 1 & 2 \end{bmatrix}.$$

Using the Gauss-Jordan method:

$$\begin{bmatrix} 1 & 1 & 2 & 1 \\ 0 & 1 & 1 & 2 \\ 0 & 3 & 1 & 2 \end{bmatrix} \longrightarrow \begin{bmatrix} 1 & 0 & 1 & -1 \\ 0 & 1 & 1 & 2 \\ 0 & 0 & -2 & -4 \end{bmatrix} \longrightarrow \begin{bmatrix} 1 & 0 & 1 & -1 \\ 0 & 1 & 1 & 2 \\ 0 & 0 & 1 & 2 \end{bmatrix}$$

$$\longrightarrow \begin{bmatrix} 1 & 0 & 1 & -1 \\ 0 & 1 & 0 & 0 \\ 0 & 0 & 1 & 2 \end{bmatrix} \longrightarrow \begin{bmatrix} 1 & 0 & 0 & -3 \\ 0 & 1 & 0 & 0 \\ 0 & 0 & 1 & 2 \end{bmatrix}.$$

While the same number (5) of row operations are performed in each method, the Gaussian elimination method allows us to perform the row operations on shorter rows. This results in fewer arithmetic operations.

Section 1.5 Exercises

In Exercises 1-6, do the following, where A represents the given matrix.

(a) Find the reduced echelon form B of A.

(b) Find the constants associated with B.

(c) Find elementary matrices $E_1, \ldots, E_k$ with $B = E_k \ldots E_1 A$.

1. $\begin{bmatrix} 1 & 2 \\ -1 & -1 \end{bmatrix}$

2. $\begin{bmatrix} -2 & 3 \\ 6 & -9 \end{bmatrix}$

3. $\begin{bmatrix} -1 & 0 & 1 & 2 \\ 1 & 2 & 1 & 1 \\ 0 & -1 & -2 & 1 \end{bmatrix}$

4. $\begin{bmatrix} 0 & 0 & 0 & 1 & 2 \\ 0 & 1 & 1 & 2 & 1 \\ 0 & 1 & 1 & 3 & 2 \end{bmatrix}$

5. $\begin{bmatrix} 1 & 0 & 1 \\ -1 & 2 & 0 \\ 1 & 1 & 0 \end{bmatrix}$

6. $\begin{bmatrix} 1 & 2 & -1 & 3 \\ 3 & 1 & 1 & 2 \\ 2 & -1 & 2 & -1 \end{bmatrix}$

In Exercises 7-12 find the complete solution of the given system of linear equations using the augmented matrix and Gauss-Jordan reduction.

7. $\begin{aligned} x_1 + x_2 &= 2 \\ x_1 - 2x_2 &= 5 \end{aligned}$

8. $\begin{aligned} x_1 + x_2 + x_3 &= 4 \\ 2x_1 + 5x_2 - 2x_3 &= 3 \end{aligned}$

9.
$$\begin{aligned} x_1 + x_2 - x_3 &= 2 \\ 2x_1 + 5x_2 - 2x_3 &= -3 \\ x_1 + 7x_2 - 7x_3 &= -12 \end{aligned}$$

11.
$$\begin{aligned} x_1 \phantom{{}+x_2} + x_3 \phantom{{}+x_4} &= 3 \\ x_1 + x_2 + 2x_3 \phantom{{}+x_4} &= 4 \\ x_1 \phantom{{}+x_2} + x_3 + x_4 &= 5 \\ 2x_1 + x_2 + 3x_3 + x_4 &= 9 \end{aligned}$$

10.
$$\begin{aligned} x_1 + 2x_2 \phantom{{}+x_3} + x_4 &= 3 \\ x_3 + 3x_4 &= 2 \\ -x_1 - 2x_2 + x_3 + 2x_4 &= -1 \end{aligned}$$

12.
$$\begin{aligned} x_1 + x_2 + x_3 &= 2 \\ x_1 - 2x_2 + 3x_3 &= -5 \\ 3x_1 + 5x_3 &= -1 \end{aligned}$$

In Exercises 13-15, assume that the $m \times (n+1)$ matrix $B = [A|H]$ is the augmented matrix of a system of linear equations and assume that B is in reduced echelon form with constants $r, j_1, \ldots, j_r$.

13. Give an example of a matrix B, as above, such that the system of equations is inconsistent; that is, has no solution.

14. Find conditions on B that will guarantee that the system of equations is consistent.

15. Assuming that the system of equations with augmented matrix B is consistent, how many of the variables may be arbitrarily chosen?

1.6 Homogeneous Systems of Equations and Solution Sets

In this section we will investigate systems of linear equations in which the constant terms are all equal to 0. In the form of a matrix equation the system would look like: $AX = 0$. Such a system of linear equations is said to be **homogeneous**. The system of equations $AX = H$ with $H \neq 0$ is said to be **nonhomogeneous**. We will see in Theorem 1.6.2 below that homogeneous systems are important in that the solution set of a nonhomogeneous system is largely determined by the solution set of the "associated homogeneous system."

For example, the matrix equation

$$\begin{bmatrix} 2 & 1 \\ -1 & 3 \end{bmatrix} \begin{bmatrix} x \\ y \end{bmatrix} = \begin{bmatrix} 0 \\ 0 \end{bmatrix},$$

which is equivalent to the system of equations

$$\begin{aligned} 2x + y &= 0 \\ -x + 3y &= 0, \end{aligned}$$

is homogeneous.

Now consider the homogeneous system of equations $AX = 0$, where A is some $m \times n$ matrix, and let S be the solution set. Unlike for nonhomogeneous systems, S is never the empty set because 0, the zero vector, is always a solution. Furthermore, if $X_1, X_2 \in S$, then $AX_1 = 0$ and $AX_2 = 0$, and so $A(X_1 + X_2) = AX_1 + AX_2 = 0 + 0 = 0$. Thus $X_1 + X_2 \in S$. Also, if $X \in S$ and r is any scalar, then $A(rX) = r(AX) = r(0) = 0$ and so $rX \in S$. Thus, we have proved the following theorem.

> **Theorem 1.6.1.** *Let S be the solution set of the homogeneous system of equations $AX = 0$. Then a) If $X_1, X_2 \in S$ then $X_1 + X_2 \in S$.*
> (The solution set of a homogeneous system is closed under addition.)
> *b) If $X \in S$ and r is any scalar then $rX \in S$.*
> (The solution set of a homogeneous system is closed under scalar multiplication.)
> *c) $0 \in S$.*   (The zero vector is a solution of a homogeneous system.)

Sets of vectors having the properties in Theorem 1.6.1 above are defined in Chapter 2 and given the name "vector spaces." The relationship between a system of equations $AX = H$ and its associated homogeneous system $AX = 0$ is very close, as the following theorem shows.

> **Theorem 1.6.2.** *Let X_p be some solution of $AX = H$. Then any solution X of $AX = H$ can be expressed in the form $X = X_p + X_h$, where X_h is a solution of the associated homogeneous equation $AX = 0$. Furthermore, any vector $X = X_p + X_h$, where $AX_h = 0$, is a solution of $AX = H$.*   (Any solution is of the form $X_p + X_h$.)

Proof. We'll do the second part first. Let $X = X_p + X_h$ with $AX_h = 0$. Then $AX = A(X_p + X_h) = AX_p + AX_h = H + 0 = H$, and so $X_p + X_h$ is a solution of $AX = H$.

Now let X be any solution of $AX = H$. Define $X_h = X + (-1)X_p$. Then $X = X_p + X_h$ and X_h is a solution of $AX = 0$ since $AX_h = AX + (-1)AX_p = H + (-1)H = 0$. It follows that the solution set S of $AX = H$ is given by $S = \{X_p + X_h | AX_h = 0\}$. $\square$

Theorem 1.6.2 says that it is important to understand or be able to characterize solution sets of homogeneous systems of equations and Theorem 1.6.1 tells us some of the properties of these solution sets. For this and other reasons, we will study the structure of these solution sets, or vector spaces, in the next chapter.

Example 1.6.1.   (a) Let us first consider the very simple system of equations $x + y = 5$. There is only one equation and we may picture the solution set as ordered pairs of real numbers lying in the xy-plane. The associated homogeneous equation is $x + y = 0$ and its solution set is the line through the origin as in Figure 1.

If we choose a particular solution of the nonhomogeneous equation, say $X_p = (1, 4)$, then the solution set of the original equation is the "translate" of the solution set of the associated homogeneous equation by this vector.

(b) We will next illustrate Theorem 1.6.2 with the example solved in Section 1.5. Recall that in Example 1.5.3, the general solution was given by $x_1 = 1 - x_3 - x_5, x_2 = 1 - 2x_3 - 2x_5, x_4 = 1 - x_5$, where x_3 and x_5 are arbitrary real numbers. If we let $x_3 = x_5 = 0$, then we obtain the specific solution $X_p = (1, 1, 0, 1, 0)^t$. It is not hard to see that had we solved the associated homogeneous equation $AX = 0$, we would have obtained the general solution

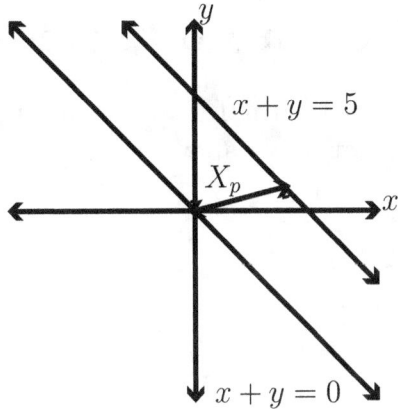

Figure 1.5: Solution sets of Equation and Associated Homogeneous Equation

$$
\begin{aligned}
x_1 &= - x_3 - x_5 \\
x_2 &= - 2x_3 - 2x_5 \\
x_4 &= - x_5
\end{aligned}
$$

with x_3, x_5 arbitrary.

If we call this solution X_h, we see that

$$X_h = (-x_3 - x_5, -2x_3 - 2x_5, x_3, -x_5, x_5)^t,$$

and so the general solution is given by

$$X = (1 - x_3 - x_5, 1 - 2x_3 - 2x_5, x_3, 1 - x_5, x_5)^t = X_p + X_h.$$

This expression for the general solution can be further expanded. We can obtain two specific solutions of the associated homogeneous equation $AX = 0$, one by choosing $x_3 = 1, x_5 = 0$ and the other by choosing $x_3 = 0$ and $x_5 = 1$. Call them X_1 and X_2. Then $X_1 = (-1, -2, 1, 0, 0)^t$ and $X_2 = (-1, -2, 0, -1, 1)^t$ and the general solution of $AX = 0$ can be written in the form $X_h = aX_1 + bX_2 = (-a - b, -2a - 2b, a, -b, b)^t$ (where a and b represent arbitrary constants. In fact, $a = x_3$ and $b = x_5$). The general solution of the nonhomogeneous equation $AX = H$ can then be written in the form

$$X = X_p + X_h = (1, 1, 0, 1, 0)^t + a(-1, -2, 1, 0, 0)^t + b(-1, -2, 0, -1, 1)^t.$$

Notice that the vectors X_1 and X_2 are "linearly independent;" a "linear combination" $aX_1 + bX_2$ of these vectors cannot be the zero vector unless both of the scalars a and b are zero. In Chapter 2, the set of all of these linear combinations $aX_1 + bX_2$ will be called the "span" of the vectors X_1 and X_2.

$\square$

Section 1.6 Exercises

In Exercises 1-6, find the general solution of each system of equations and then express it in the form $X = X_p + X_h$ as in the example above.

1. $\begin{aligned} x_1 + x_2 + x_3 &= 1 \\ x_2 - x_3 &= 2 \end{aligned}$

4. $\begin{aligned} x_1 + x_2 \qquad\quad + x_4 &= 2 \\ x_1 + x_2 + x_3 - 2x_4 &= 1 \\ x_1 + x_2 + 3x_3 \qquad\quad &= 4 \end{aligned}$

2. $\begin{aligned} x_1 \qquad\quad - x_3 &= 3 \\ x_2 + x_3 &= 6 \end{aligned}$

5. $\begin{aligned} x_1 + 2x_2 - x_3 &= -1 \\ -2x_1 + x_2 - 3x_3 &= 2 \\ -x_1 - x_2 + x_3 &= -1 \end{aligned}$

3. $\begin{aligned} x_1 \qquad\quad + x_3 &= 1 \\ x_2 + 2x_3 &= 3 \end{aligned}$

6. $\begin{aligned} x_1 + x_2 - x_3 - x_4 &= 4 \\ 2x_1 - x_2 \qquad\quad + x_4 &= 3 \\ -x_1 + x_2 - x_3 + 2x_4 &= 2 \end{aligned}$

In Exercises 7-12, express the general solution of the given homogeneous systems of equations as a sum of scalar multiples of fixed vectors as in the example above; that is, write

$$X_h = aX_1 + bX_2 + \dots$$

Note that the systems given below are the associated homogeneous systems of equations for the systems in Exercises 1-6 above.

7. $\begin{aligned} x_1 + x_2 + x_3 &= 0 \\ x_2 - x_3 &= 0 \end{aligned}$

10. $\begin{aligned} x_1 + x_2 \qquad\quad + x_4 &= 0 \\ x_1 + x_2 + x_3 - 2x_4 &= 0 \\ x_1 + x_2 + 3x_3 \qquad\quad &= 0 \end{aligned}$

8. $\begin{aligned} x_1 \qquad\quad - x_3 &= 0 \\ x_2 + x_3 &= 0 \end{aligned}$

11. $\begin{aligned} x_1 + 2x_2 - x_3 &= 0 \\ -2x_1 + x_2 - 3x_3 &= 0 \\ -x_1 - x_2 + x_3 &= 0 \end{aligned}$

9. $\begin{aligned} x_1 \qquad\quad + x_3 &= 0 \\ x_2 + 2x_3 &= 0 \end{aligned}$

12. $\begin{aligned} x_1 + x_2 - x_3 - x_4 &= 0 \\ 2x_1 - x_2 \qquad\quad + x_4 &= 0 \\ -x_1 + x_2 - x_3 + 2x_4 &= 0 \end{aligned}$

13. Let $B = [A|H]$ be the augmented matrix of a system of linear equations, and assume that B is in reduced echelon form. Prove that

    (a) A is in reduced echelon form.

    (b) $[A|0]$ is in reduced echelon form.

14. Let A be an $m \times n$ matrix. Assume that X_1 and X_2 are solutions of the homogeneous system of equations $AX = 0$, and let a_1 and a_2 be scalars. Prove that $X = a_1X_1 + a_2X_2$ is also a solution of $AX = 0$.

15. Let B be a 3×3 matrix in reduced echelon form with associated constants $r, j_1, \ldots, j_r$. Consider the homogeneous system $BX = 0$. If $r = 1$, then B is of the form

$$\begin{bmatrix} b_1 & b_2 & b_3 \\ 0 & 0 & 0 \\ 0 & 0 & 0 \end{bmatrix}$$

and so the system $BX = 0$ reduces to $b_1 x + b_2 y + b_3 z = 0$. This is the equation of a plane through the origin in 3-space. Describe geometrically the situation when

(a) $r = 0$,

(b) $r = 2$,

(c) $r = 3$.

1.7 The LU-Factorization (optional)

The use of computers gives rise to special problems. Computer scientists study the efficiency of various algorithms or methods and they are concerned most often about both "space" and "time". "Space" in the sense of how much of the computer's memory is required in the solution of the problem, and "time" in how much computer time it takes to solve the problem. The issue of time is usually studied by estimating the number of arithmetic operations required in the solution of the problem, and this is dealt with in Section 4.5. This general area of study is called the "analysis of algorithms."

In solving the system of equations in the matrix form $AX = H$, we formed the augmented matrix $[A|H]$ and reduced it to reduced echelon form using the Gauss-Jordan reduction. While this method works well for some hand computations, it is not always the best. Suppose that systems of the form $AX = H$ must be solved for several values of H. It is wasteful to perform the reduction of A each time. In this case, one could record the row operations used and then apply these operations to each of the H's thereby avoiding the computations on A for each of the solutions.

Considering space-time issues, what should be done about the problem of solving $AX = H$ for several values of H? If one records the row operations used, these operations may be applied to each of the H's. That is, if $P[A|H]$ is the reduced echelon form of the augmented matrix $[A|H]$ (the matrix P is the product of the elementary matrices corresponding to the row operations used in the reduction to reduced echelon form), then one needs to compute PA once and PH for each value of H. This is more efficient in terms of the number of arithmetic operations needed, but space could become a problem for large systems since we must store not only the matrices A and H, but also the matrix P.

There are further issues that arise when one tries to solve a system of equations using a computer. Computers do not do exact arithmetic. The number $2/3$ may be stored in a computer in a form that is related to the approximation 0.66666667. Because of this fact, roundoff error can result. The study of these and other problems is the subject of the general area of "numerical analysis" which is a part of mathematics and computer science. We cannot deal with these important areas and issues here - they should be the subject of further course work. Until then, it will suffice to remember that when solving a system of linear equations using a computer program or solving the system using a calculator where round off might occur, small errors in the coefficients can produce large errors in the solution.

While the Gauss-Jordan reduction method in theory finds the general solution of every system of m equations in n unknowns that has a solution, some systems resulting from physical situations (such as electrical networks obeying Kirchhoff's Laws) have a unique (only one) solution and are systems of n equations in n unknowns. For such systems, other methods of solution are better than the Gauss-Jordan method. Accuracy in the solution and efficiency in finding the solution are more important than generality in the method of solution.

We will describe here one of the tools which is used in the real-world solution of systems of equations- that is the factorization of a matrix A into a product LU of a "lower triangular" matrix L and an "upper triangluar matrix" U.

1.7.1   TRIANGULAR MATRICES

The efficiency gained in the process that we are about to present comes from the factorization of a matrix into the product of two "triangular" matrices. One of these matrices will have entries on or above the diagonal, and the other will have entries on or below the diagonal. So, except for the overlap on the diagonal, the two matrices will fit into the space of one matrix and save on storage space.

An $m \times n$ matrix $A = [a_{ij}]$ is called **upper triangular** if $a_{ij} = 0$ when $i > j$ and A is **lower triangular** if $a_{ij} = 0$ for $i < j$. We see that an upper triangular $n \times n$ matrix looks like

$$\begin{bmatrix} * & * & \ldots & * \\ 0 & * & \ldots & * \\ \vdots & \vdots & \ddots & \vdots \\ 0 & 0 & \ldots 0 & * \end{bmatrix}$$

and a lower triangular $n \times n$ matrix looks like

$$\begin{bmatrix} * & 0 & \ldots & 0 \\ * & * & \ldots & 0 \\ \vdots & \vdots & \ddots & \vdots \\ * & * & \ldots & * \end{bmatrix},$$

where the $*$'s denote possibly nonzero entries. If a matrix is either lower or upper triangular, we will say that it is in **triangular form**. In an $m \times n$ matrix $[a_{ij}]$, the entries a_{ij} with $i = j$ are called **diagonal entries** and the collection of all of them $\{a_{11}, a_{22}, \ldots\}$ is called the **diagonal**.

Notice that for 3×3 matrices,

$$R_{13}(a) = \begin{bmatrix} 1 & 0 & 0 \\ 0 & 1 & 0 \\ a & 0 & 1 \end{bmatrix} \text{ and } R_{31}(a) = \begin{bmatrix} 1 & 0 & a \\ 0 & 1 & 0 \\ 0 & 0 & 1 \end{bmatrix}.$$

If $i < k$ the elementary matrix $R_{ik}(a)$ is lower triangular and likewise if $i > k$, $R_{ik}(a)$ is upper triangular. In addition, these matrices have all 1's on the diagonal. Further, for 3×3 matrices,

$$R_2(a) = \begin{bmatrix} 1 & 0 & 0 \\ 0 & a & 0 \\ 0 & 0 & 1 \end{bmatrix},$$

so this matrix is both lower and upper triangular.

We will need some general results concerning triangular matrices. Since similar results hold for both lower and upper triangular matrices, we will state both results in a single statement by putting alternate assumptions in parentheses. Using this convention, the statement "the inverse of a lower (upper) triangular matrix is lower (upper) triangular" represents the two results: "the inverse of a lower triangular matrix is lower triangular" and "the inverse of an upper triangular matrix is upper triangular".

Theorem 1.7.1. *(a) A product of lower (upper) triangular matrices is lower (upper) triangular.*

(b) The diagonal entries in a product of two lower (upper) triangular matrices are the products of the corresponding diagonal entries of the two factors. So, a product of two lower (upper) triangular matrices with all ones on the diagonal has all ones on the diagonal.

(c) Each of the elementary matrices $R_i(a)$ is both lower and upper triangular, and if $i < k$ ($i > k$) then $R_{ik}(a)$ is lower (upper) triangular.

(d) If an $n \times n$ lower (upper) triangular matrix has an inverse, then it is lower (upper) triangular.

Proof. See Exercise 11. □

Let $A = [a_{ij}]$ be an $m \times n$ matrix and assume that a certain entry a_{ij} is nonzero. If $k > i$ (k denotes a row lying below row i), then performing the row operation $R_{ik}(-a_{kj}/a_{ij})$ results in a matrix with a 0 in row k and column j. That is, the matrix $R_{ik}(-a_{kj}/a_{ij})A$ has a zero in row k and column j. Notice that $R_{ik}(-a_{kj}/a_{ij})$ is a lower triangular matrix with ones on the diagonal. While it might seem silly to dignify this simple observation in the form of a theorem, the result is important and so we will do just that.

Theorem 1.7.2. *Let $A = [a_{ij}]$ be an $m \times n$ matrix and assume that $a_{ij} \neq 0$ for some i, j. Then if $k > i$, there is a lower triangular matrix L, with ones on the diagonal, such that the matrix LA has a zero in row k and column j.*

(We can make a zero below a nonzero matrix entry with a row operation corresponding to a lower triangular matrix.)

We can apply the process of the theorem above to reduce a matrix to an upper triangular matrix, provided that no zeros are encountered along the way. For example, consider the matrix

$$A = \begin{bmatrix} 2 & 1 & 3 \\ 4 & 1 & 5 \\ 6 & 4 & 7 \end{bmatrix}.$$

If we perform the row operation $R_{12}(-2)$ on the matrix A we obtain

$$\begin{bmatrix} 2 & 1 & 3 \\ 0 & -1 & -1 \\ 6 & 4 & 7 \end{bmatrix}$$

and so this matrix is $R_{12}(-2)A$ or

$$\begin{bmatrix} 2 & 1 & 3 \\ 0 & -1 & -1 \\ 6 & 4 & 7 \end{bmatrix} = \begin{bmatrix} 1 & 0 & 0 \\ -2 & 1 & 0 \\ 0 & 0 & 1 \end{bmatrix} \begin{bmatrix} 2 & 1 & 3 \\ 4 & 1 & 5 \\ 6 & 4 & 7 \end{bmatrix}.$$

Continuing, we next perform $R_{13}(-3)$ and obtain

$$\begin{bmatrix} 2 & 1 & 3 \\ 0 & -1 & -1 \\ 0 & 1 & -2 \end{bmatrix},$$

and finally we perform $R_{23}(1)$, obtaining

$$\begin{bmatrix} 2 & 1 & 3 \\ 0 & -1 & -1 \\ 0 & 0 & -3 \end{bmatrix}.$$

We see that this last matrix is the original matrix A times the elementary matrices that correspond to the row operations that we performed:

$$\begin{bmatrix} 2 & 1 & 3 \\ 0 & -1 & -1 \\ 0 & 0 & -3 \end{bmatrix} = R_{23}(1)R_{13}(-3)R_{12}(-2)A$$

$$= \begin{bmatrix} 1 & 0 & 0 \\ 0 & 1 & 0 \\ 0 & 1 & 1 \end{bmatrix} \begin{bmatrix} 1 & 0 & 0 \\ 0 & 1 & 0 \\ -3 & 0 & 1 \end{bmatrix} \begin{bmatrix} 1 & 0 & 0 \\ -2 & 1 & 0 \\ 0 & 0 & 1 \end{bmatrix} \begin{bmatrix} 2 & 1 & 3 \\ 4 & 1 & 5 \\ 6 & 4 & 7 \end{bmatrix}$$

$$= \begin{bmatrix} 1 & 0 & 0 \\ -2 & 1 & 0 \\ -5 & 1 & 1 \end{bmatrix} \begin{bmatrix} 2 & 1 & 3 \\ 4 & 1 & 5 \\ 6 & 4 & 7 \end{bmatrix}.$$

Thus, we have reduced A to an upper triangular matrix $U = KA$, where

$$K = R_{23}(1)R_{13}(-3)R_{12}(-2) = \begin{bmatrix} 1 & 0 & 0 \\ -2 & 1 & 0 \\ -5 & 1 & 1 \end{bmatrix} \text{ and } U = \begin{bmatrix} 2 & 1 & 3 \\ 0 & -1 & -1 \\ 0 & 0 & -3 \end{bmatrix}.$$

We let $L = K^{-1}$ and, solving for A, we obtain $A = K^{-1}U = LU$. Notice that L is also a lower triangular matrix with 1's on the diagonal; in fact,

$$L = \begin{bmatrix} 1 & 0 & 0 \\ 2 & 1 & 0 \\ 3 & -1 & 1 \end{bmatrix}.$$

Finally, we have A factored as

$$A = LU = \begin{bmatrix} 1 & 0 & 0 \\ 2 & 1 & 0 \\ 3 & -1 & 1 \end{bmatrix} \begin{bmatrix} 2 & 1 & 3 \\ 0 & -1 & -1 \\ 0 & 0 & -3 \end{bmatrix}.$$

We can see that this process will work as long as no zeros are encountered along the diagonal in the reduction process. So we may conclude the following:

Theorem 1.7.3. *Let A be an $m \times n$ matrix. If no zeros are encountered in the reduction process, then A can be factored in the form $A = LU$ where L is a lower triangular $m \times m$ matrix with 1's on the diagonal and U is an upper triangular $m \times n$ matrix.* *(Certain matrices can be factored as a lower times an upper.)*

The factorization described in the above theorem is called the **LU-factorization**. Notice that this factorization is essentially the same as the Gaussian elimination process if no row interchanges are needed. Notice further that it does not always work! The matrix $A = \begin{bmatrix} 0 & 1 \\ 1 & 0 \end{bmatrix}$ cannot be factored in this form. To see this, assume that $A = LU$ with L lower triangular and with 1's on the diagonal. Then $L^{-1}A = U$, but $L^{-1}A$ is the matrix L^{-1} with columns 1 and 2 interchanged. Since L^{-1} is lower triangular, $L^{-1}A$ cannot be upper triangular and so cannot be equal to U.

Corollary 1.7.4. *If A is an $n \times n$ lower triangular matrix and the diagonal entries of A are nonzero, then A has an inverse which is also lower triangular and the diagonal entries of A^{-1} are the reciprocals of the corresponding entries in A.* *(The inverse of a lower is a lower.)*

Proof. See Exercise 12. □

How can the LU-decomposition be of help? Suppose $A = LU$ with L lower triangular and U upper triangular. Because L is lower triangular, it is easy to solve a system of the form $LY = H$ by "forward" substitution. It is then similarly easy to solve the system $UX = Y$ and then we have the desired solution: $AX = LUX = L(UX) = LY = H$. So, once L and U are found, systems of the form $AX = H$ may be easily solved for any column vector H.

Example 1.7.1. Suppose that we wish to solve the system of equations $AX = H$, where

$$A = \begin{bmatrix} 2 & 1 & 3 \\ 4 & 1 & 5 \\ 6 & 4 & 7 \end{bmatrix} \text{ and } H = \begin{bmatrix} 3 \\ 7 \\ 5 \end{bmatrix}.$$

A is the matrix in the example above and we found that

$$A = LU = \begin{bmatrix} 1 & 0 & 0 \\ 2 & 1 & 0 \\ 3 & -1 & 1 \end{bmatrix} \begin{bmatrix} 2 & 1 & 3 \\ 0 & -1 & -1 \\ 0 & 0 & -3 \end{bmatrix}.$$

We first solve $LY = H$, or

$$\begin{bmatrix} 1 & 0 & 0 \\ 2 & 1 & 0 \\ 3 & -1 & 1 \end{bmatrix} \begin{bmatrix} y_1 \\ y_2 \\ y_3 \end{bmatrix} = \begin{bmatrix} 3 \\ 7 \\ 5 \end{bmatrix}.$$

We see that $y_1 = 3$, $y_2 = 7 - 2y_1 = 7 - 6 = 1$, and $y_3 = 5 - 3y_1 + y_2 = 5 - 9 + 1 = -3$. Next we solve $UX = Y$, or

$$\begin{bmatrix} 2 & 1 & 3 \\ 0 & -1 & -1 \\ 0 & 0 & -3 \end{bmatrix} \begin{bmatrix} x_1 \\ x_2 \\ x_3 \end{bmatrix} = \begin{bmatrix} 3 \\ 1 \\ -3 \end{bmatrix}.$$

We see now that $-3x_3 = -3$, so $x_3 = 1$ and $-x_2 = 1 + x_3$; thus, $x_2 = -2$ and $2x_1 = 3 - x_2 - 3x_3$, giving $x_1 = 1$.

$\square$

From the above comments we can see that the problem of efficiency from the point of view of time has been addressed, but the problem of space is also a concern. It appears that we have made matters worse by factoring A as LU. A is $m \times n$, L is $m \times m$, and U is $m \times n$, and so it appears that we must store an additional $m \times m$ matrix. However, since L has 1's on the diagonal and 0's above the diagonal, the entries below the diagonal in L may be stored in the matrix U, for U has only 0's below the diagonal. We see then that all of the essential data can be stored in a single $m \times n$ matrix. For example, with L and U as above, we can collapse L into U as pictured below:

$$\begin{bmatrix} 1 & 0 & 0 \\ 2 & 1 & 0 \\ 3 & -1 & 1 \end{bmatrix} \longrightarrow \begin{bmatrix} 2 & 1 & 3 \\ 0 & -1 & -1 \\ 0 & 0 & -3 \end{bmatrix}, \text{ giving } \begin{bmatrix} 2 & 1 & 3 \\ 2 & -1 & -1 \\ 3 & -1 & -3 \end{bmatrix}.$$

Section 1.7 Exercises

For each matrix A in Exercises 1 - 6, find a lower triangular matrix L and an upper triangular matrix U with $A = LU$.

1. $\begin{bmatrix} 2 & 1 \\ 3 & 2 \end{bmatrix}$

3. $\begin{bmatrix} 2 & 1 & 3 \\ 4 & 2 & 5 \\ 2 & 1 & 6 \end{bmatrix}$

5. $\begin{bmatrix} 2 & 1 & 3 & 4 \\ 2 & 5 & 3 & 1 \\ 6 & 3 & 4 & 2 \end{bmatrix}$

2. $\begin{bmatrix} -1 & 2 \\ 1 & 3 \end{bmatrix}$

4. $\begin{bmatrix} 1 & 0 & 2 \\ 2 & -1 & 5 \\ 1 & 2 & 1 \end{bmatrix}$

6. $\begin{bmatrix} 1 & 0 & 2 & 2 \\ -3 & 5 & 1 & 2 \end{bmatrix}$

7. Suppose that $A = LU$ and that L and U are stored together as described above. If the matrix storing L and U together is

$$\begin{bmatrix} 2 & -1 & 0 & 2 \\ 1 & 2 & -1 & 3 \\ 1 & -2 & 1 & 3 \\ 2 & 1 & 3 & 4 \end{bmatrix},$$

find L and U, and then determine the matrix A.

In Exercises 8 - 10, assume that $A = LU$, where L and U are as given, then solve the system $AX = H$ for the given value of H, using the method illustrated above.

8. $L = \begin{bmatrix} 1 & 0 \\ 2 & 1 \end{bmatrix}, U = \begin{bmatrix} 2 & 3 \\ 0 & 2 \end{bmatrix}, H = \begin{bmatrix} 2 \\ 1 \end{bmatrix}$

9. $L = \begin{bmatrix} 1 & 0 \\ -3 & 1 \end{bmatrix}, U = \begin{bmatrix} -1 & 3 \\ 0 & 2 \end{bmatrix}, H = \begin{bmatrix} -1 \\ 2 \end{bmatrix}$

10. $L = \begin{bmatrix} 1 & 0 & 0 \\ -3 & 1 & 0 \\ 2 & 1 & 1 \end{bmatrix}, U = \begin{bmatrix} -1 & 2 & 4 \\ 0 & 1 & 2 \\ 0 & 0 & 1 \end{bmatrix}, H = \begin{bmatrix} 1 \\ -2 \\ 3 \end{bmatrix}$

11. Prove Theorem 1.7.1. [Hint: consider our alternative perspective on matrix multiplication.]

12. Prove Corollary 1.7.4.

1.8 Applications - Kirchhoff's Laws (Optional)

In Section 1.1 we introduced an example of an electrical network obeying Kirchhoff's Laws. These networks are studied in beginning and intermediate level physics courses. In certain elementary cases, these problems give rise to systems of linear equations, and the methods developed in this chapter can be used to solve these systems, and so, to calculate the currents flowing in the given network. In this section, we will give a brief introduction to this area of physics.

Electrical networks made up of conductors (wires), resistors, capacitors, inductors, and power supplies obey two fundamental laws known as **Kirchhoff's laws**:

1. The algebraic sum of voltage changes around any loop in a network is zero.

2. The algebraic sum of currents traveling into and out of any branch point (or node) in a network is zero.

In these laws, the term "**algebraic**" means that the sign (+ or -) of the current or voltage drop must be considered. As examples will show, this sign arises from a chosen orientation in the loops of the network.

It is useful to make an analogy between the flow of electricity and the flow of a fluid-the charge is the amount of fluid, the voltage is the amount of pressure, the current is the rate of flow of the fluid, and the resistance is a force opposing the flow of the fluid (an obstruction in the pipe, perhaps).

The following notation is commonly used:

Symbol	Quantity Represented	Units
E	Voltage	volts
I	current	amps
R	resistance	ohms
L	inductance	henrys
C	capacitance	farads
Q	charge	coulombs

We have the following relationships:

$$
\begin{aligned}
I &= dQ/dt && \text{(the current is the rate of change of charge)}\\
E_R &= IR && \text{(the voltage drop across a resistor-Ohm's Law)}\\
E_L &= LdI/dt && \text{(the voltage drop across a inductor)}\\
E_C &= Q/C && \text{(the voltage drop across a capacitor)}
\end{aligned}
$$

For a given network, Kirchhoff's laws give a system of equations that relate the current flow in each part of the circuit. If only resistors and power supplies are involved in the circuit, the current I may be taken as the fundamental variable and the resulting system of equations will be linear and involve no derivatives. If capacitors, inductors, and resistors are all involved, the result will be a system of differential equations - some equations possibly of second order. Since we have not yet considered systems of differential equations, we will consider only networks involving resistors and constant voltage sources in this section.

Example 1.8.1.   (a) Consider the network in Figure 1. Assume that

$$
\begin{aligned}
R_1 &= 6 \text{ohms},\\
R_2 &= 3 \text{ohms, and}\\
V &= 6 \text{volts}.
\end{aligned}
$$

An orientation has been chosen for each of the loops 1 and 2. Applying Kirchhoff's first law, we get

$$
\begin{array}{llll}
\text{Loop 1:} & V - I_2 R_1 = 0 & \text{or} & 6 - 6I_2 = 0\\
\text{Loop 2:} & I_2 R_1 - I_3 R_2 = 0 & \text{or} & -6I_2 + 3I_3 = 0.
\end{array}
$$

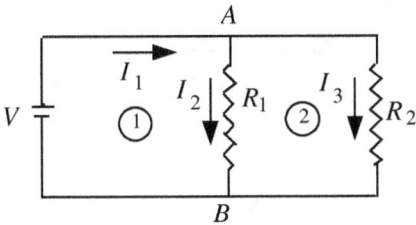

Figure 1.6: Circuit for (a)

Applying Kirchhoff's second law at point A yields

$$I_1 - I_2 - I_3 = 0.$$

(Point B gives essentially the same equation.)

We obtain the following system of linear equations:

$$\begin{array}{rcrcrcr}
6I_2 & & & = & 6 \\
-6I_2 & + & 3I_3 & = & 0 \\
I_1 & - & I_2 & - & I_3 & = & 0.
\end{array}$$

Solving, we get $I_1 = 1$ amp, $I_2 = 2$ amps, $I_3 = 3$ amps.

(b) Consider the circuit in Figure 2 below. We will make the following assumptions regarding the resistances and the voltage of the one power supply:

$$\begin{array}{rcl}
R_1 & = & 1000 \text{ ohms} \\
R_2 & = & 500 \text{ ohms} \\
R_3 & = & 750 \text{ ohms} \\
R_4 & = & 1000 \text{ ohms} \\
R_5 & = & 2000 \text{ ohms} \\
R_6 & = & 500 \text{ ohms} \\
V & = & 10 \text{ volts.}
\end{array}$$

Applying Kirchhoff's laws as in part (a), we get the following:

$$\begin{array}{llllllllllllll}
\text{Point A:} & I_1 & - & I_2 & - & I_3 & & & & & & & & = & 0 \\
\text{Point B:} & & & I_2 & & & + & I_4 & - & I_5 & & & & = & 0 \\
\text{Point C:} & & & & & I_3 & - & I_4 & & & - & I_6 & & = & 0 \\
\end{array}$$

$$\begin{array}{llllllllllllll}
\text{Loop 1:} & -I_1R_1 & - & I_2R_2 & & & & & - & I_5R_5 & & & + & V & = & 0 \\
\text{Loop 2:} & & & I_2R_2 & - & I_3R_3 & - & I_4R_4 & & & & & & & = & 0 \\
\text{Loop 3:} & & & & & & & I_4R_4 & + & I_5R_5 & - & I_6R_6 & & & = & 0.
\end{array}$$

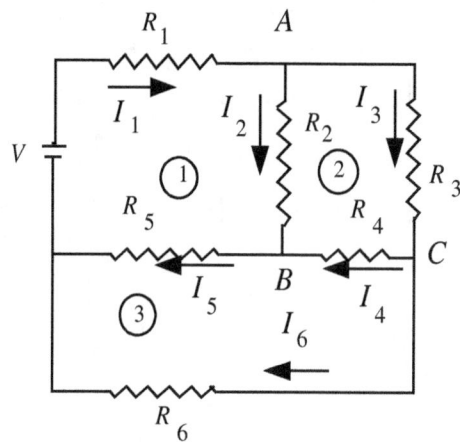

Figure 1.7: Circuit for (b)

Substituting the values for the resistances $R_1, \ldots, R_6$, rearranging, and writing the augmented matrix of this system of equations, we get

$$
\begin{bmatrix}
1 & -1 & -1 & 0 & 0 & 0 & 0 \\
0 & 1 & 0 & 1 & -1 & 0 & 0 \\
0 & 0 & 1 & -1 & 0 & -1 & 0 \\
-1000 & -500 & 0 & 0 & -2000 & 0 & -10 \\
0 & 500 & -750 & -1000 & 0 & 0 & 0 \\
0 & 0 & 0 & 1000 & 2000 & -500 & 0
\end{bmatrix}.
$$

It would be a chore to reduce this matrix to reduced echelon form and thereby solve the system of equations. We use a computer program to eliminate this drudgery and obtain the reduced echelon form below. We see that in the reduced echelon form, that the coefficient matrix has transformed into the identity matrix. It follows that the solution is unique and that the resulting currents $I_1, \ldots, I_6$ are given in the last column. The values have been expressed in scientific notation. Rounding to the nearest hundredth of a milliamp, we get: $I_1 = 5.64$ mA, $I_2 = 2.56$ mA, $I_3 = 3.08$ mA, $I_4 = -1.03$ mA, $I_5 = 1.54 mA$, and $I_6 = 4.10$ mA. (Note: mA is an abbreviation for milliampere(s), or one one-thousandth of an ampere.)

$$
\begin{bmatrix}
1 & 0 & 0 & 0 & 0 & 0 & 5.6410259E-03 \\
0 & 1 & 0 & 0 & 0 & 0 & 2.5641026E-03 \\
0 & 0 & 1 & 0 & 0 & 0 & 3.0769231E-03 \\
0 & 0 & 0 & 1 & 0 & 0 & -1.0256411E-03 \\
0 & 0 & 0 & 0 & 1 & 0 & 1.5384616E-03 \\
0 & 0 & 0 & 0 & 0 & 1 & 4.1025641E-03
\end{bmatrix}
$$

The last column gives the currents $I_1, \ldots, I_6$. we conclude that $I_1 = 5.64$ mA, $I_2 = 2.56$ mA, etc.

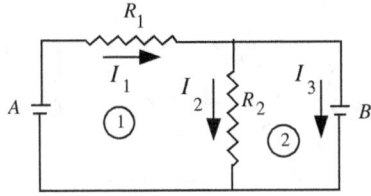

Figure 1.8: Circuit for Problem 5

□

Section 1.8 Exercises

1. In the network in Figure 1, assume $R_1 = 9$ ohms, $R_2 = 2$ ohms and $V = 10$ volts. Calculate the currents in the circuit.

2. In the network in Figure 1, assume $R_1 = 10$ ohms, $R_2 = 2$ ohms and $V = 5$ volts. Calculate the currents in the circuit.

3. In the network in Figure 2, assume that

$$
\begin{aligned}
R_1 &= 400 \text{ ohms} \\
R_2 &= 900 \text{ ohms} \\
R_3 &= 850 \text{ ohms} \\
R_4 &= 2000 \text{ ohms} \\
R_5 &= 1500 \text{ ohms} \\
R_6 &= 1500 \text{ ohms} \\
V &= 100 \text{ volts.}
\end{aligned}
$$

Write the system of equations that result from applying Kirchhoff's laws to the network.

4. In the network in Figure 2, change R_1 to 2000 ohms, leaving the other variables as in the example, and calculate the currents in the resulting circuit.

5. Find the voltages of the power cells A and B in the circuit in Figure 5, where $R_1 = 3$ ohms, $R_2 = 10$ ohms, $I_1 = 5$ amps, and $I_2 = 3$ amps.

6. Find the currents in the circuit in Figure 6, assuming that $R_1 = 10$ ohms, $R_2 = 5$ ohms, $R_3 = 15$ ohms, $R_4 = 5$ ohms, and $V = 10$ volts.

7. Find the currents in the circuit in Figure 6, assuming that $R_1 = 10$ ohms, $R_2 = 50$ ohms, $R_3 = 75$ ohms, $R_4 = 100$ ohms, and $V = 100$ volts.

8. Find the currents in the circuit in Figure 6, assuming that $R_1 = 20$ ohms, $R_2 = 40$ ohms, $R_3 = 80$ ohms, $R_4 = 10$ ohms, and $V = 50$ volts.

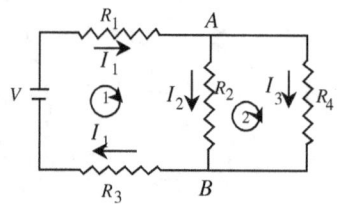

Figure 1.9: Circuit for Problem 6

1.9   Applications - Stochastic Matrices (optional)

In courses in probability, Markov chains and stochastic matrices are studied, and since they give an elementary, but very nice, application of matrices and matrix multiplication, we will present here a brief outline of the topic. The method of Markov chains and stochastic matrices gives a way of predicting future trends, given certain assumptions and known probabilities.

Assume that a certain collection $s_1, \ldots, s_n$ of **states** or **conditions** exists and that all individuals in a certain group must fall into one of these states. For example, with $n = 2$ the states may be $s_1 =$ "employed" and $s_2 =$ "unemployed," or with $n = 3$, the conditions might be $s_1 =$ "Republican," $s_2 =$ "Democrat," and $s_3 =$ "Independent." We will pursue the latter example. At a given moment, a certain proportion of all individuals satisfy the given state or condition, giving rise to a **probability** or **distribution vector**

$$X = \begin{bmatrix} x_1 \\ \vdots \\ x_n \end{bmatrix}, \tag{1.9.1}$$

where x_i is the proportion of all individuals in state s_i. Note that x_i is also the probability of an individual being in state s_i.

Now, if observations are made at discrete intervals (once a minute, once a day, once a generation, etc.), a distribution vector arises with each observation. Let these vectors be $X_1, X_2, X_3, \ldots$. We will adopt the notation

$$X_k = \begin{bmatrix} x_1^{(k)} \\ \vdots \\ x_n^{(k)} \end{bmatrix} \tag{1.9.2}$$

The chain $X_1, X_2, X_3, \ldots$ of distribution vectors is called a **Markov chain**, provided (in crude terms) that X_{k+1} is determined by X_k. We will consider a special type of Markov chain: Assume that at any given observation the probability of transition from state s_j to state s_i is p_{ij} (called the **transition probability**) and that this probability remains fixed, that is, the transition probability depends only on the states, and will not change with time. The matrix

$$P = \begin{bmatrix} p_{11} & \cdots & p_{1n} \\ \vdots & \ddots & \vdots \\ p_{n1} & \cdots & p_{nn} \end{bmatrix} \qquad (1.9.3)$$

is called the **transition matrix**. This matrix has special properties, but before we discuss them, let us pursue an example.

Assume that as before s_1, s_2, and s_3 represent the states "Republican," "Democrat," and "Independent," respectively. Assume that observations are made at intervals of one year and that currently the distribution vector is

$$X = \begin{bmatrix} 0.4 \\ 0.5 \\ 0.1 \end{bmatrix} \qquad (1.9.4)$$

that is 40% of all voters are Republican, 50% Democrats, and 10% Independent. We will assume that the transition probabilities are fixed and given by the following chart:

		Republican	Democrat	Independent
Next Year	Republican	0.8	0.3	0.3
	Democrat	0.1	0.5	0.3
	Independent	0.1	0.2	0.4

For example, given an individual who is a Democrat this year, the probability that he or she will switch to a Republican next year is 0.3, remain a Democrat is 0.5, and change to an Independent is 0.2. The transition matrix P and initial distribution vectors are thus given by:

$$P = \begin{bmatrix} 0.8 & 0.3 & 0.3 \\ 0.1 & 0.5 & 0.3 \\ 0.1 & 0.2 & 0.4 \end{bmatrix} \text{ and } X_1 = \begin{bmatrix} 0.4 \\ 0.5 \\ 0.1 \end{bmatrix}. \qquad (1.9.5)$$

The question arises as to what the distribution will be next year, that is, what will X_2 be. To determine this, let us return to the general discussion.

Recall that p_{ij} is the probability that an individual will move from state s_j to state s_i and that $x_j^{(k)}$ represents the proportion of individuals in state s_j during observation k. Thus, $p_{ij}x_j^{(k)}$ is the proportion of individuals moving to the state s_i in the $(k+1)$-th observation. To find $x_j^{(k+1)}$, we must add these proportions; that is,

$$x_i^{(k+1)} = p_{i1}x_1^{(k)} + p_{i2}x_2^{(k)} + \ldots + p_{in}x_n^{(k)}. \qquad (1.9.6)$$

This sum is exactly the result of multiplying row i of the matrix P in Equation 1.9.3 by the column vector X_k in Equation 1.9.2. We have shown that

$$X_{k+1} = PX_k. \qquad (1.9.7)$$

It follows that

$$
\begin{aligned}
X_2 &= PX_1 \\
X_3 &= PX_2 = P(PX_1) = P^2X_1 \\
X_4 &= PX_3 = P(P^2X_1) = P^3X_1 \\
&\vdots \\
X_{k+1} &= PX_k = P^kX_1.
\end{aligned}
$$

Assuming that the transition probabilities are known and remain fixed and that the initial distribution vector X_1 is known, one can calculate the distribution vector X_k for the k-th observation by using the formula

$$X_{k+1} = P^kX_1 \tag{1.9.8}$$

Returning to our example involving the distribution of voters amongst the categories Republican, Democrat, and Independent with transition matrix P and initial distribution vector X_1 given in 1.9.5 by

$$
P = \begin{bmatrix} 0.8 & 0.3 & 0.3 \\ 0.1 & 0.5 & 0.3 \\ 0.1 & 0.2 & 0.4 \end{bmatrix} \text{ and } X_1 = \begin{bmatrix} 0.4 \\ 0.5 \\ 0.1 \end{bmatrix}
$$

we see that

$$
X_2 = PX_1 = \begin{bmatrix} 0.5 \\ 0.32 \\ 0.18 \end{bmatrix},
$$

$$
X_3 = P^2X_1 = PX_2 = \begin{bmatrix} 0.55 \\ 0.264 \\ 0.186 \end{bmatrix},
$$

etc.

The matrix P in 1.9.5 has special properties since p_{ij} represents the probability of transition from state s_j to state s_i. We can see that:

(a) P is square ($n \times n$)

(b) $p_{ij} \geq 0$ for all i, j

(c) The sum of the entries in each column is 1 (since $p_{1j} + p_{2j} + \ldots + p_{nj}$ represents the probability of moving from state s_j to one of the states $s_1, s_2, \ldots, s_n$).

A matrix P with the above properties is called a **stochastic matrix**.

Section 1.9 Exercises

1. Determine which of the following matrices are stochastic matrices.

(a) $\begin{bmatrix} 0.1 & 0.3 \\ 0.9 & 0.7 \end{bmatrix}$

(b) $\begin{bmatrix} 1 & 0 \\ 0 & 1 \end{bmatrix}$

(c) $\begin{bmatrix} -0.1 & 0.2 \\ 1.1 & 0.8 \end{bmatrix}$

(d) $\begin{bmatrix} 0.2 & 0.3 & 0.1 \\ 0.8 & 0.7 & 0.9 \end{bmatrix}$

(e) $\begin{bmatrix} 0.1 & 1 \\ 0.9 & 0 \end{bmatrix}$

2. Fill in the blanks so that the following matrix is stochastic: $\begin{bmatrix} 0.1 & \_\_ & 0.7 \\ \_\_ & 0.2 & 0.3 \\ \_\_ & \_\_ & \_\_ \end{bmatrix}$

3. In the example in the text with P and X_1 as in [9.5], calculate X_4 and X_5. What is the long-term trend?

4. Let P and X_1 be given by

$$P = \begin{bmatrix} 0.8 & 0.2 \\ 0.2 & 0.8 \end{bmatrix} \text{ and } X_1 = \begin{bmatrix} 0.4 \\ 0.6 \end{bmatrix}.$$

Calculate X_2, X_3, and X_4 .

5. Let P and X_1 be given by:

$$P = \begin{bmatrix} 0.6 & 0.2 & 0.2 \\ 0.4 & 0.6 & 0.0 \\ 0.0 & 0.2 & 0.8 \end{bmatrix} \text{ and } X_1 = \begin{bmatrix} 0.1 \\ 0.6 \\ 0.3 \end{bmatrix}.$$

Calculate X_2, X_3, and X_4 . Can you find a formula for X_n?

6. Assume that currently 90% of all adults are employed and 10% are unemployed. Assume further that the child of an employed person (when reaching adulthood) has a probability of 0.8 of being employed and 0.2 of being unemployed, whereas the offspring of an unemployed person has a 0.6 probability of being employed and a 0.4 probability of being unemployed. Find the transition matrix P and the distribution vector X_1, and then find X_2 and X_3. At what intervals are observations made in this situation?

7. Let P be a stochastic matrix and X some distribution vector. Assume that $PX = X$. Prove that $P^n X = X$ for all integers $n > 0$.

8. Let $P = [p_{ij}]$ be a 2×2 stochastic matrix with $p_{12} = 0$. What can be said about the other entries in the matrix?

9. A matrix P that is stochastic and that has the property that the sum of each of the rows is 1 is called **doubly stochastic**. Give an example of a 2×2 matrix that is stochastic but not doubly stochastic.

10. Let $[p_{ij}]$ be a 2×2 doubly stochastic matrix (see Exercise 9) with $p_{12} = 0.2$. What are the other entries in the matrix?

11. Can a doubly stochastic matrix (see Exercise 9) be singular?

Chapter 2

VECTOR SPACES

2.1   Introduction: Vectors in $\mathbb{R}^2$

Vectors are usually introduced in calculus, where they are described as either directed line segments or ordered pairs of real numbers. In beginning courses in physics, vectors are described as quantities that possess both direction and magnitude. Vectors describe common elementary physical quantities such as velocity, displacement, and force. Quantities such as speed, pressure, and time are scalar rather than vector quantities; they have magnitude, but not direction.

In Chapter 1, we introduced matrices and systems of linear equations. Solutions of systems of linear equations were presented first as "n-tuples," and then as "column vectors," where column vectors were defined as special types of matrices. We saw that n-tuples and row vectors could be identified. In Section 1.6 we saw that homogeneous systems of equations have special properties - their solution sets are closed under addition and scalar multiplication. This chapter takes a deeper look at vectors. In Section 2.3 we define the term "vector space." In order to motivate our abstract definition and to gain a feel for the notion of a vector space, we begin with the familiar view of vectors as first presented in calculus.

TWO-DIMENSIONAL VECTORS

The set of real two-dimensional vectors is sometimes defined as the set of directed line segments $\overrightarrow{AB}$, where A and B are two points lying in a fixed plane (see Figure 2.1).

Vectors having the same direction and magnitude (length) are, by definition, identified, and so they are regarded as mathematically equal. The vector $\overrightarrow{AB}$ in Figure 2.1 can be identified mathematically with the vector $\overrightarrow{OC}$. We can see that all vectors with a given direction and a given nonzero magnitude can be identified with a vector extending from the origin to a unique point C. If the point C has coordinates (x_0, y_0), then the vector $\overrightarrow{OC}$ is determined by (x_0, y_0). (The reader will recall from earlier courses that points in the plane may be uniquely specified by "ordered pairs" of numbers (x_0, y_0), and that equality of ordered pairs is defined by $(x_0, y_0) = (x_1, y_1)$ if and only if $x_0 = x_1$ and $y_0 = y_1$.) We will think of two-dimensional vectors as pairs of real numbers (x_0, y_0). We then define

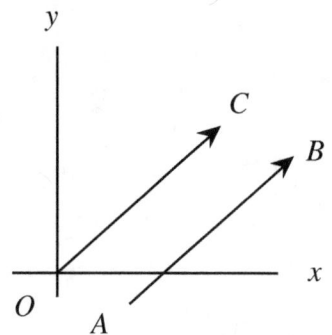

Figure 2.1: Parallel Vectors

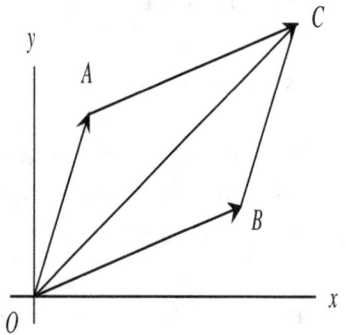

Figure 2.2: Sum of two vectors

$$\mathbb{R}^2 = \{(x,y)|x,y \in \mathbb{R}\}.$$

Operations of addition and scalar multiplication are defined on this set. Addition of vectors is defined as follows: Let $\overrightarrow{OA}$ and $\overrightarrow{OB}$ be the two-dimensional vectors with endpoints (a_1, a_2) and (b_1, b_2), respectively. The sum of the two vectors is defined to be the diagonal $\overrightarrow{OC}$ of the completed parallelogram as in Figure 2.2; that is, $\overrightarrow{OA} + \overrightarrow{OB} = \overrightarrow{OC}$. In Exercise 13, the reader is asked to verify that C has coordinates $(a_1 + b_1, a_2 + b_2)$ so that we have $(a_1, a_2) + (b_1, b_2) = (a_1 + b_1, a_2 + b_2)$.

The product of a scalar r and a vector $\overrightarrow{OA}$ is defined as follows (see Figure 2.3). If $r \geq 0$, then $r(\overrightarrow{OA})$ is defined to be the vector in the direction of $\overrightarrow{OA}$ with length equal to r times the length of $\overrightarrow{OA}$. If $r < 0$, then $r(\overrightarrow{OA})$ is defined to be the vector that has the same length as $(-r)\,\overrightarrow{OA}$ but that has opposite direction. In Exercise 2, the reader is asked to verify that if $\overrightarrow{OA}$ is the vector represented by the pair (a, b), then $r(\overrightarrow{OA})$ is the vector represented by (ra, rb); that is, $r(a, b) = (ra, rb)$.

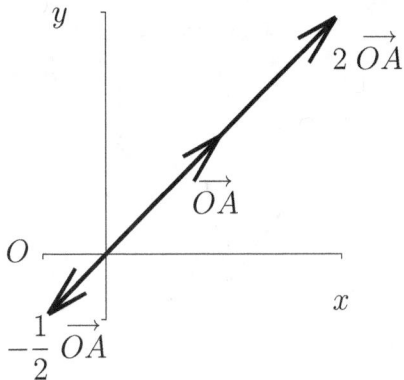

Figure 2.3: Scalar Multiples

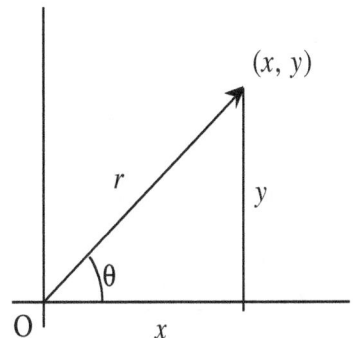

Figure 2.4: Polar Coordinates

As explained in Chapter 1, these two-dimensional vectors may be regarded as 1×2 matrices or row vectors. The operations of addition and scalar multiplication defined for matrices are identical to the above operations defined for vectors considered as ordered pairs. Further, the definition of equality of ordered pairs is identical to the definition of equality for 1×2 matrices, and so, one may regard these ordered pairs as either matrices or vectors. While it is traditional to write the vector with parentheses and a comma and the matrix with square brackets and without a comma, we will not distinguish between the ordered pair (x_0, y_0) and the 1×2 matrix $\begin{bmatrix} x_0 & y_0 \end{bmatrix}$.

The switch from coordinates (x, y) to magnitude and direction is not difficult. Let θ be the angle that the vector makes with the positive x-axis and let r denote the magnitude of the vector. Using elementary trigonometry, we see that (see Figure 2.4) $x = r \cos\theta, y = r \sin\theta, r = \sqrt{x^2 + y^2}$, and $\tan\theta = y/x$.

As an illustration of the manner in which two-dimensional vectors are used, consider the following situation: An airplane travels northwest at 150 miles per hour for one hour and then proceeds northeast at the same speed for 1.5 hours. Where is the plane at the end of this time?

The velocity vectors (speed and direction) and the time of travel determine displacement

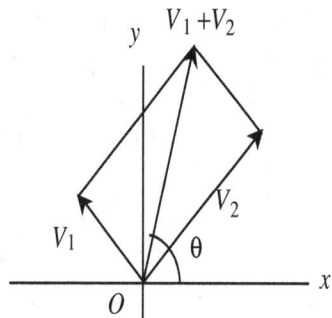

Figure 2.5: Sum of V_1 and V_2

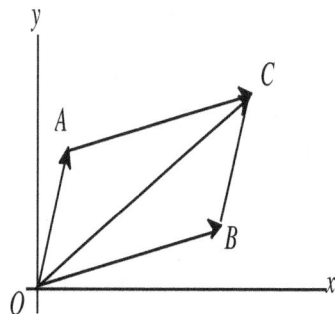

Figure 2.6: The Parallelogram Law

vectors (distance and direction) for each part of the trip. These displacement vectors may be determined and then the final location of the plane may be calculated as the sum of the two displacement vectors.

We will assume that the motion of the plane in reference to the coordinate system is as shown in Figure 2.4 and that the plane begins its trip at the origin. Using elementary trigonometry, we see that the displacement vector describing the first part of the trip is found to be $V_1 = (-150/\sqrt{2}, 150/\sqrt{2})$ and the displacement vector for the second part is given by $V_2 = (225/\sqrt{2}, 225/\sqrt{2})$. The final displacement is given by the sum $V_1 + V_2$ of the two vectors. We see that $V_1 + V_2 = (75/\sqrt{2}, 375/\sqrt{2})$. This is a vector of length $L = 75\sqrt{13} \approx 270$, subtending an angle $\theta = \arctan(375/75) \approx 79°$ with the x-axis; see Figure 2.5.

The method of addition described above is often called the **parallelogram method** (see Figure 2.2). One can also think of adding vectors head to tail. Notice that in Figure 2.2, the line segment $\overline{AC}$ is parallel to $\overline{OB}$ and the two segments have the same length. These vectors may be identified, and we can think of placing the tail of the second vector at the head of the first, the sum then being the vector $\overline{OC}$ having its tail at the tail of the first vector and its head at the head of the second vector (see Figure 2.6).

Section 2.1 Exercises

In Exercises 1-4, perform the indicated vector operations.

1. $(2, 3) + (-4, 2) =$

2. $(-5, 2) + (3, -1) =$

3. $2(-1, 3) + 4(3, -6) =$

4. $3(2, 1) + (-2)(4, 1) =$

In Exercises 5-8, find the direction and magnitude of the given vectors.

5. $(3, 2)$

6. $(-1, 3)$

7. $(1, 3)$

8. $(2, -2)$

In Exercises 9-12, θ represents the angle a given vector makes with the positive x-axis and r represents the magnitude of the given vector. Express the vector as an ordered pair.

9. $\theta = 30°, r = 2$

10. $\theta = 45°, r = 3$

11. $\theta = 210°, r = 5$

12. $\theta = 315°, r = 4$

13. As discussed in this section, verify that $\overline{OA} + \overline{OB}$ is the vector given by $(a_1 + b_1, a_2 + b_2)$ and $r(\overline{OA})$ is given by (ra_1, ra_2). (Hint: Use similar triangles.)

14. A plane flies southeast for 2 hours at 125 miles per hour and then northeast at 150 miles per hour for 1.5 hours. What direction must it fly to return home, and how long will the return flight take at 175 miles per hour?

15. Consider the vectors $A = (1, 2)$ and $B = (2, 5)$. Find the cosine of the angle between the two vectors. (Hint: Use the law of cosines.)

16. Force is a vector quantity. Assume that a given force F_1 has a magnitude of 10 pounds, and makes an angle of 30° with the positive x-axis; a force F_2 has a magnitude of 20 pounds and makes an angle of 90° with the positive x-axis; and a third force F_3 has a magnitude of 15 pounds and makes an angle of 225° with the positive x-axis. What is the resultant force? That is, what is $F_1 + F_2 + F_3$?

17. Let $(a, b), (c, d) \in \mathbb{R}^2$. Show that $(a, b) + (c, d) = (c, d) + (a, b)$.

2.2 Vector Spaces

In order to develop theories applicable to many different systems, mathematicians often consider the abstract properties of a given system and then make a general definition that includes these properties. The notion of a "vector space" is one such generalization. A precise definition of the term vector space is given in the next section-here we will present motivation for this definition.

In the previous section, the set $\mathbb{R}^2$ of two-dimensional vectors was defined along with the operations of addition and scalar multiplication. Certainly other systems have similar structures-consider $\mathbb{R}^3$ the set of 3-dimensional vectors or, for that matter, the set of $m \times n$ matrices over some field F. In each case there is an addition and a scalar multiplication.

In Section 1.6, it was shown that if one could find just one solution X_p of a system of linear equations $AX = H$, and if one could then find all solutions X_h of the homogeneous equation $AX = 0$, all solutions of $AX = H$ would be obtained in the form $X = X_p + X_h$. Because of this, it is important to be able to characterize the solution sets of homogeneous systems of linear equations. Now recall that in Section 1.6 it was shown that solution sets of such homogeneous systems were nonempty, closed under addition, and closed under scalar multiplication. Let us list the common properties of the mathematical systems mentioned above:

PROPERTIES OF VECTOR SPACES

Observation 1. Properties common to "vector spaces":

(a) In each case, there is a field F associated with the system, often the field of real numbers. The elements of the field are called scalars.

(b) There is a set $\mathbf{V}$ that is nonempty. The elements of $\mathbf{V}$ are called **vectors**.

(c) There is an addition of vectors that has the following properties:

 (i) Given two vectors X and Y in $\mathbf{V}$, there is a unique sum $X+Y$. (Sums are well defined.)

 (ii) If X and Y are in $\mathbf{V}$, then $X + Y$ is in $\mathbf{V}$.   ($\mathbf{V}$ is closed under addition.)

(d) There is a scalar multiplication that has the following properties:

 (i) Given a scalar r in F and a vector X in $\mathbf{V}$, there is a unique scalar product rX. (Scalar multiplication is well defined.)

 (ii) If r is in F and X is in $\mathbf{V}$, then rX is in $\mathbf{V}$.   ($\mathbf{V}$ is closed under scalar multiplication.)

(e) There are algebraic properties of the operations. For example, addition is associative and commutative, there is a zero vector, etc.

We saw in Chapter 1 that addition and scalar multiplication of $m \times n$ matrices are well-defined operations. Also, we proved many algebraic properties of these operations. While the definition of the term "vector space" is still to be presented, we will make the following observation:

Observation 2. Let $\mathbf{V}$ be a nonempty set of $m \times n$ matrices over some field F and assume that $\mathbf{V}$ is closed under addition and scalar multiplication; that is, assume that

(a) If X and Y are in $\mathbf{V}$, then $X + Y$ is in $\mathbf{V}$.   ($\mathbf{V}$ is closed under addition .)

(b) If X is in $\mathbf{V}$ and r in F, then rX is in $\mathbf{V}$.   ($\mathbf{V}$ is closed under scalar multiplication .)

Then $\mathbf{V}$ is a vector space over F.

This observation is quite general in that any nonempty collection of $m \times n$ matrices with the closure properties (a) and (b) forms a vector space. In most of our applications, the set of "vectors" will be a set of $1 \times n$ matrices or a set of $n \times 1$ matrices; that is, a set of row vectors or column vectors. Recall from Section 1.2 that we identified the n-tuple $(a_1, a_2, \ldots, a_n)$ and the $1 \times n$ matrix $\begin{bmatrix} a_1 & a_2 & \ldots & a_n \end{bmatrix}$.

Let us now consider examples of sets of matrices that satisfy the conditions observed above, and consequently form vector spaces.

Example 2.2.1.   (a) The most trivial example of a vector space is the set $\{0\}$ consisting of the $m \times n$ zero matrix alone. Since the zero vector is a member of this set, the set is nonempty and is closed under addition and scalar multiplication.

(b) As in Section 2.1, the set $\mathbb{R}^2$ of real two-dimensional vectors may be thought of as the set of all pairs of real numbers. $\mathbb{R}^2$ may be regarded as the set of all 1×2 matrices over the real number field $\mathbb{R}$, and so $\mathbb{R}^2$ is closed under addition and scalar multiplication. It follows that $\mathbb{R}^2$ is a vector space.

(c) As a generalization of the previous example, let $\mathbb{R}^n$ denote the set of all n-tuples or $1 \times n$ matrices over $\mathbb{R}$. As above, $\mathbb{R}^n$ is a vector space.

(d) Let F denote any field of scalars and for positive integers m and n, let $F^{m \times n}$ denote the set of all $m \times n$ matrices over F. Since the sum of two $m \times n$ matrices over F is again an $m \times n$ matrix over F, and since a scalar times an $m \times n$ matrix produces another $m \times n$ matrix, we see that $F^{m \times n}$ is closed under addition and scalar multiplication. Thus, $F^{m \times n}$ is a vector space.

Notice that with the above notation, $\mathbb{R}^n = R^{1 \times n}$, the vector space of all $1 \times n$ matrices or row vectors over the real field $\mathbb{R}$. Likewise the vector space $\mathbb{R}^{m \times 1}$ consists of all $m \times 1$ column vectors over $\mathbb{R}$.

(e) Let S be the solution set of a homogeneous system of equations $AX = 0$ as in Theorem 1.6.2. By this theorem, S is closed under addition and scalar multiplication, and so it is a vector space.

(f) Not all sets of vectors satisfy the closure properties. Let $A = \{(x, y) \in \mathbb{R}^2 | x, y > 0\}$. It is not hard to see that A is closed under addition: $(x, y), (x', y')$ in $A \implies x, x', y, y' > 0 \implies x + x', y + y' > 0 \implies (x + x', y + y') \in A$, but A is not closed under scalar multiplication $((1, 1)$ is in A but $-1(1, 1)$ is not).

(g) We can find a set closed under scalar multiplication, but not closed under vector addition. Consider $B = \{(x, y) \in \mathbb{R}^2 | x = 0 \text{ or } y = 0\}$. If $x = 0$ or $y = 0$, then the vector $r(x, y) = (rx, ry)$ has the same property. On the other hand, $(1, 0) + (0, 1) = (1, 1) \notin B$.

$\square$

We need to consider the common properties of the above systems in order to see how to formulate the abstract definition.

Vector Spaces of Matrices

In Section 2.3, we will present the general definition of the term "vector space." The definition of this "abstract" vector space will involve axioms that give properties of operations on the set of vectors. In our first observation above, we noted that the operations on vector spaces had "properties." What are these properties? In order to see that the sets described in the observation above form a vector space, we state a theorem giving the properties of vectors and their operations that correspond to the axioms in the coming section. It might be more proper to call this result a corollary rather than a theorem, since almost all of the results were proved in Section 1.3. Note that the "properties" below are also referred to as "laws," as in, "distributive law." We will use both terms freely.

Theorem 2.2.1. *Let* **V** *be nonempty set of* $m \times n$ *matrices over the field* F *and assume that* **V** *is closed under addition and scalar multiplication as in Observation 2 above. Then:*

(a) $X + Y = Y + X$ *for all* $X, Y \in \mathbf{V}$. *(Commutative property)*

(b) $X + (Y + Z) = (X + Y) + Z$ *for all* $X, Y, Z \in \mathbf{V}$. *(Associative property)*

(c) The zero vector, 0, *is in* **V** *and* $0 + X = X$ *for all* $X \in \mathbf{V}$. *(Existence of zero)*

(d) If $X \in \mathbf{V}$, *there is a* $Y \in \mathbf{V}$ *with* $X + Y = 0$. *(Existence of additive inverses)*

(e) $(r + s)X = rX + sX$ *for all* $X \in \mathbf{V}$ *and all scalars* r *and* s. *(Distributive property)*

(f) $r(X + Y) = rX + rY$ *for all* $X, Y \in \mathbf{V}$ *and all scalars* r. *(Distributive property)*

(g) $r(sX) = (rs)X$ *for all* $X \in \mathbf{V}$ *and all scalars* r *and* s. *(Associative property)*

(h) $1X = X$ *for all* $X \in \mathbf{V}$, *where* 1 *is the identity scalar.* *(Identity property)*

Proof. Parts a), b), e), f), g) and h) were proved in Theorems 1.3.2, 1.3.3, and 1.3.6. We will prove parts c) and d).

c) Since **V** is nonempty, there is some vector X in **V**. By Theorem 1.3.6, $0X = 0$, where the first 0 is the zero scalar and the second 0 is the zero matrix. Since **V** is closed under scalar multiplication, $0X = 0$ is in **V**, so the zero vector lies in **V**. Also, by the same theorem, $0 + X = X$ for all X in **V**.

d) Let X be in **V**. Then using Theorem 1.3.6 we see that $Y = (-1)X$ is in **V** and $X + Y = X + (-1)X = 1X + (-1)X = (1 + (-1))X = 0X = 0$. Hence for every vector in **V**, there is an additive inverse in **V**.

$\square$

The properties listed in Theorem 2.2.1 allow us to perform many of the familiar algebraic operations related to solving equations. Given a vector X in **V**, part d) guarantees the existence of a vector Y with $X + Y = 0$. The vector Y is called the **additive inverse** of X and is denoted by $-X$. From the proof of Theorem 1.3.6, we see that $-X = (-1)X$. Now if $Z \in V$, the **difference** of Z and X is defined by $Z - X = Z + (-X)$. By the previous comment we see that $Z - X = Z + (-1)X$.

Corollary 2.2.2. *Let* $\mathbf{V}$ *be a vector space and let* $X, Y, Z \in \mathbf{V}$. *Then:*

(a) *If* $X + Y = X + Z$ *or* $Y + X = Z + X$, *then* $Y = Z$.   *(Cancellation law)*

(b) *If* a *is a nonzero scalar and* $aX + Y = Z$, *then* $X = (1/a)(Z - Y)$. *(Linear equations can be solved.)*

Proof. Part a) is true by Corollary 1.3.7 and part b) is left as an exercise.   □

Closure under addition and scalar multiplication can be pictured geometrically. Suppose that X is a vector in $\mathbb{R}^3$. We can think of a line segment in space:

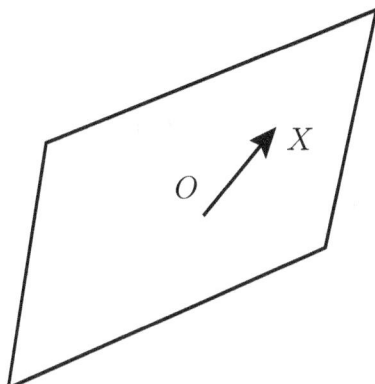

Figure 2.7: A vector in space

If we consider the set of all scalar multiples of this vector, $\{rX | r \in R\}$, we see that we have the collection of vectors lying on the line through the vector X:

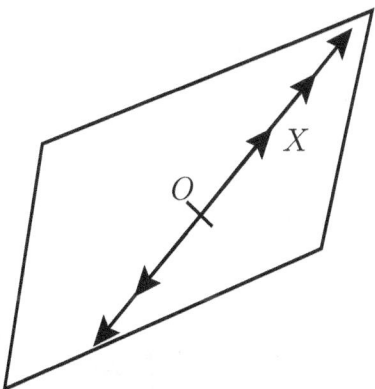

Figure 2.8: All multiples of X

Two vectors in $\mathbb{R}^3$ may be pictured as line segments from the origin. Either these line segments lie on the same line or they determine a plane. Call the vectors X and Y, and assume that they are not collinear. The picture is as in Figure 2.2.

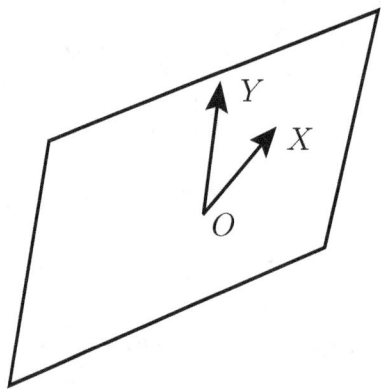

Figure 2.9: X and Y

It is not difficult to see that any vector Z lying in the plane of X and Y is a sum of two vectors lying along the lines determined by X and Y (see Figure 2.2). This means that Z is a sum of scalar multiples of X and Y.

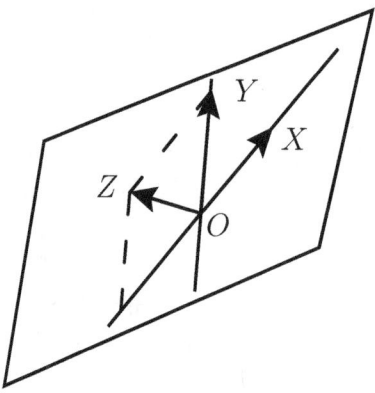

Figure 2.10: Z is a linear combination of X and Y

We see that scalar multiples of a nonzero vector give a line in space and that sums of these scalar multiples, for two non-collinear vectors, give all vectors in the plane determined by the two vectors.

Section 2.2 Exercises

1. Let **V** be the set of all vectors (x, y) in $\mathbb{R}^2$ such that $x = 2y$. Show that **V** is closed under addition and scalar multiplication.

2. Let $\mathbf{V}$ be the set of vectors (x, y, z) in $\mathbb{R}^3$ satisfying the conditions $x + y - z = 0, 2x - z = 0$. Show that $\mathbf{V}$ is closed under addition and scalar multiplication.

3. Let $\mathbf{V}$ be the set of all (x, y) in $\mathbb{R}^2$ satisfying $x + y = 1$. Show that $\mathbf{V}$ is not closed under either addition or scalar multiplication.

4. Let $\mathbf{V}$ be the set of all (x, y, z) in $\mathbb{R}^3$ with $x + y = 0$ and $x - z = 1$. Show that $\mathbf{V}$ is not closed under either addition or scalar multiplication.

5. Let $\mathbf{V}$ be a subset of $\mathbb{R}^2$ that is closed under addition and scalar multiplication, and assume that $(1, 1)$ and $(1, 2)$ are in $\mathbf{V}$. Show that $V = \mathbb{R}^2$. (Hint: Approach the problem geometrically.)

6. Find a subset $\mathbf{V}$ of $\mathbb{R}^2$ satisfying the following:

   (a) $\mathbf{V}$ is not equal to $\mathbb{R}^2$.

   (b) $\mathbf{V}$ is closed under addition and scalar multiplication.

   (c) $(-2, 1)$ is in $\mathbf{V}$.

7. Use the definition of "difference," Theorem 2.2.1, and Corollary 2.2.2 to show that for any two $m \times n$ matrices X and Y, $-(X - Y) = -X + Y$.

8. Prove part b) of Corollary 2.2.2.

9. Let $\mathbf{V}$ be a nonempty set of $m \times n$ matrices over the field F and assume that the following condition holds: If r and s are in F and X and Y in $\mathbf{V}$, then $rX + sY$ is in $\mathbf{V}$. Prove that $\mathbf{V}$ is closed under addition and scalar multiplication.

10. Let X be in $\mathbb{R}^2$ and assume $2X + 3(2, 1) = (-1, 2)$. Solve for X as in Corollary 2.2.2 b).

11. As in Exercise 10, solve for X given that $3X - (2, 3) = 3(-5, 7)$.

12. Find a subset of $\mathbb{R}^3$ that is closed under addition, but not closed under scalar multiplication.

13. Find a subset of $\mathbb{R}^3$ that is closed under scalar multiplication, but not closed under addition.

2.3 Abstract Vector Spaces

In Section 2.2, it was observed that a nonempty set of $m \times n$ matrices over some field F that is closed under addition and scalar multiplication is a vector space. Theorem 2.2.1 gives properties of such a system. These properties are consequences of the definitions of the operations and of the properties of the operations on the field of scalars. There are other mathematical structures that are very similar in nature but that do not consist of matrices. One example of such a structure is the collection of continuous functions on a given closed interval. Let us review some definitions and theorems from elementary calculus.

Let a and b be real numbers with $a < b$, and let f and g be continuous, real-valued functions defined on the closed interval $[a,b]$, and let c be some real number. The sum of the two functions f and g is the function $f + g$ defined on $[a,b]$ by

$$(f+g)(x) = f(x) + g(x),$$

and the product of c and f is the function cf defined on $[a,b]$ by

$$(cf)(x) = cf(x).$$

Now in calculus one learns that the sum of two continuous functions is continuous and that a constant times a continuous function also gives a continuous function. It follows that the set $C[a,b]$ of continuous real-valued functions defined on the closed interval $[a,b]$ is closed under addition and scalar multiplication. Furthermore, properties corresponding to those of Theorem 2.2.1 hold for $C[a,b]$. The constant function, $z(x) = 0$ for x in $[a,b]$, serves as the zero of part c), and $-f$, for f in $C[a,b]$, is defined by $(-f)(x) = -(f(x))$. We see that $C[a,b]$ has the properties that the previous examples of vector spaces had. We wish to bring this example into our list of vector spaces.

A GENERAL DEFINITION

In order to include the example above, as well as other examples, we give an abstract definition of the notion of a vector space. This definition is "abstract" in the sense that the set V of vectors is an arbitrary set (not necessarily matrices), and operations are assumed to exist with properties like those in Theorem 2.2.1. Furthermore, the possibility is included that scalars from fields other than the real number field will be permitted. For information on fields, the reader may refer to Appendix C. In practice, little harm will be done if it is assumed that the field of scalars is the field of real numbers, for the algebraic manipulations that can be performed on the real numbers (adding, subtracting, multiplying, and dividing by nonzero numbers) can also be performed on the elements of any field of scalars.

Definition 2.3.1. Let F be a field of scalars (say $\mathbb{Q}, \mathbb{R}$, or $\mathbb{C}$). A **vector space** over F is a nonempty set **V** (the elements of **V** will be called **vector**s and denoted by capital letters $X, Y, \ldots$) along with two operations: an **addition** that associates with each pair of vectors X and Y a unique vector $X + Y$, and a **scalar multiplication** that associates with each scalar r and vector X a unique vector rX. This system consisting of the set of vectors, the field of scalars, and the two operations satisfies the following conditions:

A1) $X + Y = Y + X$ for all $X, Y \in \mathbf{V}$.

A2) $X + (Y + Z) = (X + Y) + Z$ for all $X, Y, Z \in \mathbf{V}$.

A3) There is an element $0 \in \mathbf{V}$ with $0 + X = X$ for all $X \in \mathbf{V}$ (0 is called the **zero vector**).

A4) For any $X \in \mathbf{V}$, there is an element $-X \in \mathbf{V}$ with $X + (-X) = 0$ ($-X$ is the **additive inverse** of X).

M1) $(r + s)X = rX + sX$ for all $X \in \mathbf{V}$ and all scalars r and s.

M2) $r(X + Y) = rX + rY$ for all $X, Y \in \mathbf{V}$ and all scalars r.

M3) $r(sX) = (rs)X$ for all $X \in \mathbf{V}$ and all scalars r and s.

M4) $1X = X$ for all $X \in \mathbf{V}$, where 1 is the identity of F.

It follows from Theorem 2.2.1 that any nonempty set $\mathbf{V}$ of $m \times n$ matrices over a field F that is closed under addition and scalar multiplication is a vector space according to the above definition. We see then that Observation 2 is correct. In the theory that follows, we will most often consider vectors in an abstract sense. That is, we will work with the assumptions A1)-A4) and M1)-M4) from above and we will not use a particular representation for the vectors, such as assuming the vector to be a matrix. There are vector spaces, such as $C[a, b]$ above, in which the vectors are not matrices and are not representable as matrices. Before giving examples, we will state a theorem that establishes some of the elementary properties of vector spaces and their operations.

A THEOREM ABOUT VECTOR SPACES

Since we are now considering vector spaces in an abstract sense, it is important to pay attention to the proofs of theorems. While the properties cited in the theorem are known to be true for matrices, we may no longer assume that the vectors are matrices-instead we must use only the axioms for a vector space. A subtlety in the above definition should be pointed out: The phrase "an addition that associates with each pair of vectors X and Y a unique vector $X + Y$" is assumed to contain both the closure of the addition operation and the fact that it is well-defined. The closure comes from the statement that $X + Y$ is a vector, and the word "unique" implies that vector addition is well-defined. Similar remarks may be made concerning the closure and well-definedness of scalar multiplication.

Theorem 2.3.1. *Let* $\mathbf{V}$ *be a vector space with* $X, Y, Z \in \mathbf{V}$ *and* r *some scalar. Then:*

(a) If $X + Y = X + Z$ *or* $Y + X = Z + X$, *then* $Y = Z$.   *(Cancellation law)*

(b) If $X + Y = 0$ *then* $Y = -X$.   *(Additive inverses are unique)*

(c) $r0 = 0$, *where* 0 *denotes the zero vector.*   *(A scalar times zero is zero)*

(d) $0X = 0$, *where the first* 0 *is a scalar and the second, a vector.*
(Zero times a vector is zero)

(e) $-X = (-1)X$.   *(−1 times a vector is the vector's additive inverse)*

(f) $rX = 0$ *implies* $r = 0$ *or* $X = 0$.   *(There are no nonzero zero divisors)*

Proof.   (a) Assume $X + Y = X + Z$. By property A4) there is a vector $-X$ in $\mathbf{V}$ with $X + (-X) = 0$. Adding $-X$ to both sides, we obtain $-X + (X + Y) = -X + (X + Z)$. Using A1) and A3), we obtain $(X + (-X)) + Y = (X + (-X)) + Z$. It follows that $0 + Y = 0 + Z$ and so $Y = Z$.

(b) If $X + Y = 0$, then $X + Y = X + (-X)$ by A4). Thus, $Y = -X$ by a).

(c) By A3), $0 + 0 = 0$, and multiplying both sides by r, we obtain $r(0 + 0) = r(0)$, or $r0 + r0 = r0$ using M2). Adding the zero vector to the right side, we obtain $r0 + r0 = r0 + 0$, and so by a) $r0 = 0$.

Proofs of the remaining parts are left as exercises. $\square$

Example 2.3.1.   (a) The first example of a vector space can be taken to be any one of the examples of Section 2.2. By Theorem 2.2.1, any one of the closed and nonempty sets of matrices must satisfy the above definition, and so forms a vector space.

(b) As mentioned at the beginning of this section, the set $C[a, b]$ of all continuous, real-valued functions defined on the closed interval $[a, b] = \{x \in \mathbb{R} | a = x = b\}$ is a vector space over the real numbers. If f, g are elements in $C[a, b]$, then the **equality** of f and g is defined by $f = g$ if and only if $f(x) = g(x)$ for all x in $[a, b]$. Addition and scalar multiplication are defined by $(f + g)(x) = f(x) + g(x)$ and $(rf)(x) = rf(x)$ for any real number r. With these definitions, $C[a, b]$ is a vector space over the real field $\mathbb{R}$. Notice that over the complex field $\mathbb{C}, C[a, b]$ fails to be closed under scalar multiplication. For example, if $f \in C[a, b]$, then $(if)(x) = if(x)$ (where i is the imaginary unit with $i^2 = -1$) is not a real-valued function unless $f(x) = 0$ for all x, and so $C[a, b]$ is not a vector space over $\mathbb{C}$.

(c) The field of real numbers $\mathbb{R}$ may be regarded as a vector space over the field of rational numbers $\mathbb{Q}$. The set of vectors is the set of real numbers, while the rational numbers are the scalars. The "vector" addition is the ordinary addition operation on real numbers, and the scalar multiplication is defined to be the operation of ordinary multiplication of real numbers. It follows that properties A1)-A4) and M1)-M4) are satisfied, and so $\mathbb{R}$ is a vector space over $\mathbb{Q}$.

This example is really much more general than its statement suggests. If F_1 is a subfield of F_2 (that is, $F_1 \subseteq F2$ and F_1 is a field under the operations of F_2), then F_2 is a vector space over F_1. For example, the complex field $\mathbb{C}$ is a vector space over the real field $\mathbb{R}$ and over the rational field $\mathbb{Q}$.

(d) Our final example of an "abstract" vector space consists of the system of polynomials over a field F. Since these polynomials are considered over an arbitrary field, this presentation differs from the treatment of polynomials in a calculus class, where they are regarded as a special sort of function. By a **polynomial in the indeterminate x over the field** F we mean an expression of the form $a_0 + a_1 x + \ldots + a_n x^n$ where n is a non-negative integer and each $a_i \in F$. Let $p(x) = a_0 + a_1 x + \ldots + a_n x^n$ and let $q(x) = b_0 + b_1 x + \ldots + b_m x^m$ be a second polynomial over F. Assume $n \leq m$. We write $p(x) = q(x)$ if and only if $a_i = b_i$ for $0 \leq i \leq n$ and $b_{n+1} = \ldots = b_m = 0$. Let $F[x]$ be the set of all polynomials over F and define operations on $F[x]$ by

$$p(x) + q(x) = (a_0 + b_0) + (a_1 + b_1)x + \ldots + (a_n + b_n)x^n + b_{n+1}x^{n+1} + \ldots + b_m x^m$$

and

$$rp(x) = ra_0 + ra_1x + \ldots + ra_nx^n.$$

From these definitions, one can prove that $F[x]$ is a vector space over F.

$\square$

The vector spaces $C[a, b]$ and $F[x]$ described in Examples b) and d) obviously are not vector spaces consisting of matrices. We will see later that there are even more fundamental differences in these vector spaces - those vector spaces consisting of matrices are "finite dimensional," while $C[a, b]$ and $F[x]$ are "infinite dimensional."

Our definition is now sufficiently general to include all of the specific cases of vector spaces encountered in undergraduate mathematical studies. In fact, we have given the standard definition of the term vector space. We would like to understand more about the structure of these vector spaces. How can they be characterized? Are there common features of these spaces? We can answer these questions in the case of finite-dimensional vector spaces. The results will come in Section 2.7.

Section 2.3 Exercises

1. Let $f(x) = x^2 + 2$ and $g(x) = x - 1$. Then f and g are in $C[0, 1]$. Calculate the following:

    (a) $(f + g)(x)$

    (b) $(f + 3g)(a)$

    (c) $(2f + (-1)g)(.5)$

2. Let $f(x) = \sin x$ and $g(x) = \cos x$. Then f and g are in $C[0, p]$. Calculate the following:

    (a) $(2f)(p/2)$

    (b) $(f + 2g)(p/4)$

    (c) $(f + (-1)g)(p/6)$

3. For which pairs of functions may we write $f = g$ on the given interval? Explain why.

    (a) $f(x) = \sin x$ $g(x) = \cos x$ $[0, \pi]$

    (b) $f(x) = 1$ $g(x) = x^2 + 2x + 1$ $[0, 1]$

    (c) $f(x) = 1$ $g(x) = \sin^2 x + \cos^2 x$ $[0, \pi]$

    (d) $f(x) = x + 1$ $g(x) = (x^2 - 1)/(x - 1)$ $[-1, 0]$

4. In Example 2.3.1 (b), verify that property A3) holds. (Hint: Define z on $[a, b]$ by $z(x) = 0$ for all x in $[a, b]$. Show that z is in $C[a, b]$ and that $z + f = f$ for all f in $C[a, b]$.

5. Verify that property A4) holds in $C[a,b]$. (Hint: For f in $C[a,b]$ define $-f$ by $(-f)(x) = -(f(x))$. Show that $-f$ is in $C[a,b]$ and $f + (-f) = z$ (z as in Exercise 4).)

6. Verify that property M1) holds in $C[a,b]$.

7. Let $\mathbb{R}$ be the real field and let $f(x) = 1 + 3x$ and $g(x) = 1 + 2x + 3x^2$ be elements in $R[x]$. Calculate $2f(x) + 3g(x)$.

8. Let $f(x) = a_0 + a_1x + \ldots + a_nx^n$ be in $F[x]$ and assume that m is an integer with $m > n$. Let $g(x) = a_0 + a_1x + \ldots + a_nx^n + 0x^{n+1} + \ldots + 0x^m$. Show that $f(x) = g(x)$. (Hint: Apply the definition of equality of polynomials as given in Example 2.3.1 (d).)

9. Let F be a field. Show that property A3) holds in $F[x]$. (Hint: Define $z(x) = 0$. Then $z(x)$ is in $F[x]$ and $z(x) + f(x) = f(x)$ for all $f(x)$ in $F[x]$.)

10. Let F be a field. Show that property A4) holds in $F[x]$. (Hint: For $f(x) = a_0 + a_1x + \ldots + a_nx^n$ in $F[x]$, define $(-f)(x) = (-a_0) + (-a_1)x + \ldots + (-a_n)x^n$.)

11. In $F[x]$, where F is any field, verify that property M2) holds.

12. Prove the remaining parts of Theorem 2.3.1.

13. Let $\mathbf{V}$ be a vector space over the complex number field $\mathbb{C}$. Explain why $\mathbf{V}$ may be regarded as a vector space over the real field $\mathbb{R}$.

14. Does the set of rational numbers $\mathbb{Q}$ form a vector space over the system of integers $\mathbb{Z}$? Explain.

15. Consider the field Z_2 consisting of the two elements 0 and 1 (see Appendix 3). Let $f(x) = x + x^2$ and $g(x) = 0$ be two polynomials in $Z_2[x]$. Show that $f(a) = g(a)$ for all a in Z_2, but $f \neq g$.

16. Over the real numbers $\mathbb{R}$, a polynomial $f(x) = a_0 + a_1x + \ldots + a_nx^n$ may be regarded as a member of $\mathbb{R}[x]$ or $C[a,b]$, where $a \leq b$. Find and review in your calculus book the proof that equality of polynomials in $C[a,b]$ implies equality of polynomials in $\mathbb{R}[x]$.

17. Define the set of **quaternions** Q by

$$Q = \left\{ \begin{bmatrix} \alpha & \beta \\ -\bar{\beta} & \bar{\alpha} \end{bmatrix} : \alpha, \beta \in \mathbb{C} \right\}.$$

Show that Q is a vector space over $\mathbb{R}$.

2.4 Subspaces

In Chapter 1, the solution set of a system of equations in the variables $x_1, \ldots, x_n$ over the real numbers is described as a set of n-tuples of real numbers - a subset of the vector space $\mathbb{R}^n$. If the system of equations is homogeneous, this solution set is closed under addition and scalar multiplication and consequently forms a vector space. This situation is described by saying that the solution set is a "subspace" of the set of n-tuples $\mathbb{R}^n$, and we make the following definition.

Definition 2.4.1. Let $\mathbf{V}$ be a vector space over the field F and let $\mathbf{W}$ be a subset of $\mathbf{V}$. $\mathbf{W}$ is a **subspace** of $\mathbf{V}$ if and only if $\mathbf{W}$ is a vector space over F under the operations of addition and scalar multiplication in $\mathbf{V}$.

If the specific or nonabstract definition of the term vector space as given in Section 2.2 is assumed, then it is relatively easy to show whether a subset of a vector space is a subspace. One need only check that it is nonempty and closed under both of these operations. On the surface, it appears that the abstract definition is harder to deal with, but observe the following: If $\mathbf{W} \subseteq \mathbf{V}$ and $X + Y = Y + X$ for all $X, Y \in \mathbf{V}$ (so A1) holds for $\mathbf{V}$), then, trivially, $X + Y = Y + X$ holds for all $X, Y \in \mathbf{W}$, so that A1) is valid for $\mathbf{W}$. We see that, the subset $\mathbf{W}$ inherits the commutativity property from its parent set $\mathbf{V}$. The following theorem makes use of just such reasoning to simplify the proof that a subset of a vector space is a subspace.

> **Theorem 2.4.1.** *Let $\mathbf{V}$ be a vector space and let $\mathbf{W}$ be a nonempty subset of $\mathbf{V}$. Then $\mathbf{W}$ is a subspace of $\mathbf{V}$ if and only if $\mathbf{W}$ is closed under addition and scalar multiplication.* (A nonempty subset is a subspace iff it's closed.)

(Note: "iff" is often used by mathematicians as an abbreviation for "if and only if.")

Proof. Note that the conclusion of this theorem is an equivalence - a statement of the form p iff q, or symbolically $p \Leftrightarrow q$. To prove such a statement, we must prove the two implications: p implies q or $p \implies q$; and q implies p, or $q \implies p$; which we may write as $p \Leftarrow q$. The arrow in parentheses indicates which part of the proof is being given.

$(\Rightarrow)$ If $\mathbf{W}$ is a subspace of $\mathbf{V}$, then it is closed under addition and scalar multiplication, and so the first part of the proof is trivial.

$(\Leftarrow)$ Assume $\mathbf{W} \subseteq \mathbf{V}, \mathbf{W} \neq \emptyset$ and that $\mathbf{W}$ is closed under addition and scalar multiplication. We must prove that $\mathbf{W}$ satisfies conditions A1)-A4) and M1)-M4). As mentioned above, properties A1), A2), M1), M2), M3), and M4) are obvious. Only conditions A3) and A4) need to be checked. Since $\mathbf{W} \neq \emptyset$, there is some vector $X_0 \in \mathbf{W}$. Then $0X_0 = 0 \in \mathbf{W}$ by Theorem 2.3.1 and so the zero vector of $\mathbf{V}$ is in $\mathbf{W}$. Consequently, A3) holds. If $X \in \mathbf{W}$, then $-X = (-1)X \in \mathbf{W}$, also by 2.3.1. Hence A4) holds and so $\mathbf{W}$ is a vector space and so $\mathbf{W}$ is a subspace of $\mathbf{V}$. $\square$

Example 2.4.1. (a) If $\mathbf{V}$ is any vector space, then $\mathbf{V}$ is trivially a subspace of itself. Also $\{0\}$ is trivially a subspace of $\mathbf{V}$, where 0 represents the zero vector of $\mathbf{V}$.

(b) Let $\mathbf{W}$ be the solution set of a homogeneous system of linear equations in the variables $x_1, \ldots, x_n$ with real coefficients. If we regard the solutions as n-tuples, then $W \subseteq \mathbb{R}^n$ and $W \neq \emptyset$. By Theorem 1.6.1, $\mathbf{W}$ is closed under addition and scalar multiplication and so $\mathbf{W}$ is a subspace of $\mathbb{R}^n$.

If we let A be an $m \times n$ matrix and consider the homogeneous matrix equation $AX = 0$, then the solutions become $n \times 1$ column vectors and the solution set is $\{X | AX = 0\}$. This set is called the **nullspace** of A and is denoted by $N(A)$. It is, of course, a subspace of the space of $n \times 1$ column vectors.

(c) Let $\mathbf{V}$ be any vector space and let X_1 and X_2 be any two vectors in $\mathbf{V}$. Define span$\{X_1, X_2\} = \{rX_1 + sX_2 | r \text{ and } s \text{ are arbitrary scalars}\}$. Then span$\{X_1, X_2\}$ is a subspace of $\mathbf{V}$ and is called the **subspace spanned by** X_1 **and** X_2.

(d) Recall that for a real-valued function $f(x)$ of a real variable x, $f^{(n)}(x)$ denotes the n-th derivative of f. Let $C^n[a, b] = \{f \in C[a, b] | f^{(n)} \text{ exists and is continuous on } [a, b]\}$. Then $C^n[a, b]$ is a subset of $C[a, b]$ and is nonempty since, for example, the zero function has derivatives of all orders. Now let $f, g \in C^n[a, b]$ and let r be some scalar. Then $f^{(n)}$ and $g^{(n)}$ exist and are continuous, and by a theorem of basic calculus we have that $(f + g)^{(n)} = f^{(n)} + g^{(n)}$ and $(rf)^{(n)} = r(f)^{(n)}$. Also the fact that $f^{(n)}$ and $g^{(n)}$ are continuous implies that $f^{(n)} + g^{(n)}$ and $(rf)^{(n)}$ are continuous. We see that $f + g$ and rf are also elements of $C^n[a, b]$. It follows that $C^n[a, b]$ is a subspace of $C[a, b]$.

(e) Let $\mathbf{W}$ be the set of all functions $y \in C^1[a, b]$ satisfying the homogeneous linear differential equation $y' - y = 0$. It is not hard to prove that $\mathbf{W}$ is a subspace of $C^1[a, b]$.

(f) Let $p(x) = a_0 + a_1 x + \ldots + a_n x^n$ be a nonzero polynomial over the field F. If $a_n \neq 0$, we say that $p(x)$ has degree n. For example, $2x + 1$ has degree 1, $x^2 + x^5$ has degree 5, etc. The zero polynomial is not assigned a degree. Let $F[x]_n$ be the set consisting of the zero polynomial along with all polynomials $p(x)$ in $F[x]$ that have degree less than or equal to n. Then $F[x]_n$ is a subspace of $F[x]$.

$\square$

Section 2.4 Exercises

In most of the exercises below, Theorem 2.4.1 may be used to simplify the proof that a subset is in fact a subspace.

1. Show that the subset $\mathbf{V} = \{(x, y) \in \mathbb{R}^2 | x = 2y\}$ is a subspace of $\mathbb{R}^2$.

2. Show that the subset $\mathbf{V} = \{(x, y, z) \in \mathbb{R}^3 | x = 2y\}$ is a subspace of $\mathbb{R}^3$.

3. Show that the subset $\mathbf{V} = \{(x, y, z) \in \mathbb{R}^3 | x = 2y \text{ or } x = z\}$ is not a subspace of $\mathbb{R}^3$.

4. Show that the subset $\mathbf{V} = \{(x, y) \in \mathbb{R}^2 | x - y = 1\}$ is not a subspace of $\mathbb{R}^2$.

5. Let $\mathbf{V}$ be a subspace of $\mathbb{R}^2$ and let $\mathbf{W} = \{(x, y, z) \in \mathbb{R}^3 | (x, y) \in \mathbf{V} \text{ and } z = 0\}$. Show that $\mathbf{W}$ is a subspace of $\mathbb{R}^3$.

6. If $\mathbf{W}$ is a subspace of $\mathbb{R}^3$ and $\mathbf{V} = \{(x, y) \in \mathbb{R}^2 | (x, y, z) \in \mathbf{W} \text{ for some } z\}$, is $\mathbf{V}$ a subspace of $\mathbb{R}^2$? Explain.

7. If $\mathbf{U}$ and $\mathbf{W}$ are subspaces of a vector space $\mathbf{V}$, is it possible that $\mathbf{U} \cap \mathbf{W} = \emptyset$? (See Appendix A for definition of the intersection ($\cap$) of two sets and the definition of the empty set $\emptyset$.

8. If $\mathbf{U}$ and $\mathbf{W}$ are subspaces of a vector space $\mathbf{V}$, what can be said about $\mathbf{U} \cap \mathbf{W}$ and $\mathbf{U} \cup \mathbf{W}$? That is, can one prove that these sets form subspaces? ($\cap$ and $\cup$ are explained in Appendix A.)

9. Show that the set $\text{span}\{X_1, X_2\}$ in Example 2.4.1 (c) is a subspace of $\mathbf{V}$.

10. Prove that the set $\mathbf{W}$ in Example 2.4.1 (e) is a subspace of $C^1[a, b]$. Can you find a nonzero function in $\mathbf{W}$? Can you determine all functions in $\mathbf{W}$?

11. Let $\mathbf{U}$ be the set of all functions in $C^2[a, b]$ satisfying the differential equation $y'' + y = 0$.

    (a) Show that $\mathbf{U}$ is a subspace of $C^2[a, b]$.

    (b) Find two functions $f_1, f_2 \in \mathbf{U}$ such that $f_1 \neq c f_2$ for any constant c.

    (c) Show that $\mathbf{U} = \text{span}\{f_1, f_2\}$. (This requires some knowledge of differential equations.)

12. Consider the vector space $\mathbb{R}^2$. Show that the set of all vectors lying on a given straight line through the origin is a subspace of $\mathbb{R}^2$.

13. Show that the only subspaces of $\mathbb{R}^2$ (considered as a vector space over $\mathbb{R}$) are $\{0\}, \mathbb{R}^2$, and straight lines through the origin. What can be said about $\mathbb{R}^3$?

14. Show that any subspace $\mathbf{U}$ of $\mathbb{R}^2$ can be expressed in the form $\mathbf{U} = \text{span}\{X, Y\}$ for some not necessarily distinct vectors $X, Y \in \mathbb{R}^2$.

15. If $\mathbf{U}$ and $\mathbf{W}$ are subspaces of a vector space $\mathbf{V}$ and $\mathbf{U} \subseteq \mathbf{W}$, is $\mathbf{U}$ a subspace of $\mathbf{W}$? Explain.

16. Let $\mathbf{V}$ be the vector space of all n-dimensional column vectors and let $\mathbf{U}$ be a subspace of $\mathbf{V}$. Let P be an $n \times n$ matrix. Show that $\mathbf{W} = \{PX | X \in \mathbf{U}\}$ is also a subspace of V.

2.5 Linear Combinations and the Span of a Set of Vectors

In the examples in the previous section, we considered a subspace formed by taking all vectors of the form $rX_1 + sX_2$ where r and s are scalars and X_1 and X_2 are vectors. Because they are useful in describing subspaces, we will now consider more general sums of this form. We make the following definition.

Definition 2.5.1. Let $\mathbf{V}$ be a vector space and let $X_1, \ldots, X_n \in \mathbf{V}$ for some integer n. A **linear combination** of $X_1, \ldots, X_n$ is a vector X of the form $X = a_1 X_1 + \ldots + a_n X_n$ for some scalars $a_1, \ldots, a_n$.

Notice that a linear combination is a sum of finitely many scalar products. We will not have occasion to consider infinite sums or infinite series. We often say that "X is a linear combination of $X_1, \ldots, X_n$," meaning $X = a_1 X_1 + \ldots + a_n X_n$ for some $a_1, \ldots, a_n$.

We have seen these before: our alternative perspectives on matrix multiplication give us a product of two matrices as a matrix whose columns are linear combinations of the columns in the left matrix, or whose rows are linear combinations of the rows in the right matrix. That perspective will be helpful again in this section and beyond.

Example 2.5.1. $(-1, -16, 0) = 2(1, 1, 3) + (-3)(1, 6, 2)$ is a linear combination of $(1, 1, 3)$ and $(1, 6, 2)$. The vector $(-1, 4, -4)$ is also a linear combination of $(1, 1, 3)$ and $(1, 6, 2)$ since $(-1, 4, -4) = (-2)(1, 1, 3) + (1)(1, 6, 2)$. Notice that the vector $(1, 1, 4)$ is not a linear combination of $(1, 1, 3)$ and $(1, 6, 2)$, for $(1, 1, 4) = r(1, 1, 3) + s(1, 6, 2) = (r + s, r + 6s, 3r + 2s)$ implies that $r + s = 1 = r + 6s$. This means that $s = 0$ and $r = 1$, but this fails the condition $4 = 3r + 2s$.

$\square$

Example 2.5.2. If we consider the solution set $S = \{(x, y, z)^t | x + y + z = 0\}$ of the linear equation $x + y + z = 0$, we see that both of the vectors $(1, -1, 0)^t$ and $(0, 1, -1)^t$ are in S. Since S is closed under addition and scalar multiplication, any linear combination $r(1, -1, 0)^t + s(0, 1, -1)^t$ is also in S. We will see that S is the set of all of these linear combinations.

$\square$

THE SPAN

Definition 2.5.2. Let $\mathbf{V}$ be a vector space and let S be a nonempty subset of $\mathbf{V}$. By the **span** of S we mean the set of all linear combinations of elements of S; we denote the span of S by span(S). If $S = \{X_1, \ldots, X_n\}$ is finite, we will write span$\{X_1, \ldots, X_n\}$ for span(S), and by convention we define span $(\emptyset) = \{0\}$.

For a single vector, the span is the set of all scalar multiples of the vector. For two vectors, the span is the set of all sums of scalar multiples of the two vectors, etc. In $\mathbb{R}^3$, let $X_1 = (1, 2, 1)$, $X_2 = (-1, 0, 2)$, and $X_3 = (0, 2, 3)$. Let us calculate the spans of some of these vectors.

$$\text{span}\{X_1\} = \{rX_1 | r \text{ a scalar}\} = \{r(1, 2, 1) | r \text{ a scalar}\}$$

$$\text{span}\{X_2\} = \{rX_2 | r \text{ a scalar}\} = \{r(-1, 0, 2) | r \text{ a scalar}\}$$

$$\text{span}\{X_1, X_2\} = \{rX_1 + sX_2 | r, s \text{ scalars}\} = \{(r - s, 2r, r + 2s) | r, s \text{ scalars}\}.$$

Notice that since $X_3 = X_1 + X_2, X_3 \in \text{span}\{X_1, X_2\}$, and it follows that

$$\text{span}\{X_1, X_2, X_3\} = \text{span}\{X_1, X_2\}.$$

Notice that $\text{span}\{X_1, X_2\} \neq \mathbb{R}^3$ since $(0, 0, 1) \notin \text{span}\{X_1, X_2\}$.

The span of a collection of vectors may be difficult to determine. One can think of the span of a set of vectors geometrically. The scalar multiples of a single nonzero vector in, say, $\mathbb{R}^3$ form a straight line in space.

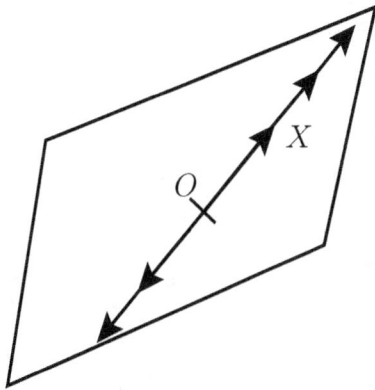

Figure 2.11: A line in space

Another vector along the same line adds nothing more to the span. The span of two nonzero, non-collinear vectors contains the plane determined by the two vectors. So, in Figure 2.12, the vectors X and Y generate all of the vectors lying in the plane.

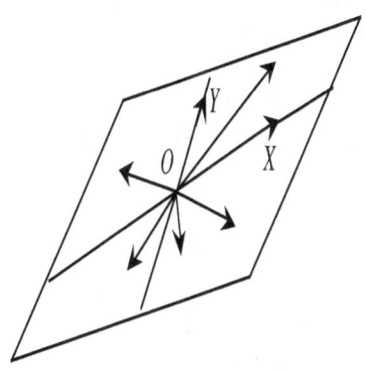

Figure 2.12: A plane spanned by X and Y

We can see that the span of a set of vectors is always defined, even though it may be difficult to determine. What are the properties of the span, how does it behave? In Example 2.4.1 (c), we saw that the span of a set of vectors is always a subspace of the vector space from which the set is taken. This result, along with other properties of spans and linear combinations are listed in the following:

Theorem 2.5.1. *Let* **V** *be a vector space.*

(a) *If* $S \subseteq \mathbf{V}$, *then* $span(S)$ *is a subspace of* **V** *and* $S \subseteq span(S)$.
   (The span of a set is a subspace containing the set.)

(b) *If* $S \subseteq S' \subseteq \mathbf{V}$, *then* $span(S) \subseteq span(S')$.
   (The span of a bigger set is bigger.)

(c) $X \in span(S)$ *if and only if* $span(S \cup \{X\}) = span(S)$.
   (A vector is in the span of a set iff it doesn't increase the span.)

(d) *If* **W** *is a subspace of* **V**, *and* $S \subseteq \mathbf{W}$, *then* $span(S) \subseteq \mathbf{W}$.
   (The span of a subset of a subspace lies in the subspace.)

Proof. (a) If $S = \emptyset$, then $span(S) = \{0\}$ and this set is a subspace of **V**. Assume $S \neq \emptyset$, and say $X \in S$. Then $0 = 0X \in span(S)$ and so $span(S) \neq \emptyset$. If $X, Y \in span(S)$, then

$$X = a_1 X_1 + \ldots + a_n X_n \text{ and } Y = b_1 Y_1 + \ldots + b_m Y_m$$

for some vectors $X_1, \ldots, X_n, Y_1, \ldots, Y_m \in S$ and some scalars $a_1, \ldots, a_n, b_1, \ldots, b_m$. Now

$$X + Y = a_1 X_1 + \ldots + a_n X_n + b_1 Y_1 + \ldots + b_m Y_m \in span(S);$$

that is, a sum of linear combinations of elements of S is a linear combination of elements of S. Also, if r is some scalar, $rX = (ra_1)X_1 + \ldots + (ra_n)X_n \in span(S)$. It follows that $span(S)$ is a subspace of **V**.

Part (b) is straightforward and Parts (c) and (d) are left as exercises.

$\square$

The rows of a matrix form a set of vectors. One wonders whether the span of these vectors provides some interesting characterization of the matrix.

THE ROW SPACE

Definition 2.5.3. Let $A = [a_{ij}]$ be an $m \times n$ matrix. By **row** i of A, we mean the n-tuple (or $1 \times n$ matrix) $(a_{i1}, a_{i2}, \ldots, a_{in})$. We will denote row i of A by A_i. The span of the rows of A is the **row space** of A and is denoted by $\boldsymbol{R}(A) = span\{A_1, A_2, \ldots, A_m\}$.

Let A be the 3×3 matrix

$$A = \begin{bmatrix} 1 & -1 & 0 \\ 2 & 1 & 2 \\ 0 & 2 & 1 \end{bmatrix}.$$

Then with the above notation, $A_1 = (1, -1, 0)$, $A_2 = (2, 1, 2)$, and $A_3 = (0, 2, 1)$. The row space of A is $\boldsymbol{R}(A) = span\{A_1, A_2, A_3\}$ and it can be shown that $\boldsymbol{R}(A) = \mathbb{R}^3$ for this particular A. Notice also that according to our alternative perspective on matrix multiplication,

$$\boldsymbol{R}\left(A\right) = \left\{\begin{bmatrix} x_1 & x_2 & \cdots & x_m \end{bmatrix} A : x_1, x_2, \ldots, x_m \in \mathbb{R}\right\}.$$

Theorem 2.5.2. *Let A and B be $m \times n$ matrices and assume that A is row equivalent to B. Then $\boldsymbol{R}\left(A\right) = \boldsymbol{R}\left(B\right)$.*   (Equivalent matrices have the same row space.)

Proof. Recall that if A is row equivalent to B, then B may be obtained from A by a finite sequence of elementary row operations. So, to prove $\boldsymbol{R}\left(A\right) = \boldsymbol{R}\left(B\right)$ it is sufficient to prove that the row space of A does not change when a single elementary row operation is performed on it. Recall that there are three types of row operations and three types of elementary row matrices corresponding to these operations: $R_i(a), R_{ij}$, and $R_{ij}(a)$. We take three cases (recall that A_i denotes the row i of A).

$\boldsymbol{R}(A) = \boldsymbol{R}(R_{ik}A)$: This part is straightforward: any linear combination of $A_1, \ldots, A_i, \ldots, A_k, \ldots, A_m$ is also a linear combination of $A_1, \ldots, A_k, \ldots, A_i, \ldots, A_m$.

$\boldsymbol{R}\left(A\right) = \boldsymbol{R}\left(R_i(a)A\right)$: Here we assume $a \neq 0$, so a^{-1} exists. If $X \in \boldsymbol{R}\left(A\right)$, then

$$\begin{aligned} X &= a_1 A_1 + \ldots + a_i A_i + \ldots + a_m A_m \\ &= a_1 A_1 + \ldots + a_i a^{-1}(aA_i) + \ldots + a_m A_m \end{aligned}$$

and so $X \in \boldsymbol{R}\left(R_i(a)A\right)$. So we see that, $\boldsymbol{R}\left(A\right) \subseteq \boldsymbol{R}\left(R_i(a)A\right)$. Also, $Y \in \boldsymbol{R}\left(R_i(a)A\right)$ implies

$$\begin{aligned} Y &= b_1 A_1 + \ldots + b_i(aA_i) + \ldots + b_m A_m \\ &= b_1 A_1 + \ldots + (b_i a)A_i + \ldots + b_m A_m \in \boldsymbol{R}\left(A\right). \end{aligned}$$

It follows that $\boldsymbol{R}\left(A\right) = \boldsymbol{R}\left(R_i(a)A\right)$.

$\boldsymbol{R}\left(A\right) = \boldsymbol{R}\left(R_{ik}(a)A\right)$: Let us assume that $i < k$. As above, let

$$X = a_1 A_1 + \ldots + a_m A_m \in \boldsymbol{R}\left(A\right).$$

Then

$$X = a_1 A_1 + \ldots + (a_i - aa_k)A_i + \ldots + a_k(A_k + aA_i) + \ldots + a_m A_m$$

and so $X \in \boldsymbol{R}\left(R_{ik}(a)A\right)$.

The proof that $\boldsymbol{R}\left(R_{ik}(a)A\right) \subseteq \boldsymbol{R}\left(A\right)$ is left as an exercise.   □

We have considered the rows of a matrix, what about the columns? If $A = [a_{ij}]$ is an $m \times n$ matrix, column j of A is the $m \times 1$ column vector

$$\begin{bmatrix} a_{1j} \\ a_{2j} \\ \vdots \\ a_{mj} \end{bmatrix}.$$

If we let $A_1, \ldots, A_n$ denote the columns of A, then $\mathrm{span}\{A_1, \ldots, A_n\}$ is the **column space** of A and it is denoted by $\boldsymbol{C}(A)$. Of course, $\boldsymbol{C}(A)$ is a subspace of the vector space of all m-dimensional column vectors. One naturally wonders whether there is some relationship between the row space of a matrix and its column space.

One final note:

$$\boldsymbol{C}(A) = \left\{ A \begin{bmatrix} x_1 \\ x_2 \\ \vdots \\ x_n \end{bmatrix} : x_1, x_2, \ldots, x_n \in \mathbb{R} \right\}$$

is the set of all linear combinations of the columns of A by our alternative perspective on matrix multiplication.

Section 2.5 Exercises

1. Show that $\mathbb{R}^2 = \mathrm{span}\{(1, -1), (2, 2)\}$.

2. Find a vector $X \in \mathbb{R}^3$ with $X \notin \mathrm{span}\{(1, -1, 2), (2, 2, 1)\}$.

3. Determine whether $(1, -2, 1) \in \mathrm{span}\{(1, 3, 1), (1, 0, -1)\}$. Give reasons.

4. Show that $\mathrm{span}\{(2, -4), (-1, 2)\} = \mathrm{span}\{(1, -2)\}$.

5. Determine whether $(5, 1, 6) \in \mathrm{span}\{(1, 2, 3), (-1, 1, 0)\}$. Give reasons.

6. Let $A = \begin{bmatrix} 1 & -1 \\ 2 & 0 \end{bmatrix}$. Show that $\boldsymbol{R}(A) = \mathbb{R}^2$.

7. Let A be an $m \times n$ matrix and let B be the reduced echelon form of A. What is the relationship between $\boldsymbol{R}(A)$ and $\boldsymbol{R}(B)$?

8. Prove parts (c) and (d) of Theorem 2.5.1.

9. Complete the proof of Theorem 2.5.2.

10. Explain what the following statement should mean and then prove it:

    If $\mathbf{V}$ is a vector space and $S \subseteq \mathbf{V}$, then $\mathrm{span}(S)$ is the smallest subspace of $\mathbf{V}$ containing S.

11. Show that $\mathbb{R}^2 \neq \mathrm{span}\{X\}$ for any vector $X \in \mathbb{R}^2$.

12. Show that $\mathbb{R}^3 \neq \mathrm{span}\{X, Y\}$ for any vectors $X, Y \in \mathbb{R}^3$.

13. Let A be an $m \times n$ matrix. How may one determine the least number k such that $\boldsymbol{R}(A) = \mathrm{span}\{X_1, X_2, \ldots, X_k\}$ for some vectors $X_1, X_2, \ldots, X_k$.

14. Let A and B be 2×2 matrices in reduced echelon form. Assume $\boldsymbol{R}(A) = \boldsymbol{R}(B)$. Prove that $A = B$.

15. Let A and B be $m \times n$ matrices in reduced echelon form. Assume $\boldsymbol{R}(A) = \boldsymbol{R}(B)$. Prove that $A = B$.

2.6 Linear Independence

Given a set of vectors, there is a certain set spanned by these vectors. In some cases a smaller set will span the same set, in other cases, no smaller set has the same span. When the latter case occurs, the vectors are said to be "linearly independent."

Let us consider the reduction of an $m \times n$ matrix A to its reduced echelon form B. As in Theorem 1.5.1, this can be accomplished by a finite sequence of elementary row operations, and as a consequence, it is not hard to see that every row of B is a linear combination of the rows of A. We will see that if a row of zeros appears in B then the rows of A must have been "linearly dependent." If no row of zeros appears, the rows of A must have been "linearly independent."

Example 2.6.1. If

$$A = \begin{bmatrix} 1 & 2 & 0 \\ 2 & 4 & 1 \\ 3 & 6 & 1 \end{bmatrix},$$

then B, the reduced echelon form, is given by

$$B = \begin{bmatrix} 1 & 2 & 0 \\ 0 & 0 & 1 \\ 0 & 0 & 0 \end{bmatrix}$$

and B is obtained by the following sequence of row operations: $R_{12}(-2)R_{13}(-3)R_{23}(-1)$. The rows of B and A are related by $B_1 = A_1, B_2 = A_2 + (-2)A_1$, and

$$B_3 = 0 = A_3 + (-3)A_1 + (-1)B_2 = A_3 + (-3)A_1 + (-1)[A_2 + (-2)A_1] = A_3 - A_1 - A_2.$$

$\square$

From the above example, it is seen that if B, the reduced echelon form of A, has a row of zeros, then it must be the case that a linear combination of some of the rows of A was equal to one of the other rows of A, say $A_m = a_1 A_1 + \ldots + a_{m-1} A_{m-1}$, so that, when the linear combination was subtracted, the zero vector was obtained:

$$-a_1 A_1 - a_2 A_2 - \ldots - a_{m-1} A_{m-1} + A_m = 0.$$

The zero vector was obtained as a linear combination of vectors and at least one of the coefficients in the linear combination was nonzero. This last observation must be made if the statement is to have any significant meaning, since the zero vector is clearly a linear combination of any collection of vectors-take all of the coefficients to be the zero scalar. This condition is known as "linear dependence."

LINEAR DEPENDENCE

Definition 2.6.1. Let $\mathbf{V}$ be a vector space and let $X_1, \ldots, X_n \in \mathbf{V}$. The n vectors $X_1, \ldots, X_n$ are **linearly dependent** if and only if there are scalars $a_1, \ldots, a_n$, not all of which are zero, such that $a_1 X_1 + \ldots + a_n X_n = 0$. We also say that the set of vectors $\{X_1, \ldots, X_n\}$ is linearly dependent.

Consider the following vectors in $\mathbb{R}^3$: $X_1 = (1, 2, -1)$, $X_2 = (0, 1, 1)$, and $X_3 = (2, 6, 0)$. These vectors are linearly dependent. To see this, we must find a nonzero solution of $aX_1 + bX_2 + cX_3 = 0$, or $a(1, 2, -1) + b(0, 1, 1) + c(2, 6, 0) = 0$. Evaluating the left-hand side of this equation we get $(a + 2c, 2a + b + 6c, -a + b) = (0, 0, 0)$, and so we must find a nonzero solution of the homogeneous system of equations:

$$
\begin{array}{rcrcrcl}
a & + & & & 2c & = & 0 \\
2a & + & b & + & 6c & = & 0 \\
-a & + & b & & & = & 0.
\end{array}
$$

We can solve this system using the methods of Section 1.6, or we can solve it by "inspection." Note from the last equation that $a = b$. Let $a = b = 1$. From the first equation, we see then that $c = 1/2$, and we check that these values satisfy the second equation. Thus, $X_1 + X_2 + (-1/2)X_3 = (1, 2, -1) + (0, 1, 1) + (-1, -3, 0) = (0, 0, 0)$. Notice that because of this relationship, we can "solve" for one of these vectors as a linear combination of the others, for example, $X_3 = 2X_1 + 2X_2$. This happens in general:

> **Theorem 2.6.1.** *Let $X_1, \ldots, X_n$ be elements in the vector space $V, n > 1$. $X_1, \ldots, X_n$ are linearly dependent if and only if one of the vectors may be expressed as a linear combination of the remaining vectors.*
> (In a set of dependent vectors, one vector is contained in the span of the others.)

Proof. ($\Rightarrow$) Assume $X_1, \ldots, X_n$ are linearly dependent with $a_1 X_1 + \ldots + a_n X_n = 0$ and some $a_i \neq 0$. For convenience assume $a_1 \neq 0$. Then we see that $X_1 = (-a_2/a_1)X_2 + \ldots + (-a_n/a_1)X_n$.

($\Leftarrow$) Now assume that some vector in the list is a linear combination of the others. For convenience we may assume that X_1 is a linear combination of the vectors $X_2, \ldots, X_n$; say $X_1 = a_2 X_2 + \ldots + a_n X_n$. But then we see that $1X_1 - a_2 X_2 - \ldots - a_n X_n = 0$, and so $X_1, \ldots, X_n$ are linearly dependent.

$\square$

If three vectors X, Y, and Z in $\mathbb{R}^3$ are linearly dependent, then the above theorem tells us what must happen. One of the vectors, say X, must lie in the span of the other two, Y and Z. The vectors Y and Z are either collinear or they determine a plane, and so X lies along the line given by X and Y or X lies in the plane of the two vectors.

> **Theorem 2.6.2.** *Let A and B be $m \times n$ matrices and assume that A is row equivalent to B. Then the rows of A are linearly dependent if and only if the rows of B are linearly dependent. Thus, if the reduced echelon form B of A has a row of zeros, then the rows of A are linearly dependent.* (If a matrix has dependent rows, so does any equivalent matrix.)

Proof. Assume that A is row equivalent to B. Recall that this means that B may be obtained from A by a sequence of elementary row operations. Also recall that elementary

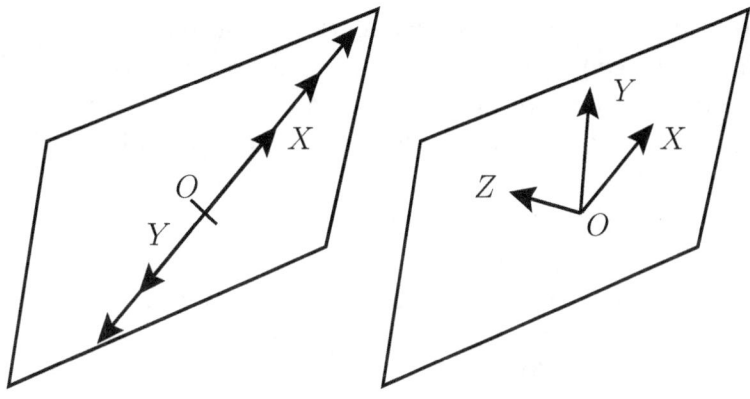

Figure 2.13: The span of X and Y in space

row operations are reversible so that the matrix A may be obtained from B by a sequence of elementary row operations. Because of these facts, it is sufficient to consider the effect of a single elementary row operation performed on the matrix A. We assume that the rows of A are linearly dependent and we consider B to be the result of a single elementary operation.

Since the rows of A are dependent, there is a linear combination equal to 0 with a nonzero coefficient; say $0 = a_1 A_1 + \ldots + a_m A_m$, with some a_i nonzero. Assume that $i < k$, and consider the following:

$$B = R_{ik}A :$$
$$0 = a_1 A_1 + \ldots + a_i A_i + \ldots + a_k A_k + \ldots + a_m A_m$$
$$0 = a_1 A_1 + \ldots + a_k A_k + \ldots + a_i A_i + \ldots + a_m A_m$$
$$0 = a_1 B_1 + \ldots + a_k B_i + \ldots + a_i B_k + \ldots + a_m B_m$$

$$B = R_i(c)A :$$
$$0 = a_1 A_1 + \ldots + a_i A_i + \ldots + a_m A_m$$
$$0 = a_1 A_1 + \ldots + (a_i/c)cA_i + \ldots + a_m A_m$$
$$0 = a_1 B_1 + \ldots + (a_i/c)B_i + \ldots + a_m B_m$$

$$B = R_{ik}(c)A :$$
$$0 = a_1 A_1 + \ldots + a_i A_i + \ldots + a_k A_k + \ldots + a_m A_m$$
$$0 = a_1 A_1 + \ldots + (a_i - ca_k)A_i + \ldots + a_k(A_k + cA_i) + \ldots + a_m A_m$$
$$0 = a_1 B_1 + \ldots + (a_i - ca_k)B_i + \ldots + a_k B_k + \ldots + a_m B_m$$

It is not hard to see (but it does need to be checked) that the assumption that there is a nonzero coefficient in the linear combination of the rows of A implies that a nonzero coefficient will appear in the linear combination of the rows of the matrix B. $\square$

Some sets of vectors are clearly linearly dependent. The zero vector by itself forms a linearly dependent set $(1 \cdot 0 = 0)$, and similarly, any set of vectors that contains the zero vector must be linearly dependent. It follows that if a matrix contains a row of 0's, then the rows of that matrix must be linearly dependent. These remarks give us an easy way to determine whether a set of vectors are dependent: form a matrix with the rows and reduce the matrix to reduced echelon form. If the echelon form has a row of 0's then the rows of the echelon form, and so also the original vectors, must be dependent.

Some sets of vectors are not linearly dependent. For example, if X is a single nonzero vector and we assume that a linear combination aX of X is zero, then, using Theorem 2.3.1, we see that $a = 0$ since $X \neq 0$. We will see that the nonzero rows of a matrix in reduced echelon form are not linearly dependent.

For a better example, consider $(1,0), (0,1) \in \mathbb{R}^2$. If $a(1,0) + b(0,1) = (0,0)$, then $(a,b) = (0,0)$ so that $a = b = 0$. We see that the zero vector may not be expressed as a linear combination of $(1,0)$ and $(0,1)$ unless both coefficients are zero. It follows that the single nonzero vector X is not linearly dependent and the two vectors $(1,0)$ and $(0,1)$ in $\mathbb{R}^2$ are not linearly dependent - they are "linearly independent."

LINEAR INDEPENDENCE

Definition 2.6.2. Let **V** be a vector space and let $X_1, \ldots, X_n \in$ **V**. The vectors $X_1, \ldots, X_n$ are **linearly independent** if and only if $X_1, \ldots, X_n$ are not linearly dependent.

Following the terminology used above we also say that the set of vectors $\{X_1, \ldots, X_n\}$ is linearly independent. This opens the question of whether the empty set should be regarded as linearly independent or linearly dependent. If $\emptyset$ is linearly dependent then there exist vectors $X_1, \ldots, X_n$ in $\emptyset$ with the appropriate nonzero linear combination. But no vectors exist in $\emptyset$, so $\emptyset$ is not linearly dependent and so it must be linearly independent.

What does it mean for a collection of vectors to be linearly independent? Using Theorem 2.6.1, we see that in order for a collection of vectors to be independent, no one of the vectors may lie in the span of the others. Figure 2.14 shows the situation in $\mathbb{R}^3$ when three vectors X, Y, and Z are linearly independent. X cannot lie in the span of Y and so these vectors cannot be collinear. The vector Z cannot lie in the span of X and Y and so it lies outside the plane of X and Y.

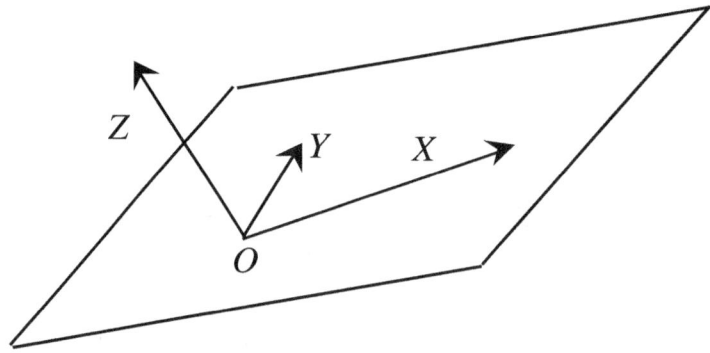

Figure 2.14: Z is not in the span of X and Y

The definition of linear independence can cause confusion. The problem arises in trying to prove that a collection of vectors is not linearly dependent. For example, consider the vectors

$E_1 = (1, 0, \ldots, 0), E_2 = (0, 1, 0, \ldots, 0), \ldots, E_n = (0, \ldots, 0, 1)$ in $\mathbb{R}^n$. We will establish that these vectors are linearly independent. We argue as follows: Assume that $a_1, \ldots, a_n$ are some scalars and $a_1 E_1 + \ldots + a_n E_n = (0, \ldots, 0)$. Then, collecting terms, we get $a_1 E_1 + \ldots + a_n E_n = (a_1, a_2, \ldots, a_n) = (0, 0, \ldots, 0)$. Hence, $a_1 = 0, a_2 = 0, \ldots, a_n = 0$. We can now conclude that the vectors $E_1, \ldots, E_n$ are not linearly dependent (that is, they are linearly independent), for if they were, there would be a linear combination that was equal to zero, but that had a nonzero coefficient. The above argument shows this to be impossible.

Theorem 2.6.3. *Let* $\mathbf{V}$ *be a vector space and let* $X_1, \ldots, X_n \in \mathbf{V}$. *Then* $X_1, \ldots, X_n$ *are linearly independent if and only if the following condition holds:*

(*) $a_1 X_1 + \ldots + a_n X_n = 0$ *for some scalars* $a_1, \ldots, a_n$ *implies* $a_1 = \ldots = a_n = 0$.

(Linear independence means that if a combination is 0, all of the coefficients must be 0.)

Proof. Notice that the statement of the theorem involves an equivalence - if and only if. We must prove two implications: linear independence implies (*) and (*) implies linear independence. The symbol $\implies$ is used for implication. The two parts of the proof are indicated by ($\implies$) and ($\impliedby$).

($\implies$) Assume $X_1, \ldots, X_n$ are linearly independent and $a_1 X_1 + \ldots + a_n X_n = 0$. Then $a_1 = a_2 = \ldots = a_n = 0$, for otherwise $X_1, \ldots, X_n$ would be linearly dependent and we would have a contradiction.

($\impliedby$) Assume that the condition (*) holds. If the vectors $X_1, \ldots, X_n$ were linearly dependent, there would be a linear combination $a_1 X_1 + \ldots + a_n X_n = 0$ with some coefficient nonzero. The condition (*) makes this impossible. It follows that $X_1, \ldots, X_n$ are not linearly dependent and so $X_1, \ldots, X_n$ are linearly independent. $\square$

Example 2.6.2. The vectors $(1, 1)$ and $(1, -1)$ in $\mathbb{R}^2$ are linearly independent, for $a(1, 1) + b(1, -1) = (0, 0)$ implies $(a + b, a - b) = (0, 0)$ so that $a + b = 0$ and $a - b = 0$. Adding these equations, we see that $2a = 0$, so $a = 0$ and $a + b = b = 0$. We have shown that $(1, 1)$ and $(1, -1)$ are linearly independent.

$\square$

Example 2.6.3. Consider $(1, 0, 0), (1, 1, 0), (1, 1, 1) \in \mathbb{R}^3$. These vectors are linearly independent, for $a(1, 0, 0) + b(1, 1, 0) + c(1, 1, 1) = (0, 0, 0)$ implies that $a + b + c = 0, b + c = 0$, and $c = 0$. Now $c = 0$ and $b + c = 0$ imply $b = 0$, and then $a + b + c = 0$ together with $b = c = 0$ imply $a = 0$. It follows that $a = b = c = 0$ and that the vectors are linearly independent.

$\square$

From Section 2.5, we know that a matrix and its reduced echelon form have the same row space. What can be said concerning the reduced echelon form and linear independence?

INDEPENDENCE AND THE REDUCED ECHELON FORM

The next result tells us that the nonzero rows in a matrix in echelon form are linearly independent. We will see in the next section that the number of these rows is the same as the maximum number of linearly independent rows in the original matrix.

Theorem 2.6.4. *Let* $B = [b_{ij}]$ *be an* $m \times n$ *matrix in echelon form and let* $r, j_1, j_2, \ldots, j_r$ *be the constants as in the definition of echelon form.   Then the nonzero rows* $B_1, \ldots, B_r$ *of* B *are linearly independent.*

(The nonzero rows in a matrix in echelon form are independent.)

Proof. Recall that r is the number of nonzero rows and j_i is the column in which the first nonzero entry in row i occurs. Assume $a_1 B_1 + \ldots + a_r B_r = 0$. Then in column j_1 of this linear combination (remember that the linear combination is just a $1 \times n$ row vector), we find a_1 since in column j_1 of B there is only one nonzero entry and it is a 1 in row 1. It follows that $a_1 = 0$. Now we have $a_2 B_2 + \ldots + a_r B_r = 0$, and similar reasoning shows that $a_2 = 0$. Continuing, we see that $a_1 = \ldots = a_r = 0$ and so we have shown that $B_1, \ldots, B_r$ are linearly independent. $\square$

The following result is very important and will be the basis for the work in the next section. This theorem will help us to see that the number of vectors in a linearly independent spanning set is an invariant quantity.

Theorem 2.6.5. *Let* $\mathbf{V}$ *be a vector space with* $X_1, \ldots, X_n \in \mathbf{V}$ *and assume*

$$\mathbf{V} = span\{X_1, \ldots, X_n\}.$$

If $Y_1, \ldots, Y_{n+1} \in \mathbf{V}$, *then* $Y_1, \ldots, Y_{n+1}$ *are linearly dependent.*

(In a vector space spanned by n vectors, any n + 1 vectors are dependent.)

Proof. Since the vectors $X_1, \ldots, X_n$ span $\mathbf{V}$, each of the vectors $Y_1, \ldots, Y_{n+1}$ may be expressed as a linear combination of the vectors $X_1, \ldots, X_n$, say

$$
\begin{aligned}
Y_1 &= a_{11}X_1 &+ \ldots + & a_{1n}X_n \\
Y_2 &= a_{21}X_1 &+ \ldots + & a_{2n}X_n \\
&\ \ \vdots & \vdots \qquad & \quad \vdots \\
Y_{n+1} &= a_{n+1,1}X_1 &+ \ldots + & a_{n+1,n}X_n.
\end{aligned}
\tag{2.6.1}
$$

Form a matrix from the coefficients and call it A. Then

$$
A = \begin{bmatrix}
a_{11} & \cdots & a_{1n} \\
a_{21} & \cdots & a_{2n} \\
\vdots & \ddots & \vdots \\
a_{n+1,n} & \cdots & a_{n+1,n}
\end{bmatrix}.
\tag{2.6.2}
$$

Notice that A is $(n+1) \times n$. Now reduce the matrix A to its reduced echelon form B (possible by Theorem 1.5.1). Let $r, j_1, \ldots, j_r$ be the constants associated with B. Recall that r is the number of nonzero rows, j_i is the column containing the first nonzero entry in row i and $j_1 < j_2 < \ldots < j_r$. Now A, and so B, has n columns and since $1 < j_1 < j_2 < \ldots < j_r \leq n$, it must be the case that $r \leq n$. But B has $n+1$ rows and so (since $r < n+1$) at least one row of B must be zero. Since B has a row of zeros, its rows are linearly dependent and so by Theorem 2.6.2, the rows of A are linearly dependent. Thus we have a linear combination of the form

$$b_1 A_1 + \ldots + b_{n+1} A_{n+1} = 0$$

with some $b_i \neq 0$. Note that the components in this linear combination are

$$b_1 a_{11} + b_2 a_{21} + \ldots + b_n a_{n1} \text{ in the first coordinate,}$$

$$b_1 a_{12} + b_2 a_{22} + \ldots + b_n a_{n2} \text{ in the second, etc.}$$

These expressions are all equal to 0 and they are the coefficients of $X_1, \ldots, X_n$ in the expansion of the linear combination $b_1 Y_1 + \ldots + b_{n+1} Y_{n+1}$. Thus, $b_1 Y_1 + \ldots + b_{n+1} Y_{n+1} = 0$ with one of the coefficients not equal to 0. In other words, since $b_1 A_1 + \ldots + b_{n+1} A_{n+1} = 0$, we see that $b_1 Y_1 + \ldots + b_{n+1} Y_{n+1} = 0$.

It follows that $Y_1, \ldots, Y_{n+1}$ are linearly dependent. $\qquad \square$

Corollary 2.6.6. *If* $\mathbf{V}$ *is a vector space,* $\mathbf{V} = span\{X_1, \ldots, X_n\}$, *and* $Y_1, \ldots, Y_m$ *are linearly independent in* $\mathbf{V}$, *then* $m \leq n$. *(Independent sets must contain no more vectors than spanning sets.)*

Proof. If $m > n$, then $m \geq (n+1)$. By assumption, $Y_1, \ldots, Y_m$ are linearly independent, and so $Y_1, \ldots, Y_{n+1}$ are linearly independent, since a subset of a linearly independent set is linearly independent by Exercise 10. This is impossible by Theorem 2.6.5 and so $m \leq n$. $\quad \square$

The following result is useful in extending linearly independent sets of vectors.

Theorem 2.6.7. *Let* $\mathbf{V}$ *be a vector space and let* $X_1, \ldots, X_n$ *be linearly independent vectors in* $\mathbf{V}$. *Then* $X_1, \ldots, X_n, X_{n+1}$ *are linearly independent for some* $X_{n+1} \in \mathbf{V}$ *if and only if* $X_{n+1} \notin span\{X_1, \ldots, X_n\}$. *(Choosing vectors outside the span expands an independent set.)*

Proof. We will prove the first part and leave the remainder as an exercise. Assume

$$X_1, \ldots, X_n, X_{n+1}$$

are linearly independent, but that $X_{n+1} \in span\{X_1, \ldots, X_n\}$. Then $X_{n+1} = a_1 X_1 + \ldots + a_n X_n$ and so $a_1 X_1 + \ldots + a_n X_n + (-1)X_{n+1} = 0$. This linear combination being zero with the coefficient of X_{n+1} nonzero implies that $X_1, \ldots, X_n, X_{n+1}$ are linearly dependent, a contradiction. It follows that $X_{n+1} \notin span\{X_1, \ldots, X_n\}$. $\qquad \square$

This theorem has an extremely useful application that relates independence and spanning. It is contained in the following corollary, the proof of which is left as Exercise 17.

Corollary 2.6.8. *Let* **V** *be a vector space with* $V = span\{X_1, \ldots, X_n\}$. *If* $Y_1, \ldots, Y_n$ *are linearly independent in* **V** *then* $V = span\{Y_1, \ldots, Y_n\}$.

We defined the terms linear dependence and linear independence for a finite set $X_1, \ldots, X_n$ of vectors. If **V** is a vector space and $S \subseteq V$ (the set S may be infinite), we say that S is **linearly dependent** if and only if some finite subset of S is linearly dependent. It follows that S is linearly dependent if and only if there exist vectors $X_1, \ldots, X_n \in S$ and scalars $a_1, \ldots, a_n$ not all of which are zero, with $a_1 X_1 + \ldots + a_n X_n = 0$. The set S is **linearly independent** if and only if it is not linearly dependent; that is, S is linearly independent if and only if every finite subset of S is linearly independent.

Section 2.6 Exercises

1. Show that the vectors $(1, 2), (1, 1)$ and $(3, 2)$ in $\mathbb{R}^2$ are linearly dependent by using the definition. Which theorem also applies?

2. Show that the vectors $(1, 0, 0), (1, 1, 0)$ and $(1, 0, 1)$ in $\mathbb{R}^3$ are linearly independent.

3. For the vectors $E_1, \ldots, E_n$ in $\mathbb{R}^n$ defined in this section, prove that $span\{E_1, \ldots, E_n\} = \mathbb{R}^n$.

4. Show that the vectors $(1, 2, -1, 0), (1, 0, -1, 0), (0, 1, 0, 1)$ in $\mathbb{R}^4$ are linearly independent.

5. Determine whether the vectors $(1, 1, 2), (2, -1, 1), (0, 3, 3)$ in $\mathbb{R}^3$ are linearly independent or linearly dependent.

6. Determine whether the vectors $(1, 2, -1, 2), (3, 1, 1, 1)$, and $(-4, 2, -4, 2)$ in $\mathbb{R}^4$ are linearly independent or linearly dependent.

7. Let $X_1, \ldots, X_4$ be four vectors in $\mathbb{R}^3$ with the pair of vectors X_1, X_2 linearly independent and the pair X_3, X_4 also linearly independent. Show that there is a nonzero vector in $span\{X_1, X_2\} \cap span\{X_3, X_4\}$.

8. Let Z be the zero vector. Prove that $\{Z\}$ is linearly dependent.

9. Let $X_1, \ldots, X_n$ be a set of vectors and assume one of them is the zero vector. Prove that $X_1, \ldots, X_n$ are linearly dependent.

10. Let **V** be a vector space and let $S \subseteq S' \subseteq V$. Prove that:

    (a) If S' is linearly independent, then S is linearly independent.

    (b) If S is linearly dependent, then S' is linearly dependent.

11. Complete the proof of Theorem 2.6.7.

12. Let $X_1, \ldots, X_n$ be vectors in some vector space $\mathbf{V}$. Assume that two of the vectors in the list are equal. Prove that the vectors are linearly dependent. (To simplify the proof it is convenient to assume that $X_1 = X_2$.) Note that this makes our terminology referring to an independent set of vectors inconsistent, for $\{X_1, \ldots, X_n\}$ is a dependent set of vectors, but $\{X_2, \ldots, X_n\}$ may be linearly independent and the two sets may be equal.

13. Can one find vectors $X_1, X_2, X_3 \in \mathbb{R}^2$ with $X_1 \neq 0, X_2 \notin \text{span}\{X_1\}$, and $X_3 \notin \text{span}\{X_1, X_2\}$?

14. Let X_1 and X_2 be linearly independent vectors in $\mathbb{R}^2$ and let X be any other vector in $\mathbb{R}^2$. Show that $X \in \text{span}\{X_1, X_2\}$.

15. Let X_1 and X_2 be vectors in $\mathbb{R}^2$ with $\mathbb{R}^2 = \text{span}\{X_1, X_2\}$. Show that X_1 and X_2 are linearly independent.

16. Let $X_1, \ldots, X_n$ be $n \times 1$ column vectors, and let P be an $n \times n$ nonsingular matrix. Prove that $X_1, \ldots, X_n$ are linearly independent if and only if $PX_1, \ldots, PX_n$ are linearly independent. (Can a more general theorem be stated?)

17. Prove Corollary 2.6.8.

2.7 Basis and Dimension

We are now in a position to characterize certain vector spaces in two nice ways - the first is by finding a "basis" and the second is by its "dimension." Our notion of dimension coincides with the usual notion of dimension, and the notion of a basis allows us to characterize a vector space (that is usually infinite) in terms of finitely many of its elements.

Definition 2.7.1. Let $\mathbf{V}$ be a vector space and let $B \subseteq \mathbf{V}$. We say that B is a **basis** for $\mathbf{V}$ if and only if B is a linearly independent set and $\text{span}(B) = \mathbf{V}$. We assume as a convention that $\emptyset$ is a basis for the zero vector space. *A basis is an independent set that spans.*

For example, the vectors $(1, 0), (0, 1)$ in $\mathbb{R}^2$ are linearly independent, as we saw in Section 2.6, and the span of the two vectors is $\mathbb{R}^2$, since

$$(a, b) = a(1, 0) + b(0, 1) \in \text{span}\{(1, 0), (0, 1)\}.$$

Thus, the set $\{(1, 0), (0, 1)\}$ is a basis for $\mathbb{R}^2$, or as we will often say, "the vectors $(1, 0)$ and $(0, 1)$ form a basis for $\mathbb{R}^2$." We also have observed that the vectors $(1, 1)$ and $(1, -1)$ are linearly independent in $\mathbb{R}^2$. It is not hard to show that $\text{span}\{(1, 1), (1, -1)\} = \mathbb{R}^2$ and so the vectors $(1, 1)$ and $(1, -1)$ form a basis for $\mathbb{R}^2$. The basis $\{(1, 0), (0, 1)\}$ seems more natural, or standard.

THE STANDARD BASIS

The vectors $E_1, \ldots, E_n \in \mathbb{R}^n$ were defined in Section 2.6 and shown to be linearly independent; recall that E_i is the vector with a 1 in the i-th coordinate and 0's in the other $n-1$ coordinates. It is not hard to see that $(a_1, a_2, \ldots, a_n) = a_1 E_1 + a_2 E_2 + \ldots + a_n E_n$ and so $\text{span}\{E_1, \ldots, E_n\} = \mathbb{R}^n$. It follows that $\{E_1, \ldots, E_n\}$ is a basis for $\mathbb{R}^n$. This basis is called the **standard basis** for $\mathbb{R}^n$.

Definition 2.7.2. A vector space $\mathbf{V}$ is said to be **finite dimensional** if and only if there is a finite set $X_1, \ldots, X_n$ of vectors in $\mathbf{V}$ with $\text{span}\{X_1, \ldots, X_n\} = \mathbf{V}$. If $\mathbf{V}$ is not finite dimensional, we say that $\mathbf{V}$ is **infinite dimensional**.

Of course, by the above remarks $\mathbb{R}^n$ is finite dimensional, but not every vector space is. For example, consider the vector space $\mathbb{R}[x]$ of all polynomials over the real numbers $\mathbb{R}$ (see Example 2.3.1, part (d)). If $p_1(x), \ldots, p_n(x)$ are any polynomials in $\mathbb{R}[x]$, then we can prove that $\text{span}\{p_1(x), \ldots, p_n(x)\} \neq \mathbb{R}[x]$. Let m be the largest of the degrees of the polynomials $p_1(x), \ldots, p_n(x)$, and observe that any linear combination $a_1 p_1(x) + \ldots + a_n p_n(x)$, for scalars $a_1, \ldots, a_n$, has degree less than or equal to m. But $\mathbb{R}[x]$ has polynomials of degree greater than m and so $\text{span}\{p_1(x), \ldots, p_n(x)\} \neq \mathbb{R}[x]$. From this we see that $\mathbb{R}[x]$ is infinite dimensional.

> **Theorem 2.7.1.** *Let $\mathbf{V}$ be a finite-dimensional vector space and let $\{X_1, \ldots, X_n\}$ and $\{Y_1, \ldots, Y_m\}$ be bases for $\mathbf{V}$. Then $n = m$.*   *(Any two bases have the same number of elements.)*

Proof. We make use of Theorem 2.6.5 and its Corollary 2.6.6. The Corollary states that if $\mathbf{V}$ is spanned by n vectors, then any linearly independent set of vectors in $\mathbf{V}$ has n or fewer elements.

Now by our assumptions $\mathbf{V} = \text{span}\{X_1, \ldots, X_n\}$ and $Y_1, \ldots, Y_m$ are linearly independent. It follows from Corollary 2.6.6 that $n \geq m$. But also $\mathbf{V} = \text{span}\{Y_1, \ldots, Y_m\}$ and $X_1, \ldots, X_n$ are linearly independent and so $n \leq m$. It follows that $n = m$. $\qquad\square$

Definition 2.7.3. Let $\mathbf{V}$ be a finite-dimensional vector space. The **dimension** of $\mathbf{V}$ is the number of distinct vectors in any one basis of $\mathbf{V}$. If $\mathbf{V}$ has dimension n, we will write $\dim \mathbf{V} = n$.   *(The dimension is the number of elements in a basis.)*

By the convention adopted in Definition 2.7.1, $\emptyset$ is a basis for the zero vector space and so $\dim\{0\} = 0$. Since the standard basis for $\mathbb{R}^n$ has n vectors, $E_1, \ldots, E_n$, we see that $\dim \mathbb{R}^n = n$. Also by the above discussion, $\mathbb{R}[x]$ has infinite dimension.

The study of infinite-dimensional vector spaces involves some aspects of set theory that the reader has probably not encountered before and that are too complicated for us to present here. Because of this, much of the discussion that follows applies only to finite-dimensional vector spaces. Of course, results applicable to either case will be presented in their most general form.

> **Theorem 2.7.2.** *Let $\mathbf{V}$ be a finite-dimensional vector space and let $\mathbf{U}$ be a subspace of $\mathbf{V}$. Then $\mathbf{U}$ is finite dimensional.*   *(Finite-dimensional spaces have only finite-dimensional subspaces.)*

Proof. Since **V** is finite dimensional, $\mathbf{V} = \text{span}\{X_1, \ldots, X_n\}$ for some vectors $X_1, \ldots, X_n$. If $\mathbf{U} = \{0\}$, then $\emptyset$ is a basis by definition, and so $\mathbf{U} = \text{span}\{\emptyset\}$. Assume $\mathbf{U} \neq \{0\}$. Then **U** contains a nonzero vector; call it Y_1. If $\text{span}\{Y_1\} = \mathbf{U}$, we are done, and if not, there is a vector $Y_2 \in \mathbf{U}$ with $Y_2 \notin \text{span}\{Y_1\}$. Then by Theorem 2.6.7, Y_1 and Y_2 are linearly independent and $\text{span}\{Y_1, Y_2\} \subseteq \mathbf{U}$. Continue this process; that is, assume $Y_1, \ldots, Y_k$ are linearly independent vectors in **U**. If $\text{span}\{Y_1, \ldots, Y_k\} = \mathbf{U}$, we're done; if not, choose $Y_{k+1} \in \mathbf{U}$, with $Y_{k+1} \notin \text{span}\{Y_1, \ldots, Y_k\}$.

By Corollary 2.6.6, this process must stop, for since **V** is spanned by n vectors, any linearly independent subset must have n or fewer vectors. It follows that $\mathbf{U} = \text{span}\{Y_1, \ldots, Y_m\}$ for some integer m and so **U** is finite dimensional. $\quad\square$

Corollary 2.7.3. *Every subspace of a finite-dimensional vector space has a basis.*

Proof. By the proof of Theorem 2.7.2, the subspace of **U** of the finite-dimensional vector space **V** is spanned by vectors $Y_1, \ldots, Y_m$ for some integer m. By construction, $Y_1, \ldots, Y_m$ are linearly independent, and so these vectors form a basis for **U**. $\quad\square$

Since every vector space is a subspace of itself, the preceding corollary immediately gives the following:

Corollary 2.7.4. *Every finite-dimensional vector space has a basis.*

The term "dimension" was defined for finite-dimensional vector spaces with reference to a basis for the space, but prior to Corollary 2.7.4, the existence of a basis for any given finite-dimensional vector space had not been established. Now we know that every finite-dimensional vector space has a basis and so its dimension is defined.

Corollary 2.7.5. *Let* **V** *be a finite-dimensional vector space and let* **U** *be a subspace of* **V**. *Then* $dim\mathbf{U} \leq dim\mathbf{V}$.   *(Subspaces have the same dimension or smaller dimension.)*

Proof. By the previous corollaries, both **U** and **V** have bases, and so dim **U** and dim **V** are both defined. If $\{X_1, \ldots, X_m\}$ is a basis for **U** and $\{Y_1, \ldots, Y_n\}$ is a basis for **V**, then the vectors $X_1, \ldots, X_n$ are independent in **V** and so $m \leq n$ (by Corollary 2.6.6). It follows that dim $\mathbf{U} \leq$ dim$\mathbf{V}$. $\quad\square$

Theorem 2.7.2 and its corollaries tell us that there is a limit on the size of independent sets in a finite-dimensional vector space. Suppose that we have a basis for a subspace of a finite-dimensional vector space. Is this basis contained in some basis for the whole space? The following theorem says "yes" and the proof of the theorem tells us how to find the basis.

EXTENSION OF A BASIS

Theorem 2.7.6. *Let* **V** *be a finite-dimensional vector space and let* **U** *be a subspace of* **V** *with* $\{X_1, \ldots, X_k\}$ *a basis for* **U**. *Then there exist vectors* $X_{k+1}, \ldots, X_n \in \mathbf{V}$ *such that* $\{X_1, \ldots, X_k, X_{k+1}, \ldots, X_n\}$ *is a basis for* **V**.

(A basis for a subspace can be extended to a basis for the whole space.)

Proof. The method here is much like Theorem 2.7.2 If $\mathbf{U} = \mathbf{V}$, we're done and $X_1, \ldots, X_k$ form a basis for $\mathbf{V}$. If $\mathbf{U} \neq \mathbf{V}$, choose $X_{k+1} \in \mathbf{V}, X_{k+1} \notin \mathbf{U}$. Then $X_1, \ldots, X_k, X_{k+1}$ are linearly independent. If we continue this process, eventually we see that for some n, $\text{span}\{X_1, \ldots, X_k, \ldots, X_n\} = \mathbf{V}$ and a basis has been found. $\qquad\square$

To illustrate the method of the theorem above, let us consider the subspace

$$\mathbf{U} = \text{span}\{(1, 1, -1, 0), (2, 1, 1, 1)\}$$

of $\mathbb{R}^4$. The two vectors $X_1 = (1, 1, -1, 0)$ and $X_2 = (2, 1, 1, 1)$ form a basis for $\mathbf{U}$. We wish to "extend" this basis $\{X_1, X_2\}$ to a basis for $\mathbb{R}^4$. Choose X_3 satisfying $X_3 \in \mathbb{R}^4$, but $X_3 \notin \mathbf{U}$. To do this, observe that

$$\mathbf{U} = \{a(1, 1, -1, 0) + b(2, 1, 1, 1) | a, b \text{ scalars}\}.$$

We can see that $(0, 0, 0, 1) \notin U$ for $a(1, 1, -1, 0) + b(2, 1, 1, 1) = (0, 0, 0, 1)$ implies $(a + 2b, a + b, -a + b, b) = (0, 0, 0, 1)$. From this, we see that $b = 1, -a + b = 0$ and $a + b = 0$. This implies $b = a = -a = 1$, and so no such a and b can exist. So $(0, 0, 0, 1) \notin U$. Now let $X_3 = (0, 0, 0, 1)$ and observe that X_1, X_2, and X_3 are linearly independent. We know that $\text{span}\{X_1, X_2, X_3\} \neq \mathbb{R}^4$ and so we can choose $X_4 \notin \text{span}\{X_1, X_2, X_3\}$.

To do this, we will reason as follows: One of the standard basis vectors must lie outside $\text{span}\{X_1, X_2, X_3\}$, for otherwise $\text{span}\{X_1, X_2, X_3\} = \mathbb{R}^4$ and this would prove $\dim \mathbb{R}^4 = 3$, a contradiction. Now an arbitrary element of $\text{span}\{X_1, X_2, X_3\}$ is of the form $a(1, 1, -1, 0) + b(2, 1, 1, 1) + c(0, 0, 0, 1)$. This could not equal $(1, 0, 0, 0)$, for this equality would imply $a + b = 0, -a + b = 0$ so that $a = b = 0$, a contradiction. Hence $X_4 = (1, 0, 0, 0) \notin \text{span}\{X_1, X_2, X_3\}$ and so X_1, X_2, X_3, and X_4 are linearly independent. Now $\text{span}\{X_1, X_2, X_3, X_4\} = \mathbb{R}^4$, for otherwise we could find a vector $X_5 \in \mathbb{R}^4, X_5 \notin \text{span}\{X_1, X_2, X_3, X_4\}$. This would imply that $X_1, \ldots, X_5$ are linearly independent vectors in $\mathbb{R}^4$, but this is impossible. It follows that $\{X_1, X_2, X_3, X_4\}$ is a basis for $\mathbb{R}^4$ and it contains the two original vectors X_1 and X_2.

Section 2.7 Exercises

1. Show that the vectors $(1, 0, 0), (1, 1, 0)$, and $(1, 1, 1)$ form a basis for $\mathbb{R}^3$.

2. Find a basis for $\mathbb{R}^3$ that contains the vector $(1, -1, 2)$.

3. Show that the vectors $(1, 0, 1), (1, 1, 0)$, and $(2, 1, 1)$ do not form a basis for $\mathbb{R}^3$.

4. Find a basis for $\mathbb{R}^2$ that contains the vector $(1, 3)$.

5. Let X and Y be vectors in $\mathbb{R}^2$ with $Y \neq 0$ and $X \neq kY$ for any scalar k. Show that $\{X, Y\}$ is a basis for $\mathbb{R}^2$.

6. Prove: If $\mathbf{V}$ is a vector space with $\dim \mathbf{V} = n$ and $X_1, \ldots, X_n$ are n distinct, linearly independent vectors in $\mathbf{V}$, then $X_1, \ldots, X_n$ forms a basis for $\mathbf{V}$.

7. Let X be a nonzero vector in $\mathbb{R}^n$, say $X = (a_1, \ldots, a_n)$ with $a_k \neq 0$. Show that $E_1, \ldots, E_{k-1}, X, E_k + 1, \ldots, E_n$ are linearly independent in $\mathbb{R}^n$ and so form a basis for $\mathbb{R}^n$. (Recall that the E_i's are the standard basis vectors.)

8. Let $\mathbf{W} = \{(x, y, z) | x = -y\}$. Then $\mathbf{W}$ is a subspace of $\mathbb{R}^3$. Find a basis for $\mathbf{W}$. What is the dimension of $\mathbf{W}$?

9. Find a basis for the subspace $\mathbf{U} = \{(x, y, z, t) | x = y, z = 2t\}$ of $\mathbb{R}^4$ and extend it to a basis for $\mathbb{R}^4$. What is the dimension of $\mathbf{U}$?

10. Prove: If $\mathbf{V}$ is a vector space and dim $\mathbf{V} = n$ and $\mathbf{V} = \mathrm{span}\{X_1, \ldots, X_n\}$, then $X_1, \ldots, X_n$ are linearly independent and so form a basis for $\mathbf{V}$.

11. Let $\mathbf{V}$ be a finite-dimensional vector space and let $\mathbf{U}$ be a subspace of $\mathbf{V}$ with $\mathbf{V} \neq \mathbf{U}$. Prove that there exists a basis $X_1, \ldots, X_n$ of $\mathbf{V}$ such that $X_i \notin \mathbf{U}$ for all $i = 1, 2, \ldots, n$.

12. Find a basis for $\mathbb{R}[x]$.

13. Show that $\sin x$ and $\cos x$ are linearly independent vectors in $C[0, 2\pi]$ so that dim $\mathrm{span}\{\sin x, \cos x\} = 2$.

14. Let $\mathbf{U}$ be a subspace of the space of n-dimensional column vectors and let P be a nonsingular $n \times n$ matrix. Let $\mathbf{V} = \{PX | X \in U\}$. Show that dim $\mathbf{U} = \mathrm{dim}\mathbf{V}$. (See Exercises 2.4.16 and 2.6.16.)

15. Prove: Let $\mathbf{V}$ be a vector space and let $X_1, \ldots, X_n \in \mathbf{V}$. Then $X_1, \ldots, X_n$ forms a basis for $\mathbf{V}$ if and only if every vector $X \in \mathbf{V}$ can be expressed as a linear combination of $X_1, \ldots, X_n$ in one and only one way.

16. Let $\mathbf{V} = \mathrm{span}\{Y_1, \ldots, Y_m\}$ be a subspace of $\mathbb{R}^n$. Let A be the $m \times n$ matrix with the i-th row vector A_i of A being Y_i; that is, $A_i = Y_i$. Let B be the reduced echelon form of A. Prove that the nonzero row vectors of B ($B_1, \ldots, B_r$) form a basis for $\mathbf{V}$.

17. Let B be a matrix in reduced echelon form with r being the number of nonzero rows. Let $\boldsymbol{C}$ (B) be the column space of B (see Section 2.5). Prove that $\boldsymbol{C}$ (B) has dimension r.

2.8 The Rank of a Matrix and Consistency in Systems of Equations

We can use the notion of dimension to give a clear condition on the coefficient and augmented matrices that will guarantee that a system of linear equations is consistent. This result will be of theoretical interest, though not of importance computationally. We will first need to make some general observations about matrices.

Let $A = [a_{ij}]$ be an $m \times n$ matrix. As before, we denote row i of A by A_i; that is, $A_i = (a_{i1}, a_{i2}, \ldots, a_{in})$ for $i = 1, \ldots, m$. In Section 2.5, the **row space** of A was defined as the span of the rows of A and was denoted $\boldsymbol{R}$ (A). So we have $\boldsymbol{R}$ $(A) = \mathrm{span}\{A_1, A_2, \ldots, A_m\}$. The dimension of $\boldsymbol{R}$ (A) is a useful quantity and we give it a name.

Definition 2.8.1. The **rank** of a matrix A is the dimension of $\boldsymbol{R}\,(A)$, the row space of A, and is denoted by rank(A). $(\mathrm{rank}(A) = \dim \boldsymbol{R}\,(A))$

The definition is straightforward. For example, consider the matrices

$$A = \begin{bmatrix} 1 & 0 \\ 1 & 1 \end{bmatrix} \text{ and } B = \begin{bmatrix} 1 & 1 \\ 1 & 1 \end{bmatrix}.$$

The matrix A has rank two (rank(A) = 2), while B has rank 1 (rank(B) = 1).

By Theorem 2.5.2, row-equivalent matrices have the same row space; that is, A row equivalent to B implies $\boldsymbol{R}\,(A) = \boldsymbol{R}\,(B)$. We see that dim $\boldsymbol{R}\,(A) = \dim\boldsymbol{R}\,(B)$ and so the following result is established:

Theorem 2.8.1. *If A and B are row-equivalent matrices, then A and B have the same rank, that is, rank(A) = rank(B).*   *(Equivalent matrices have the same rank.)*

Now let B be an $m \times n$ matrix in reduced echelon form and assume $B_1, \ldots, B_r$ are the nonzero rows and that the other $m-r$ rows are all zeros. Then $\boldsymbol{R}\,(B) = \mathrm{span}\{B_1, \ldots, B_m\} = \mathrm{span}\{B_1, \ldots, B_r\}$ since the rows of zeros add nothing to the span. By Theorem 2.6.3, the nonzero rows $B_1, \ldots, B_r$ are linearly independent since B is in reduced echelon form. We see that the vectors $B_1, \ldots, B_r$ form a basis for $\boldsymbol{R}\,(B)$. Thus, $\boldsymbol{R}\,(B)$ has dimension r and so B has rank r; that is, rank(B) = r. By the above theorem, row-equivalent matrices have the same rank, and we know that every matrix is row-equivalent to a matrix that is in reduced echelon form. By the comments above we have proved:

Theorem 2.8.2. *Let A be an $m \times n$ matrix and let B be the reduced echelon form of A.   The rank of A is the number of nonzero rows in B.*
(The rank is the number r in the reduced echelon form.)

Example 2.8.1. Consider the matrix $A = \begin{bmatrix} 1 & -1 & 2 \\ 0 & 1 & -1 \\ 2 & 1 & 1 \end{bmatrix}$. The row space of A is the space spanned by the three rows A_1, A_2, and A_3; that is, $\boldsymbol{R}\,(A) = \mathrm{span}\{A_1, A_2, A_3\}$. The row vectors A_1, A_2, and A_3 are not linearly independent since $A_3 = 2A_1 + 3A_2$, and it follows that $\boldsymbol{R}\,(A) = \mathrm{span}\{A_1, A_2\}$. Now A_1 and A_2 are linearly independent and so dim $\boldsymbol{R}\,(A) = 2 = \mathrm{rank}(A)$. Thus, A has rank 2.

$\square$

CONSISTENCY

The notion of rank provides a convenient means for characterizing consistency in systems of linear equations. Recall that a system of linear equations can be expressed more compactly in the form of a matrix equation, $AX = H$, where, for some integers m and n, A is an $m \times n$ matrix, X is $n \times 1$, and H is $m \times 1$. The system $AX = H$ is said to be **consistent** provided that there exists at least one $n \times 1$ column vector X_0, with $AX_0 = H$. That is, the solution set is nonempty. Not all systems are consistent. For example, the system of linear equations $x + y = 1, x + y = 2$ has no simultaneous solution. The augmented matrix of this inconsistent system is

$$\begin{bmatrix} 1 & 1 & 1 \\ 1 & 1 & 2 \end{bmatrix},$$

and reducing the matrix to its reduced echelon form we get

$$\begin{bmatrix} 1 & 1 & 0 \\ 0 & 0 & 1 \end{bmatrix}.$$

This is the augmented matrix of the system of equations $x + y = 0, 0x + 0y = 1$. Clearly, there is no solution. The coefficient matrix is

$$\begin{bmatrix} 1 & 1 \\ 1 & 1 \end{bmatrix}$$

and it has

$$\begin{bmatrix} 1 & 1 \\ 0 & 0 \end{bmatrix}$$

as its reduced echelon form, and so has rank one. The augmented matrix of the system has rank two. The following theorem states that if the ranks of the coefficient and augmented matrices are equal, then the system is consistent.

Theorem 2.8.3. *Consider a system of equations $AX = H$, where A is an $m \times n$ matrix and X and H are $n \times 1$ and $m \times 1$ column vectors, respectively. The system $AX = H$ is consistent if and only if the coefficient matrix A and the augmented matrix $[A|H]$ have the same rank.*

(A system is consistent if and only if the coefficient and augmented matrices have the same rank.)

Proof. Let $[B|K]$ be the reduced echelon form of the augmented matrix $[A|H]$. Then $BX = K$ is equivalent to $AX = H$ by Theorems 1.4.2 and 1.5.1, and so $AX = H$ is consistent if and only if $BX = K$ is consistent. By the comments above, A and B have the same rank and $[A|H]$ and $[B|K]$ have the same rank. From this we see that it is sufficient to prove: $BX = K$ is consistent if and only if B and $[B|K]$ have the same ranks.

($\Rightarrow$) If $BX = K$ is consistent and the rank of B is r, then $[B|K]$ has rank at least r. If $[B|K]$ has rank $r + 1$, then the $(r + 1)$-th row of $[B|K]$ is $(0, \ldots, 0, 1)$. The equation in the system $BX = K$ corresponding to this row is $0x_1 + \ldots + 0x_n = 1$. This equation has no solution, but the system is assumed to be consistent and so the rank of $[B|K]$ must be r.

($\Leftarrow$) Now assume that B and $[B|K]$ have the same rank, say r. Let $j_1, \ldots, j_r$ be the columns of the first nonzero entries in the first r rows - as in the definition of reduced echelon form. Let $B = [b_{ij}]$ and

$$K = \begin{bmatrix} k_1 \\ k_2 \\ \vdots \\ k_m \end{bmatrix}.$$

Define a solution by

$$x_{j_1} = k_1, x_{j_2} = k_2, \ldots, x_{j_r} = k_r$$

and $x_j = 0$ for $j \neq j_1, \ldots, j_r$. It is not hard to see that this defines a solution and so $BX = K$ is consistent. $\square$

The above theorem, while it seems impressive, is of little use in solving systems of equations. Given a system of equations $AX = H$, we determine the ranks of A and $[A|H]$ by reducing to reduced echelon form. Once the reduced echelon form of $[A|H]$ is in hand, the solution of the system is not hard to obtain. So establishing the existence of a solution of the system involves nearly the same effort as finding the solution. We will see in the coming chapters that the notion of rank is helpful in a variety of practical situations.

Section 2.8 Exercises

1. Find the rank of $A = \begin{bmatrix} 2 & 0 & 4 \\ 2 & 1 & 3 \\ -1 & 0 & -2 \end{bmatrix}$.

2. Find the rank of $B = \begin{bmatrix} 1 & 2 & 0 & -1 \\ 1 & 3 & 1 & 0 \\ 2 & 4 & 0 & -2 \end{bmatrix}$.

3. Find the rank of $C = \begin{bmatrix} 1 & -2 \\ 1 & 3 \\ -2 & 4 \end{bmatrix}$.

In problems 4 - 6, determine whether the system of equations is consistent or inconsistent: (If it is consistent, find the general solution.)

4.
$$\begin{array}{rcrcrcrcr}
x_1 & - & 2x_2 & + & x_3 & - & x_4 & = & -1 \\
3x_1 & & & - & 2x_3 & + & 3x_4 & = & -4 \\
5x_1 & - & 4x_2 & & & + & x_4 & = & -3
\end{array}$$

5.
$$\begin{array}{rcrcrcrcr}
2x_1 & & & & & + & 3x_4 & = & -1 \\
x_1 & + & x_2 & - & x_3 & - & 2x_4 & = & 1 \\
3x_1 & - & x_2 & + & x_3 & + & 4x_4 & = & -4
\end{array}$$

6. $\begin{aligned} 2x_1 \; + \; 2x_2 \; + \; 2x_3 \qquad\qquad &= \; 3 \\ x_1 \; + \; x_2 \; + \; x_3 \; - \; x_4 &= \; 2 \\ x_1 \; + \; x_2 \; - \; x_3 \; - \; x_4 &= \; 3 \\ x_1 \; - \; x_2 \; - \; x_3 \; - \; x_4 &= \; 4 \end{aligned}$

7. Find all values of α for which the following system of equations has a solution:

$$\begin{aligned} x \; - \; 3y \; + \; 2z &= \; 4 \\ 2x \; + \; y \; - \; z &= \; 1 \\ 3x \; - \; 2y \; + \; z &= \; \alpha \end{aligned}$$

8. Let

$$A = \begin{bmatrix} 4 & -1 & 2 & 6 \\ -1 & 5 & -1 & -3 \\ 3 & 4 & 1 & 3 \end{bmatrix}.$$

   Find all vectors H such that $AX = H$ has a solution.

9. Is the following matrix row equivalent to the matrix B in Exercise 2?

$$A = \begin{bmatrix} 1 & 0 & 0 & 1 \\ 2 & 1 & 0 & 3 \\ 0 & 2 & 1 & -1 \end{bmatrix}$$

10. If A and B are two $m \times n$ matrices and A and B have the same rank, must A be row equivalent to B? Prove your answer.

11. We know that row-equivalent matrices have the same row space. Is the converse true? That is, is the following statement true: If A and B are two $m \times n$ matrices and $R(A) = R(B)$, then A is row equivalent to B. Prove your conclusion.

12. Can a 3×2 matrix have rank 3? Explain your answer.

13. If A is an $m \times n$ matrix, what can be said about the rank of A? That is, what are the limits on the possible values of the rank of A?

14. Can a condition be found on the rank of a matrix A that will guarantee that the system of equations $AX = H$ has a unique solution? (Assume that A is $m \times n$. Find a condition involving m, n, and the rank of A.)

15. Assume that the system of equations $AX = H$ has a unique solution. Describe the reduced echelon form B of A.

2.9   The Dimension of the Solution Space of a Homogeneous System

Recall that a homogeneous system of m equations in n unknowns may be regarded as a matrix equation of the form $AX = 0$, where A is an $m \times n$ matrix and X is an $n \times 1$ column vector. In Section 2.4, we observed that the solution set of a homogeneous system was a subspace of the space of n-dimensional column vectors and as such, it is natural to inquire about the dimension of this subspace and to try to determine a basis. Recall that this solution set is called the **nullspace** of A and is denoted by null(A).

Theorem 2.9.1. *Let A be an $m \times n$ matrix and assume that A has rank r. Then null(A) has dimension $n - r$.*   *(The nullspace has dimension $n - r$.)*

Proof. By previous results, the solution space of $AX = 0$ is exactly the same as the solution space of $BX = 0$, where $B = [b_{ij}]$ is the reduced echelon form of A. Let $r, j_1, \dots, j_r$ be the constants associated with B, as in the definition of echelon form. (Recall that by Section 2.8, the rank of A and the number of nonzero rows in B are the same.) Recall that the first nonzero entry in row i of B is a one in column j_i, for $i = 1, \dots, r$, and it is the only nonzero entry in column j_i.

We will establish that the solution space of $BX = 0$ (and so also $AX = 0$) has dimension $n - r$ by constructing a basis. Now observe that for $i = 1, \dots, r$, row i of the matrix equation $BX = 0$ gives an equation of the form:

$$x_{j_i} + b_{i j_{r+1}} x_{j_{r+1}} + \dots + b_{i j_n} x_{j_n} = 0,$$

where $j_{r+1}, \dots, j_n$ is some numbering of the columns other than $j_1, \dots, j_r$. Solving the equations for the variables $x_{j_1}, \dots, x_{j_r}$ in terms of the "other" variables – that is, those with the subscripts $j_{r+1}, \dots, j_n$ – we obtain the following equivalent system of equations:

$$
\begin{aligned}
x_{j_1} &= & -\ b_{1 j_{r+1}} x_{j_{r+1}} & &-\ \dots\ - & & b_{1 j_n} x_{j_n} \\
&\vdots & \vdots & & & \vdots & \\
x_{j_r} &= & -\ b_{r j_{r+1}} x_{j_{r+1}} & &-\ \dots\ - & & b_{r j_n} x_{j_n}.
\end{aligned}
$$

Now we can see that every choice of values for the variables

$$x_{j_{r+1}}, \dots, x_{j_n}$$

uniquely determines a solution and, conversely, every solution is obtained in this manner.

For $r < i \le n$, let X_i be the solution obtained by setting

$$x_{j_i} = 1 \text{ and } x_{j_k} = 0 \text{ for } k \neq i, r < k \le n.$$

Then the vectors $X_{r+1}, \dots, X_n$ are linearly independent, and since a solution with any given values for the variables

$$x_{j_{r+1}}, \dots, x_{j_n}$$

can be obtained as a linear combination of the solutions $X_{r+1}, \ldots, X_n$, we see that

$$\{X_{r+1}, \ldots, X_n\}$$

is a basis for the solution space. It follows that the solution space has dimension $n - r$.

Here is an alternate proof:

Begin as before, with B the reduced echelon form of A and $r, j_1, \ldots, j_r$ the constants associated with B. By relabeling the variables if necessary, we may assume that $j_1 = 1, j_2 = 2, \ldots, j_r = r$. That is, if, for example, $j_1 = 3$, then let $y_1 = x_3$ and rewrite the equations with y_1 replacing x_3. Continue by replacing each x_{j_i} with y_i, and then let $y_{r+1}, \ldots, y_n$ replace the remaining x's. This will give I_r for the upper left $r \times r$ block of B', the reduced row echelon form of A', the coefficient matrix of the system after it has been rewritten with y's. (This is what is meant by the expression, "without loss of generality, we may assume that $j_1 = 1, \ldots, j_r = r$.)"

The system corresponding to $B'X = 0$ is

$$
\begin{aligned}
y_1 &= - b_{1,r+1}y_{r+1} - \cdots - b_{1,n}y_n \\
&\ \ \vdots \qquad\qquad \vdots \qquad\qquad \vdots \\
y_r &= - b_{r,r+1}y_{r+1} - \cdots - b_{r,n}y_n.
\end{aligned}
$$

Each choice of $y_{r+1}, \ldots, y_n$ yields a solution, and every solution corresponds to such a choice. We may now finish the proof as before. $\qquad\square$

The above method of proof seems complicated, but it is not really too unwieldy. In effect, the solutions $X_1, \ldots, X_{n-r}$ that make up the basis are chosen so that the standard basis vectors are embedded in the solutions. This has the effect of guaranteeing that the vectors are linearly independent and also guaranteeing that any solution may be expressed as a linear combination of these vectors.

Example 2.9.1. Consider the following system of linear equations:

$$
\begin{aligned}
x_1 + x_2 + 2x_3 &= 0 \\
-x_1 - x_2 - 2x_3 + x_4 &= 0 \\
2x_1 + 2x_2 + 4x_3 + 3x_4 &= 0.
\end{aligned}
$$

The coefficient matrix is

$$
\begin{bmatrix}
1 & 1 & 2 & 0 \\
-1 & -1 & -2 & 1 \\
2 & 2 & 4 & 3
\end{bmatrix}
$$

and the reduced echelon form of this matrix is

$$
\begin{bmatrix}
1 & 1 & 2 & 0 \\
0 & 0 & 0 & 1 \\
0 & 0 & 0 & 0
\end{bmatrix}.
$$

The constants associated with this matrix are $r = 2, j_1 = 1, j_2 = 4$ (first and fourth columns are where the first nonzero entry, a 1, appears in rows 1 and 2, respectively),

$j_3 = 2$, and $j_4 = 3$. The general solution is $x_1 = -x_2 - 2x_3, x_4 = 0; x_2, x_3$ arbitrary. The solution space has dimension $n - r = 4 - 2 = 2$ and a basis is formed by calculating the two solutions where the first has $x_2 = 1, x_3 = 0$ and the second has $x_2 = 0, x_3 = 1$. Calculating these solutions and expressing them as row vectors, we see that the basis is $X_1 = (-1, 1, 0, 0), X_2 = (-2, 0, 1, 0)$, and so every solution of the original system may be expressed uniquely in the form

$$X = a(-1, 1, 0, 0) + b(-2, 0, 1, 0).$$

$\square$

Let us summarize: Let A be an $m \times n$ matrix of rank r. We have discussed three spaces associated with A: the row space of A denoted by $\boldsymbol{R}(A)$, the column space of A denoted by $\boldsymbol{C}(A)$, and the nullspace of A denoted by null(A). By results we have established, $\boldsymbol{R}(A)$ has dimension r (by Theorem 2.8.2), $\boldsymbol{C}(A)$ has dimension r (by Exercises 2.7.14 and 2.7.17), and null(A) has dimension $n - r$ by the above theorem.

Consideration of symmetry would imply that there should be one more space to consider in addition to those above. The missing space is the set of all m-dimensional row vectors X with $XA = 0$ (sometimes called the left nullspace of A). It is not too difficult to see that this space has dimension $m - r$.

Section 2.9 Exercises

1. If the matrix A is 3×4 and has rank 2, what is the dimension of the solution set of $AX = 0$?

2. If the matrix A is 3×3 and has rank 3, what is the solution set of $AX = 0$?

3. Find the general solution of the following system of linear equations and find a basis for the solution space. What is the dimension of the solution space?

$$\begin{aligned} x - 2y &= 0 \\ -2x + 4y &= 0. \end{aligned}$$

4. Find the general solution of the following system of linear equations and find a basis for the solution space:

$$\begin{aligned} x + y &= 0 \\ x - y &= 0. \end{aligned}$$

5. If the general solution of a system of equations is given by

$$\begin{aligned} x_1 &= -3x_3 \\ x_2 &= 2x_3, \end{aligned}$$

where x_3 is arbitrary, find a basis for the solution space of $AX = 0$. What is the dimension of the solution space?

6. If the general solution of a system of equations is given by

$$
\begin{aligned}
x_1 &= x_2 &&- 3x_5 \\
x_3 &= && 2x_5 &&- x_6 \\
x_4 &= &&- 3x_5 &&+ 2x_6,
\end{aligned}
$$

with x_2, x_5, and x_6 arbitrary, find the dimension of the solution space and find a basis for the solution space.

7. Find the general solution and a basis for the solution space of the following system of equations in the variables x_1, x_2, x_3, x_4, x_5 :

$$
\begin{aligned}
x_2 + 2x_3 \quad + 2x_5 &= 0 \\
x_4 + x_5 &= 0.
\end{aligned}
$$

8. Find the general solution and a basis for the solution space of the following system:

$$
\begin{aligned}
x_1 - x_2 \quad + 2x_4 &= 0 \\
3x_1 + x_2 - x_3 + x_4 &= 0 \\
2x_2 + x_3 + 2x_4 &= 0.
\end{aligned}
$$

9. Let A be a 3×4 matrix. Can a nonzero solution of $AX = 0$ be found? Why?

10. Let A be a 4×7 matrix. What can be said about the dimension of the solution space of $AX = 0$?

11. Consider a homogeneous system of four linear equations in six unknowns. Show that two linearly independent, nonzero solutions may be found.

12. What may be said about the solution space of a homogeneous system of six linear equations in four unknowns?

13. What is the relationship between the dimension of the solution space of a homogeneous system of linear equations and the number of arbitrarily chosen variables in the general solution of the system?

14. Let (a, b) and (c, d) be linearly independent vectors. Prove that for any numbers e and f the vectors (a, e, b) and (c, f, d) are linearly independent.

15. Let (a, b) and (c, d) be linearly independent vectors. Determine whether the vectors (a, b, e) and (f, c, d) are linearly independent for values e and f. What about the vectors (a, b, e) and (c, d, f)?

Chapter 3

LINEAR TRANSFORMATIONS AND MATRICES

3.1 Definitions and Examples

While calculus may be thought of as the study of the real number system, it is perhaps better to consider it the study of functions defined on the real number system. For many of the fundamental structures in mathematics a similar statement holds - it is the study of the functions on the structures that proves to be of interest. Most often the collection of all functions is too broad and functions with special properties are studied. In calculus the sets of continuous functions and differentiable functions are of most interest. What functions defined on a vector space are of interest?

In Chapter 1, we studied systems of linear equations in the form of a matrix equation $AX = H$ where for some integers m and n, A is an $m \times n$ matrix and X and H are respectively $n \times 1$ and $m \times 1$ column vectors. One may think of this matrix multiplication operation as a function or transformation that associates with the vector X, another vector AX or H. Recall further that for homogeneous systems, $AX = 0$, we showed that the solution set was a subspace of the vector space of all $n \times 1$ column vectors and that in Chapter 2 this subspace was characterized by finding a basis for the subspace. The important properties needed in proving that the solution set is a subspace are

$$A(X + Y) = AX + AY \text{ and}$$

$$A(aX) = a(AX).$$

We define now a class of functions having the above properties and in this chapter we will study these functions. These functions, called **linear transformations**, occur in many different applications. In calculus, rotations are used to simplify certain quadratic equations. In physics, angular momentum is an example of a linear transformation (see Section 5.8). We will see that linear transformations defined on finite dimensional vector spaces are associated with matrices and we will investigate this association. In Chapters Five and Six we will find that linear transformations are associated with matrices of a relatively simple form. For information on functions the reader is referred to Appendix B.

Definition 3.1.1. Let $\mathbf{U}$ and $\mathbf{V}$ be vector spaces over the same field F. A function $T :$ $\mathbf{U} \to \mathbf{V}$ is called a **linear transformation** if and only if the following two conditions are satisfied:

(a) $T(X + Y) = T(X) + T(Y)$ for all $X, Y \in \mathbf{U}$.

(b) $T(rX) = rT(X)$ for all $X \in \mathbf{U}, r \in F$.

The notation $T : \mathbf{U} \to \mathbf{V}$ is read "T maps $\mathbf{U}$ into $\mathbf{V}$" or "T mapping $\mathbf{U}$ into $\mathbf{V}$." We express the first part of the definition by saying that T "preserves addition" and the second by saying that T "preserves scalar multiplication". Notice that combining these two properties we can see that $T(rX + sY) = T(rX) + T(sY) = rT(X) + sT(Y)$. Also note that $T(X + Y + Z) = T(X + Y) + T(Z) = T(X) + T(Y) + T(Z)$. The most general version of these observations (see Exercise 13) states that

$$T\left(\sum_{i=1}^{n} a_i X_i\right) = \sum_{i=1}^{n} a_i T(X_i).$$

Because of this, the action of a linear transformation is completely determined by its action on a basis. That is, if we know $T(X_i)$ for all elements X_i in some basis, then $T(X)$ can be computed (using the above observation) for any vector X in the vector space. In Example 3.1.1 j) below, we see that this observation may also be used to define a linear transformation which maps basis elements to arbitrary elements of another vector space.

Properties of linear transformations are explored in the next section. Here we present a rather lengthy list of examples of linear transformations and constructions that result in such transformations.

Example 3.1.1. (a) As with most definitions, there are trivial examples. Let $\mathbf{V}$ be any vector space. Define $I_v : \mathbf{V} \to \mathbf{V}$ by $I_v(X) = X$ for all $X \in \mathbf{V}$. Then I_v is a linear transformation called the **identity transformation**. The function I_v satisfies (a) and (b) trivially. To simplify the notation, I_v will be denoted by I when no confusion will result.

Define $O_v : \mathbf{V} \to \mathbf{V}$ by $O_v(X) = 0$ (the zero vector) for all $X \in \mathbf{V}$. O_v is a linear transformation called the **zero transformation** on $\mathbf{V}$. As above, we will most often use O to denote O_v.

(b) For a vector $X \in \mathbb{R}^2$, let $T(X)$ be the vector obtained from X by rotating X in the counterclockwise direction through an angle θ. (See Figure 3.1.) Then T is a linear transformation and in fact:

$$T(x, y) = (x \cos \theta - y \sin \theta, x \sin \theta + y \cos \theta).$$

(c) Define $\pi_i : \mathbb{R}^n \to \mathbb{R}^n$ by $\pi_i((x_1, \ldots, x_n)) = (0, \ldots, 0, x_i, 0, \ldots, 0)$. π_i is a linear transformation called the **projection onto the ith coordinate**. See Figure 3.2.

(d) Define $D : C^1[a, b] \to C[a, b]$ by $D(f) = f' = df/dx$. Then, since $(f + g)' = f' + g'$ and $(rf)' = rf'$, we have that D is a linear transformation.

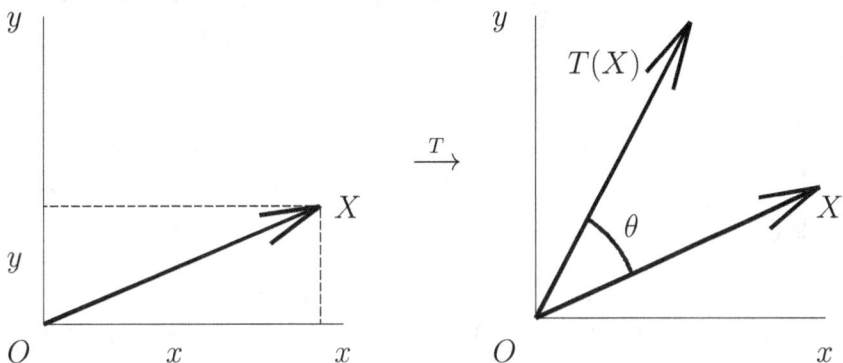

Figure 3.1: Rotation by θ

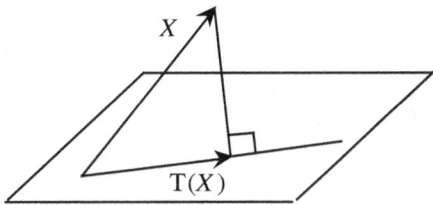

Figure 3.2: Projection onto the ith coordinate

(e) As in (d), we can think of D as a linear transformation defined on $\mathbb{R}[x]$, the vector space of all polynomials over the real field $\mathbb{R}$. If $p(x) = a_0 + a_1 x + \ldots + a_n x^n$, then $p'(x) = a_1 + 2a_2 x + \ldots + na_n x^{n-1} \in \mathbb{R}[x]$ so $D : \mathbb{R}[x] \to \mathbb{R}[x]$ defined by $D(p(x)) = p'(x)$ is a linear transformation.

(f) Consider $\mathbb{R}$ as a vector space over itself and define $S : C[a,b] \to \mathbb{R}$ by

$$S(f) = \int_a^b f(x)dx.$$

Then since

$$\int_a^b [f(x) + g(x)]dx = \int_a^b f(x)dx + \int_a^b g(x)dx \text{ and}$$

$$\int_a^b rf(x)dx = r \int_a^b f(x)dx,$$

we see that S is a linear transformation.

(g) Let $S : \mathbf{U} \to \mathbf{V}$ and $T : \mathbf{V} \to \mathbf{W}$ be linear transformations, where $\mathbf{U}, \mathbf{V}$, and $\mathbf{W}$ are vector spaces over a given field F. The **composite** TS of S and T is the function $TS : \mathbf{U} \to \mathbf{W}$ defined by $TS(X) = T[S(X)]$. TS is a linear transformation since $TS(X + Y) = T(S(X + Y)) = T(S(X) + S(Y)) = T(S(X)) + T(S(Y)) = TS(X) + TS(Y)$ and $TS(rX) = T(S(rX)) = T(rS(X)) = rT(S(X)) = rTS(X)$.

(h) If $S, T : \mathbf{U} \to \mathbf{V}$ are linear transformations, then $T + S : \mathbf{U} \to \mathbf{V}$ is defined by $(T + S)(X) = T(X) + S(X)$. $T + S$ is also a linear transformation and is called the **sum** of T and S.

(i) If $T : \mathbf{U} \to \mathbf{V}$ is a linear transformation and r is a scalar, then $rT : \mathbf{U} \to \mathbf{V}$ is the linear transformation defined by: $(rT)(X) = rT(X)$. The transformation rT is called the **scalar product** of r and T.

(j) Let $\mathbf{U}$ and $\mathbf{V}$ be vector spaces with $X_1, \ldots, X_n$ a basis for $\mathbf{U}$. Let $Y_1, \ldots, Y_n$ be any (not necessarily distinct) vectors in $\mathbf{V}$. Any vector X can be expressed as:

$$X = \sum_{i=1}^{n} a_i X_i = a_1 X_1 + \ldots + a_n X_n,$$

where $a_1, \ldots, a_n$ are scalars. Define $T : \mathbf{U} \to \mathbf{V}$ by

$$T(X) = T\left(\sum_{i=1}^{n} a_i X_i\right) = \sum_{i=1}^{n} a_i Y_i = a_1 Y_1 + \ldots + a_n Y_n$$

Then T is a linear transformation.

(k) Finally, the example with which we began the discussion. Let A be an $m \times n$ matrix over some field and for an $n \times 1$ column vector X, define $T_A(X)$ by $T_A(X) = AX$. Then as we saw before T_A is a linear transformation from the space of $n \times 1$ column vectors to the spaces of $m \times 1$ column vectors. We will see in Section 3.5 that many linear transformations can be characterized by a matrix in the above manner.

$\square$

Section 3.1 Exercises

1. Define $T : \mathbb{R}^2 \to \mathbb{R}^2$ by $T(x, y) = (x - y, x + 3y)$. Calculate $T(1, 2), T(1, 0), T(0, 2)$, and $T(1, 0) + T(0, 2)$.

2. Define $T : \mathbb{R}^2 \to \mathbb{R}^2$ by $T(x, y) = (x + 2y, 2x - y)$. Calculate $T(2, 3), T(2, 0), T(0, 3)$, and $T(2, 0) + T(0, 3)$.

3. Define $T : \mathbb{R}^3 \to \mathbb{R}^2$ by $T(x, y, z) = (x + y, 2y - x)$. Calculate $T(2, 1, 1)$ and $T(2, 3, 1)$.

4. Define $T : \mathbb{R}^3 \to \mathbb{R}^2$ by $T(x, y, z) = (x + z, y - x)$. Calculate $T(3, 2, 1)$ and $T(1, 3, 1)$.

5. With T as in Exercise 3, find a nonzero vector (x, y, z) with $T(x, y, z) = (0, 0)$.

6. With T as in Exercise 1, determine whether one can find a nonzero vector (x, y) with $T(x, y) = (0, 0)$. Give a reason for your answer.

7. As in Example 3.1.1 (j), there is a linear transformation $T : \mathbb{R}^3 \to \mathbb{R}^3$ that is determined by $T(1,0,0) = (0,1,0), T(0,1,0) = (1,2,1)$, and $T(0,0,1) = (-1,-3,-1)$. Compute $T(1,2,3)$.

8. Is there a linear transformation $T : \mathbb{R}^2 \to \mathbb{R}^2$ with $T(1,0) = (2,3)$ and $T(0,1) = (-1,2)$? Give a reason for your answer.

9. Is there a linear transformation $T : \mathbb{R}^2 \to \mathbb{R}^2$ with $T(1,0) = (2,3), T(0,1) = (-1,2)$, and $T(1,1) = (2,0)$? Give a reason for your answer.

10. Let S and T be linear transformations of $\mathbf{U}$ into $\mathbf{V}$ and let $X_1, \ldots, X_n$ be a basis for $\mathbf{U}$. Prove that if $S(X_i) = T(X_i)$ for $i = 1, \ldots, n$ then $S = T$.

11. Let $T : \mathbf{U} \to \mathbf{V}$ be a linear transformation. Show that $T(0) = 0$ and $T(-X) = -T(X)$.

12. Let $T : \mathbf{U} \to \mathbf{V}$ be a linear transformation and let $I_{\mathbf{U}} : \mathbf{U} \to \mathbf{U}$ and $I_{\mathbf{V}} : \mathbf{V} \to \mathbf{V}$ be the identity transformations on $\mathbf{U}$ and $\mathbf{V}$, respectively. Show that $TI_{\mathbf{U}} = T$ and $I_{\mathbf{V}}T = T$.

13. Let $T : \mathbf{U} \to \mathbf{V}$ be a linear transformation and let $X_1, \ldots, X_n \in \mathbf{U}$ and let $a_1, \ldots, a_n$ be scalars. Using mathematical induction (see Appendix E) prove that

$$T\left(\sum_{i=1}^{n} a_i X_i\right) = \sum_{i=1}^{n} a_i T(X_i).$$

14. Let $T : \mathbf{U} \to \mathbf{V}$ be a linear transformation and assume that $T(X) = T(Y)$ for some distinct $X, Y \in \mathbf{U}$. Prove that there is a nonzero vector $Z \in \mathbf{U}$ with $T(Z) = 0$.

15. Let $T : \mathbb{R}^2 \to \mathbb{R}^3$ be a linear transformation. Prove that there is a vector $Y \in \mathbb{R}^3$ with $T(X) \neq Y$ for all $X \in \mathbb{R}^2$.

16. Show that T in Example 3.1.1 (j) is a linear transformation.

3.2 Properties of Linear Transformations

In calculus, the idea of a function of a real variable is introduced, but it is the functions with special properties, such as continuity and differentiability, that are of greatest interest. The same holds true of linear transformations - nonsingular linear transformations are of considerable importance. In this section we will investigate nonsingular transformations and some special sets that are naturally associated with these transformations.

Associated with linear transformations there are two subspaces that, along with their associated dimensions, help to give information about the transformation. They are called the **image** and **nullspace**.

Definition 3.2.1. Let $T : \mathbf{U} \to \mathbf{V}$ be a linear transformation. The **image** of T is the set $\text{Im}(T) = \{Y \in \mathbf{V} | Y = T(X) \text{ for some } X \in \mathbf{U}\}$. The **nullspace** of T is the set $\text{null}(T) = \{X \in \mathbf{U} | T(X) = 0\}$, also denoted $N(T)$.

For example, suppose that $T : \mathbb{R}^3 \to \mathbb{R}^3$ is defined by $T(x, y, z) = (x - y, y - x, y)$. Then $T(x, y, z) = (0, 0, 0)$ implies that $x - y = 0, y - x = 0$, and $y = 0$. From this we see that $x = y = 0$ and it follows that $\text{null}(T) = \{(0, 0, z) | z \in \mathbb{R}\}$. Likewise,

$$
\begin{aligned}
\text{Im}(T) &= \{T(x, y, z) | (x, y, z) \in \mathbb{R}^3\} \\
&= \{(x - y, y - x, y) | (x, y, z) \in \mathbb{R}^3\} \\
&= \{(a, -a, b) | a, b \in \mathbb{R}\}.
\end{aligned}
$$

One may consider the nullspace of T to be the solution set of the equation $T(X) = 0$. Just as with the homogeneous systems of linear equations, this solution set forms a subspace.

Theorem 3.2.1. *If $T : \mathbf{U} \to \mathbf{V}$ is a linear transformation, then $\text{Im}(T)$ is a subspace of $\mathbf{V}$ and $\text{null}(T)$ is a subspace of $\mathbf{U}$.*   *(The image and nullspace of a transformation are subspaces.)*

Proof. See Exercise 8.   □

Functions that are one-to-one and onto are often of importance in mathematics. For linear transformations, the nullspace is related to whether the transformation is one-to-one, and of course, the image of the transformation is related to whether the function is onto. As a consequence, the dimensions of these subspaces are useful quantities.

RANK AND NULLITY

Definition 3.2.2. Let $T : \mathbf{U} \to \mathbf{V}$ be a linear transformation. The **nullity** of T is the dimension of $\text{null}(T)$ and is denoted by $\text{nullity}(T)$. The **rank** of T is the dimension of $\text{Im}(T)$ and is denoted by $\text{rank}(T)$.

If T is the linear transformation defined above ($T : \mathbb{R}^3 \to \mathbb{R}^3$ is defined by $T(x, y, z) = (x - y, y - x, y)$), we see that the nullspace of T has dimension 1 so that T has nullity 1, and the image of T has dimension 2 so that T has rank 2.

The following theorem relates the rank and nullity of a linear transformation to the dimension of the domain of the transformation. We will see that it is a useful result.

Theorem 3.2.2. *Let $T : \mathbf{U} \to \mathbf{V}$ be a linear transformation and suppose $\dim \mathbf{U} = n$. Then $rank(T) + nullity(T) = n$.*   *(The rank plus the nullity is the dimension of the domain.)*

Proof. Let $\text{nullity}(T) = k$ and choose a basis $X_1, \ldots, X_k$ for $\text{null}(T)$. Extend this basis to a basis $X_1, \ldots, X_k, X_{k+1}, \ldots, X_n$ for the entire vector space $\mathbf{U}$. We claim that $T(X_{k+1}), \ldots, T(X_n)$ is a basis for $\text{Im}(T)$. To show this we must prove that the vectors $T(X_{k+1}), \ldots, T(X_n)$ are linearly independent and that they span the image of T.

To show that the vectors are linearly independent, let $a_{k+1}T(X_{k+1}) + \ldots + a_n T(X_n) = 0$. Then $T(a_{k+1}X_{k+1} + \ldots + a_n X_n) = 0$ so $a_{k+1}X_{k+1} + \ldots + a_n X_n \in \text{null}(T)$. Since $X_1, \ldots, X_k$

is a basis for null(T), $a_{k+1}X_{k+1} + \ldots + a_nX_n = a_1X_1 + \ldots + a_kX_k$ for some scalars $a_1, \ldots, a_k$. Now rearranging we get $a_1X_1 + \ldots + a_kX_k - a_{k+1}X_{k+1} - \ldots - a_nX_n = 0$. Since $X_1, \ldots, X_n$ are linearly independent, we must have $a_1 = \ldots = a_k = \ldots = a_n = 0$. It follows that $T(X_{k+1}), \ldots, T(X_n)$ are linearly independent.

We next show that the vectors span the image. Let $Y \in \text{Im}(T)$. Then $Y = T(X)$ for some $X \in \mathbf{U}$. Let $X = b_1X_1 + \ldots + b_nX_n$. Then

$$
\begin{aligned}
Y &= T(X) \\
&= T(b_1X_1 + \ldots + b_nX_n) \\
&= b_1T(X_1) + \ldots + b_kT(X_k) + b_{k+1}T(X_{k+1}) + \ldots + b_nT(X_n) \\
&= b_{k+1}T(X_{k+1}) + \ldots + b_nT(X_n)
\end{aligned}
$$

since $T(X_1) = \ldots = T(X_k) = 0$, and so $T(X_{k+1}), \ldots, T(X_n)$ span $\text{Im}(T)$.

It follows that $T(X_{k+1}), \ldots, T(X_n)$ is a basis for $\text{Im}(T)$ and so $\text{rank}(T) + \text{nullity}(T) = (n - k) + k = n$. $\qquad\square$

By Appendix B, functions which are one-to-one and onto have inverses. Linear transformations which have inverses are called **nonsingular**.

NONSINGULARITY

A linear transformation is a correspondence that associates with a given vector X another vector Y — we may represent this correspondence by $X \to Y$. Under certain circumstances the "reverse" correspondence, $Y \to X$ is also a linear transformation. When this reverse correspondence satisfies the definition of a function, it is called the "inverse" function of the original function. The reader has no doubt encountered this situation in previous mathematics courses: the natural logarithm function $\ln(x)$ is the inverse of the exponential function e^x, and the inverses of the trigonometric functions are studied. The following definition and theorem characterize this situation for linear transformations.

Definition 3.2.3. A linear transformation $T : \mathbf{U} \to \mathbf{V}$ is **nonsingular** if and only if null$(T) = \{0\}$ and $\text{Im}(T) = \mathbf{V}$. Linear transformations that fail to satisfy this condition are called **singular**.

Example 3.2.1. Define $T : \mathbb{R}^2 \to \mathbb{R}^2$ by $T(x, y) = (x - y, 2y)$. Then T is a linear transformation and we can see that T is nonsingular since $T(x, y) = (0, 0)$ implies $x - y = 0$ and $2y = 0$, so $x = y = 0$. Further, if $(a, b) \in \mathbb{R}^2$, then $T(a + b/2, b/2) = (a, b)$ and so $\text{Im}(T) = \mathbb{R}^2$. It follows that T satisfies the two parts of the definition of nonsingularity.

$\qquad\square$

The following theorem relates the definition of nonsingularity to the existence of an inverse correspondence.

> **Theorem 3.2.3.** *Let* $T : \mathbf{U} \to \mathbf{V}$ *be a linear transformation. T is nonsingular if and only if there is a linear transformation* $S : \mathbf{V} \to \mathbf{U}$ *such that* $ST = I_{\mathbf{U}}$ *(the identity transformation on* $\mathbf{U}$*) and* $TS = I_{\mathbf{V}}$ *(the identity transformation on* $\mathbf{V}$*).*
>
> *(A transformation is nonsingular iff it has an inverse.)*

Proof. Assume a transformation $S : \mathbf{V} \to \mathbf{U}$ exists as in the statement of the theorem. If $X \in \text{null}(T)$, then $T(X) = 0$, so $ST(X) = S(T(X)) = S(0) = 0 = I(X) = X$. It follows that $\text{null}(T) = \{0\}$. If $Y \in \mathbf{V}, Y = I(Y) = TS(Y) = T(S(Y))$ and so $Y \in \text{Im}(T)$. By the definition of nonsingularity, we see that T is nonsingular.

Now assume T is nonsingular. We must define a linear transformation $S : \mathbf{V} \to \mathbf{U}$ as in the statement of the theorem and prove that it has the given properties. For a vector $Y \in \mathbf{V}$, there is a vector $X \in \mathbf{U}$ with $T(X) = Y$ since $\text{Im}(T) = \mathbf{V}$. Define $S(Y) = X$.

To complete the proof we need to show that S is well-defined, that S is a linear transformation and that S satisfies the "inverse" relationships as stated in the theorem. Assume that $S(Y) = X$ and $S(Y) = X'$. Then $T(X) = Y$ and $T(X') = Y$. It follows that (using Exercise 12) $T(X) - T(X') = T(X - X') = 0$ so that $X - X' \in \text{null}(T)$. But $\text{null}(T) = \{0\}$, so $X - X' = 0$ and hence $X = X'$. It follows that S is well defined.

The remaining parts of the proof are left as an exercise. $\qquad\square$

The linear transformation S is called the **inverse** of T and will be denoted by T^{-1}. The word "the" implies uniqueness, as does the use of the T^{-1} notation for the transformation S. What if there were several transformations that had the same properties as S?

Let us suppose that two transformations S_1 and S_2 behave like the transformation S in the above theorem. Then $S_1 T = S_2 T = I_{\mathbf{U}}$ and $TS_1 = TS_2 = I_{\mathbf{V}}$. Computing, we get $S_1(TS_2) = (S_1 T)S_2 = S_1 I_{\mathbf{V}} = I_{\mathbf{V}} S_2$. We see that $S_1 = S_2$, and so we have proved the following:

> **Theorem 3.2.4.** *The inverse of a linear transformation is unique.*

Section 3.2 Exercises

1. Define $T : \mathbb{R}^2 \to \mathbb{R}^2$ by $T(x, y) = (x + 2y, 2x - y)$. Show that $(2, -1) \in \text{Im}(T)$.

2. Define $T : \mathbb{R}^2 \to \mathbb{R}^3$ by $T(x, y) = (x + 2y, x, 2x - y)$. Show that $(1, 1, 2) \in \text{Im}(T)$.

3. Define $T : \mathbb{R}^3 \to \mathbb{R}^2$ by $T(x, y, z) = (x + z, y - x)$. Find a nonzero vector $X \in \text{null}(T)$.

4. Define $T : \mathbb{R}^3 \to \mathbb{R}^2$ by $T(x, y, z) = (2x + 3z, 2y + x)$. Find a nonzero vector $X \in \text{null}(T)$.

5. Determine the image, nullspace, rank and nullity of the transformation T of Exercise 1.

6. Determine the image, nullspace, rank and nullity of the transformation T of Exercise 2.

7. Show that $T : \mathbb{R}^3 \to \mathbb{R}^3$ defined by $T(x, y, z) = (2x + y, z, 2y)$ is nonsingular and determine T^{-1}.

8. Prove Theorem 3.2.1.

9. Let $T : \mathbf{U} \to \mathbf{U}$ be a linear transformation where $\mathbf{U}$ is a finite-dimensional vector space. Assume that $\mathrm{null}(T) = \{0\}$. Prove that T is nonsingular.

10. Complete the remaining parts to the proof of Theorem 3.2.3.

11. Show that any linear transformation $T : \mathbb{R}^3 \to \mathbb{R}^2$ is singular.

12. Show that any linear transformation $T : \mathbb{R}^2 \to \mathbb{R}^3$ is singular.

13. Let $T : \mathbf{U} \to \mathbf{V}$ be a nonsingular linear transformation with $\mathbf{U}$ and $\mathbf{V}$ finite-dimensional vector spaces. Prove that $\dim(\mathbf{U}) = \dim(\mathbf{V})$.

14. Use Theorem 3.2.3 and Theorem 3.2.4 to prove that if $T : \mathbf{U} \to \mathbf{V}$ is a nonsingular linear transformation, then T^{-1} is nonsingular, and $(T^{-1})^{-1} = T$.

15. Use Theorem 3.2.3 and Theorem 3.2.4 to prove that if $T : \mathbf{U} \to \mathbf{V}$ and $S : \mathbf{V} \to \mathbf{W}$ are nonsingular linear transformations, then ST is nonsingular, and $(ST)^{-1} = T^{-1}S^{-1}$.

3.3 The Matrix of a Linear Transformation

In some of the examples and exercises of Section 3.1, we saw that many linear transformations are closely related to matrices and that the action of a transformation is determined by its action on a basis. In this section, we define the matrix of a linear transformation that is defined on a finite-dimensional vector space and we discover that the matrix of the composite of two linear transformations is the product of the corresponding matrices.

Let $T : \mathbf{U} \to \mathbf{V}$ be a linear transformation and let $\mathbf{B} = \{X_1, \ldots, X_n\}$ be a basis for $\mathbf{U}$ and let $\mathbf{C} = \{Y_1, \ldots, Y_m\}$ be a basis for $\mathbf{V}$. We are assuming, of course, that $\mathbf{U}$ and $\mathbf{V}$ are finite-dimensional vector spaces. Since $\mathbf{C}$ is a basis for $\mathbf{V}$ and since $T(X_i) \in \mathbf{V}$, each $T(X_i)$ can be written as a linear combination of $Y_1, \ldots, Y_m$. Expanding the images of the elements in the basis B in terms of the basis $\mathbf{C}$ we have the following relationships:

$$
\begin{aligned}
T(X_1) &= a_{11}Y_1 + a_{21}Y_2 + \ldots + a_{m1}Y_m \\
T(X_2) &= a_{12}Y_1 + a_{22}Y_2 + \ldots + a_{m2}Y_m \\
&\ \vdots \qquad\quad \vdots \qquad\qquad\quad \vdots \\
T(X_n) &= a_{1n}Y_1 + a_{2n}Y_2 + \ldots + a_{mn}Y_m.
\end{aligned}
\tag{3.3.1}
$$

for some scalars a_{ij}.

Note that using this notation,

$$
T(X_j) = \sum_{k=1}^{m} a_{kj}Y_k
\tag{3.3.2}
$$

The **matrix of T relative to the bases B and C** (with the elements of **B** and **C** listed in a fixed order) is defined to be the $m \times n$ matrix

$$\begin{bmatrix} a_{11} & a_{12} & \cdots & a_{1n} \\ a_{21} & a_{22} & \cdots & a_{2n} \\ \vdots & \vdots & \ddots & \vdots \\ a_{m1} & a_{m2} & \cdots & a_{mn} \end{bmatrix}. \tag{3.3.3}$$

This matrix is denoted by $[T : \mathbf{B}, \mathbf{C}]$ or $T_{\mathbf{B}}^{\mathbf{C}}$. Notice that the scalars in the first equation of 3.3.1 are $a_{11}, a_{21}, \ldots, a_{m1}$, and that these scalars become the entries in column one of the matrix of the transformation 3.3.3. In effect, the "rows" of the equations become the "columns" in the matrix. This switch is made because it "works" and is consistent with other choices of notation. The underlying reason that the switch is necessary comes from the old habit of writing the function on the left of the variable or argument; that is, $T(X)$ with T on the left as opposed to $(X)T$, with T on the right. The author has decided that it is best to use the traditional notation, in spite of the problems it carries, and so the reader must remember to make the "switch."

A further warning concerning notation is needed: We have used **B** and **C** (in bold face type) to denote the two bases in the above discussion. In other places in this text, we have used $A, B,$ and C (italic, but not boldface) to represent matrices. It is important that the two not be confused! In particular, do not confuse the basis **B** with the matrix $[b_{ij}]$.

Example 3.3.1.   (a) Let $T : \mathbb{R}^3 \to \mathbb{R}^2$ be defined by $T(x, y, z) = (2x + y, y + z)$. Let $\mathbf{B} = \{(1, 0, 0), (1, 1, 0), (1, 1, 1)\}$ and $\mathbf{C} = \{(1, 0), (0, 1)\}$. Then:

$$\begin{array}{lllll} T(1, 0, 0) & = & (2, 0) & = & 2(1, 0) + 0(0, 1) \\ T(1, 1, 0) & = & (3, 1) & = & 3(1, 0) + 1(0, 1) \\ T(1, 1, 1) & = & (3, 2) & = & 3(1, 0) + 2(0, 1) \end{array}$$

Thus, the matrix of T relative to $\mathbf{B}, \mathbf{C}$ is

$$[T : \mathbf{B}, \mathbf{C}] = \begin{bmatrix} 2 & 3 & 3 \\ 0 & 1 & 2 \end{bmatrix}.$$

(b) Let $I : \mathbb{R}^2 \to \mathbb{R}^2$ be the identity transformation and let $\mathbf{D} = \{(1, 0), (1, 1)\}$. We would like to compute $[I : \mathbf{D}, \mathbf{D}]$ and $[I : \mathbf{D}, \mathbf{C}]$. We compute the action of the transformation I on the basis $\mathbf{D}$:

$$\begin{array}{lllllll} I(1, 0) & = & (1, 0) & = & 1(1, 0) + 0(1, 1) & = & 1(1, 0) + 0(0, 1) \\ I(1, 1) & = & (1, 1) & = & 0(1, 0) + 1(1, 1) & = & 1(1, 0) + 1(0, 1) \end{array}$$

From this we see that

$$[I : \mathbf{D}, \mathbf{D}] = \begin{bmatrix} 1 & 0 \\ 0 & 1 \end{bmatrix} \text{ and } [I : \mathbf{D}, \mathbf{C}] = \begin{bmatrix} 1 & 1 \\ 0 & 1 \end{bmatrix}.$$

The matrices come by taking the coefficients of the basis vectors in the image and entering them into the columns of the matrix.

□

Notice that since **C** was the standard basis in part (a), it was easy to express the images of the basis vectors in **B** in terms of this basis. If another basis were involved, it would possibly be necessary to solve equations in order to find the proper expression.

THE MATRIX OF A COMPOSITE

We want now to determine the matrix of the composite of two linear transformations T and S.

Let $S : \mathbf{U} \to \mathbf{V}$ and $T : \mathbf{V} \to \mathbf{W}$ be linear transformations and let $\mathbf{B} = \{X_1, \ldots, X_r\}$, $\mathbf{C} = \{Y_1, \ldots, Y_n\}$, and $\mathbf{D} = \{Z_1, \ldots, Z_m\}$ be bases for $\mathbf{U}$, $\mathbf{V}$, and $\mathbf{W}$, respectively. Let $[S : \mathbf{B}, \mathbf{C}]$ be the $n \times r$ matrix $[b_{ij}]$ and let $[T : \mathbf{C}, \mathbf{D}]$ be the $m \times n$ matrix $[a_{ij}]$. Then we have (using Equation 3.3.2):

$$
\begin{aligned}
TS(X_j) &= T(S(X_j)) \\
&= T\left(\sum_{k=1}^{n} b_{kj} Y_k\right) \\
&= \sum_{k=1}^{n} b_{kj} T(Y_k) \\
&= \sum_{k=1}^{n} b_{kj} \left(\sum_{i=1}^{m} a_{ik} Z_i\right) \\
&= \sum_{i=1}^{m} \left(\sum_{k=1}^{n} b_{kj} a_{ik} Z_i\right) \\
&= \sum_{i=1}^{m} \left(\sum_{k=1}^{n} a_{ik} b_{kj}\right) Z_i.
\end{aligned}
$$

It follows that $[TS : \mathbf{B}, \mathbf{D}]$ is the $m \times r$ matrix $[c_{ij}]$, where

$$
c_{ij} = \sum_{k=1}^{n} a_{ik} b_{kj}.
$$

Recalling the definition of the product of two matrices as given in Section 1.2, we see that the matrix $[c_{ij}]$ is the product of the two matrices $[a_{ij}]$ and $[b_{ij}]$ in that order, and so we have proved the following theorem.

Theorem 3.3.1. *If $T, S, \mathbf{U}, \mathbf{V}, \mathbf{W}, \mathbf{B}, \mathbf{C}$, and $\mathbf{D}$ are as defined above, then*

$$
[TS : \mathbf{B}, \mathbf{D}] = [T : \mathbf{C}, \mathbf{D}][S : \mathbf{B}, \mathbf{C}],
$$

or

$$
(TS)_{\mathbf{B}}^{\mathbf{D}} = T_{\mathbf{C}}^{\mathbf{D}} S_{\mathbf{B}}^{\mathbf{C}}.
$$

(The matrix of a composite is the product of the matrices of the transformations and in the same order.)

The above theorem can be difficult to remember because of the "unnatural" order of the transformations and bases. To make it easier to remember, consider with a picture like the following:

$$
\begin{array}{ccccc}
 & S & & T & \\
\mathbf{U} & \to & \mathbf{V} & \to & \mathbf{W} \\
| & & | & & | \\
\mathbf{B} & & \mathbf{C} & & \mathbf{D}
\end{array}
\qquad (3.3.4)
$$

where S and T are the transformation and $\mathbf{U}, \mathbf{V}$, and $\mathbf{W}$ are the vector spaces with $\mathbf{B}, \mathbf{C}$, and $\mathbf{D}$ the corresponding bases. Now recall that the order of the composite transformation is TS. To see this note that for $X \in \mathbf{U}, T(S(X))$ is always defined since $S(X)$ is defined for all $X \in \mathbf{U}$ and the result is an element of $\mathbf{V}$ on which T is defined. Notice also that $S(T(X))$ is not defined, and remember that the matrix of the composite is the product of the matrices of the individual transformations in the same order as the transformations occur in the composite. Considering this, we now know that:

$$[TS: \ , \ ] = [T: \ , \ ][S: \ , \ ] \qquad (3.3.5)$$

Now refer to the diagram 3.3.4 and fill in the bases in the natural order of occurrence. The composite TS is the transformation from $\mathbf{U}$ to $\mathbf{W}$ and so the bases are $\mathbf{B}$ and $\mathbf{D}$. We get $[TS : \mathbf{B}, \mathbf{D}]$. Likewise T maps $\mathbf{V}$ into $\mathbf{W}$ and so the bases are $\mathbf{C}$ and $\mathbf{D}$, from which $[T : \mathbf{C}, \mathbf{D}]$ results; similarly, we obtain $[S : \mathbf{B}, \mathbf{C}]$. Thus, the missing bases in 3.3.5 can be filled in, and you can recall the result of Theorem 3.3.1.

On the other hand, the alternative notation recalls the structure of a definite integral: we go *from* the lower index *to* the upper index. The nice thing about this notation is that the upper and lower indices "cancel" in the sense that if the map goes from $\mathbf{B}$ to $\mathbf{C}$ and then $\mathbf{C}$ to $\mathbf{D}$, then we might as well skip the middle step (pausing at $\mathbf{C}$) and go straight from $\mathbf{B}$ to $\mathbf{D}$.

Section 3.3 Exercises

1. Define $T : \mathbb{R}^2 \to \mathbb{R}^2$ by $T(x, y) = (x + y, x - y)$ and let $\mathbf{E} = \{(1, 0), (0, 1)\}$ be the standard basis for $\mathbb{R}^2$. Calculate $[T : \mathbf{E}, \mathbf{E}]$.

2. Define $T : \mathbb{R}^2 \to \mathbb{R}^2$ by $T(x, y) = (2x - y, x + 3y)$ and let $\mathbf{E}$ be the standard basis for $\mathbb{R}^2$. Calculate $[T : \mathbf{E}, \mathbf{E}]$.

3. Define $T : \mathbb{R}^3 \to \mathbb{R}^3$ by $T(x, y, z) = (x + y + z, x - y, 2z)$ and let $\mathbf{E} = \{(1, 0, 0), (0, 1, 0), (0, 0, 1)\}$ be the standard basis for $\mathbb{R}^3$. Calculate $[T : \mathbf{E}, \mathbf{E}]$.

4. Define $T : \mathbb{R}^3 \to \mathbb{R}^2$ by $T(x, y, z) = (x + y, x + y - 2z)$ and let $\mathbf{E}_3$ and $\mathbf{E}_2$ be the standard bases for $\mathbb{R}^3$ and $\mathbb{R}^2$, respectively. Calculate $[T : \mathbf{E}_3, \mathbf{E}_2]$.

5. Define $T : \mathbb{R}^2 \to \mathbb{R}^2$ by $T(x, y) = (x + y, 2x - y)$ and let $\mathbf{B} = \{(1, 0), (1, 1)\}, \mathbf{C} = \{(1, -1), (1, 1)\}$, and $\mathbf{E} = \{(1, 0), (0, 1)\}$ be bases for $\mathbb{R}^2$. Calculate the following:

   (a) $[T : \mathbf{B}, \mathbf{E}]$

   (b) $[T : \mathbf{E}, \mathbf{B}]$

   (c) $[T : \mathbf{B}, \mathbf{C}]$

6. Define $T : \mathbb{R}^2 \to \mathbb{R}^2$ by $T(x, y) = (x - y, x + 2y)$ and let $\mathbf{E} = \{(1, 0), (0, 1)\}$, $\mathbf{E}_1 = \{(0, 1), (1, 0)\}$, $\mathbf{E}_2 = \{(1, 0), (0, 2)\}$, and $\mathbf{E}_3 = \{(1, 0), (1, 1)\}$ be bases for $\mathbb{R}^2$. Notice that $\mathbf{E}$ is the standard basis for $\mathbb{R}^2$ and that the other bases are obtained from $\mathbf{E}$ by "elementary operations." Calculate the following:

   (a) $T_{\mathbf{E}}^{\mathbf{E}}$

   (b) $T_{\mathbf{E}}^{\mathbf{E}_1}$

   (c) $T_{\mathbf{E}}^{\mathbf{E}_2}$

   (d) $T_{\mathbf{E}}^{\mathbf{E}_3}$

How are the matrices in parts b) - d) related to the matrix in part a)?

7. Let $T, S : \mathbb{R}^2 \to \mathbb{R}^2$ be the linear transformations defined by $T(x, y) = (2x + y, 3y)$, $S(x, y) = (x + y, x - y)$. Let $\mathbf{E} = \{(1, 0), (0, 1)\}$ be the standard basis for $\mathbb{R}^2$. Compute $[T : \mathbf{E}, \mathbf{E}]$, $[S : \mathbf{E}, \mathbf{E}]$, and $[TS : \mathbf{E}, \mathbf{E}]$. Also compute $[T : \mathbf{E}, \mathbf{E}][S : \mathbf{E}, \mathbf{E}]$.

8. Let $T : \mathbb{R}^2 \to \mathbb{R}^2$ be a linear transformation and let $\mathbf{E} = \{(1, 0), (0, 1)\}$ be the standard basis for $\mathbb{R}^2$. Assume that

$$[T : \mathbf{E}, \mathbf{E}] = \begin{bmatrix} 1 & -1 \\ 2 & 3 \end{bmatrix}.$$

Find the following:

   (a) $T(1, 0)$

   (b) $T(0, 1)$

   (c) $T(-2, 3)$

9. Let $T : \mathbb{R}^3 \to \mathbb{R}^2$ be defined by $T(x, y, z) = (x - z, y + z)$ and assume that

$$[T : \mathbf{E}, \mathbf{B}] = \begin{bmatrix} -1 & -1 & 0 \\ 0 & 1 & 1 \end{bmatrix}.$$

where $\mathbf{E}$ is the standard basis for $\mathbb{R}^3$ and $\mathbf{B}$ is some basis for $\mathbb{R}^2$. Find the basis $\mathbf{B}$.

10. Let $I : \mathbb{R}^2 \to \mathbb{R}^2$ be the identity transformation and let $\mathbf{B}$ be some basis for $\mathbb{R}^2$. If $\mathbf{E}$ is the standard basis for $\mathbb{R}^2$, describe the matrix $[I : \mathbf{B}, \mathbf{E}]$ in terms of the basis $\mathbf{B}$. (Assume that the vectors in it are known and choose some representation for these vectors.)

11. Let $T : \mathbb{R}^2 \to \mathbb{R}^2$ be a linear transformation and let $\mathbf{B} = \{(1,1), (-1,2)\}$ be a basis for $\mathbb{R}^2$. Use Theorem 3.3.1 to find the relationship between $[T : \mathbf{B}, \mathbf{B}]$ and $[T : \mathbf{B}, \mathbf{E}]$. (Hint: consider the "picture" below:

$$
\begin{array}{ccccc}
& T & & I & \\
\mathbb{R}^2 & \to & \mathbb{R}^2 & \to & \mathbb{R}^2 \\
| & & | & & | \\
\mathbf{B} & & \mathbf{B} & & \mathbf{E}
\end{array}
$$

Apply Theorem 3.3.1 along with the fact that $IT = T$.)

12. Let $\mathbf{V}$ be a finite-dimensional vector space with the basis $\mathbf{B}$. Find the matrix of the identity transformation $I_{\mathbf{V}}$ relative to $\mathbf{B}$; that is, determine $[I_{\mathbf{V}} : \mathbf{B}, \mathbf{B}]$.

13. Let $T : \mathbf{U} \to \mathbf{V}$ be a linear transformation defined on the finite-dimensional vector spaces $\mathbf{U}$ and $\mathbf{V}$. Assume that $\mathbf{B}$ and $\mathbf{C}$ are bases for $\mathbf{U}$ and $\mathbf{V}$, respectively. Assume that T is a nonsingular linear transformation with inverse T^{-1}. Express the inverse of the matrix $[T : \mathbf{B}, \mathbf{C}]$ in terms of $T^{-1}, \mathbf{B}$, and $\mathbf{C}$.

14. Let $\mathbf{B}$ and $\mathbf{C}$ be any two bases for the n-dimensional real vector space $\mathbb{R}^n$, and let I be the identity transformation on $\mathbb{R}^n$. Show that $[I : \mathbf{B}, \mathbf{C}]$ is a nonsingular matrix and express its inverse in terms of $\mathbf{B}$ and $\mathbf{C}$.

3.4 Properties of Matrices and Matrix Multiplication

In the previous section, matrices once again arose - this time in relation to a linear transformation. Also, products of matrices became important in describing the matrix of the composite of two linear transformations. In this section (for want of a better place) we present some topics related to matrices and matrix multiplication that will be needed and used in the coming sections and chapters.

BLOCK MULTIPLICATION

The first topic is **block multiplication**. A matrix is made up of scalar entries, but in certain situations, it is useful to imagine a matrix **partitioned** into **submatrices** or **blocks**. In the discussion that follows we will show that the product of two partitioned matrices can be calculated by calculating the sums of products of the blocks — just as one does with the scalar entries in a matrix that is not partitioned — assuming all sums and products are defined.

Let $X = \begin{bmatrix} x_1 & \cdots & x_n \end{bmatrix}$ be a $1 \times n$ matrix (or row vector) and let

$$
Y = \begin{bmatrix} y_1 \\ \vdots \\ y_n \end{bmatrix}
$$

be an $n \times 1$ matrix (or a column vector). The matrix product XY is the 1×1 matrix $[x_1y_1 + \ldots + x_ny_n]$. We will identify a 1×1 matrix $[a]$ with its scalar entry a. Now if we write

$$x_1y_1 + \ldots + x_ny_n = (x_1y_1 + \ldots + x_ky_k) + (x_{k+1}y_{k+1} + \ldots + x_ny_n),$$

we see that we can think of XY as a sum of two matrix products $X_1Y_1 + X_2Y_2$ where $X_1 = [x_1 \ldots x_k], X_2 = [x_{k+1} \ldots x_n]$, and

$$Y_1 = \begin{bmatrix} y_1 \\ \vdots \\ y_k \end{bmatrix} \text{ and } Y_2 = \begin{bmatrix} y_{k+1} \\ \vdots \\ y_n \end{bmatrix}.$$

Think of X and Y "partitioned" as $X = \begin{bmatrix} X_1 & X_2 \end{bmatrix}$ and $Y = \begin{bmatrix} Y_1 \\ Y_2 \end{bmatrix}$. The computation above shows that

$$XY = \begin{bmatrix} X_1 & X_2 \end{bmatrix} \begin{bmatrix} Y_1 \\ Y_2 \end{bmatrix} = X_1Y_1 + X_2Y_2.$$

It is not hard to see that the same sort of thing would work if X and Y were partitioned into more "submatrices."

Now let A and B be matrices, A $m \times n$, B $n \times r$. Each entry in the product AB is a product of a "row vector" of A times a "column vector" of B, so if $A = [A_{ij}]$ and $B = [B_{ij}]$ are partitioned into submatrices or blocks A_{ij} and B_{ij} such that all the appropriate sums and products are defined, then the product AB is given by $AB = [A_{ij}][B_{ij}] = [C_{ij}]$, where

$$C_{ij} = \sum_{k=1}^{h} A_{ik}B_{kj}$$

and h is the number of blocks along a row in the partition of A or along a column of B.

In applications, we will need only the following special cases of this block multiplication.

1. $\begin{bmatrix} A & B \\ C & D \end{bmatrix} \begin{bmatrix} E & F \\ G & H \end{bmatrix} = \begin{bmatrix} AE + BG & AF + BH \\ CE + DG & CF + DH \end{bmatrix}$

2. $\begin{bmatrix} X_1 & \ldots & X_n \end{bmatrix} \begin{bmatrix} a_1 & 0 & \ldots & 0 \\ 0 & a_2 & \ldots & 0 \\ \vdots & \vdots & \ddots & \vdots \\ 0 & 0 & \ldots & a_n \end{bmatrix} = \begin{bmatrix} a_1X_1 & \ldots & a_nX_n \end{bmatrix}$

3. $A \begin{bmatrix} X_1 & \ldots & X_n \end{bmatrix} = \begin{bmatrix} AX_1 & \ldots & AX_n \end{bmatrix}$

The above rules are only valid, of course, when all sums and products are defined. In 2) and 3), the vectors $X_1, \ldots, X_n$, may be thought to represent the columns of the matrix, and in our use of these special cases, this will always be the case.

Notice also that our alternative perspective on matrix multiplication is a special case of block matrix multiplication.

Example 3.4.1. The following computations illustrate the three cases above. The lines form the partitions in the matrices.

1. In the following product of a 3×3 matrix and a 3×2 matrix, each matrix is divided into blocks making 2×2 matrices of blocks:

$$
\begin{bmatrix} \left. \begin{array}{cc|c} 2 & 1 & -1 \\ 1 & -1 & 2 \\ \hline 0 & 1 & 3 \end{array} \right. \end{bmatrix} \begin{bmatrix} \left. \begin{array}{c|c} 1 & 4 \\ 1 & 1 \\ \hline 0 & 2 \end{array} \right. \end{bmatrix}
$$

$$
= \begin{bmatrix} \begin{bmatrix} 2 & 1 \\ 1 & -1 \end{bmatrix} \begin{bmatrix} 1 \\ 1 \end{bmatrix} + \begin{bmatrix} -1 \\ 2 \end{bmatrix} [\,0\,] & \begin{bmatrix} 2 & 1 \\ 1 & -1 \end{bmatrix} \begin{bmatrix} 4 \\ 1 \end{bmatrix} + \begin{bmatrix} -1 \\ 2 \end{bmatrix} [\,2\,] \\ [\,0\ 1\,] \begin{bmatrix} 1 \\ 1 \end{bmatrix} + [\,3\,][\,0\,] & [\,0\ 1\,] \begin{bmatrix} 4 \\ 1 \end{bmatrix} + [\,3\,][\,2\,] \end{bmatrix}
$$

$$
= \begin{bmatrix} \begin{array}{c|c} 3 & 7 \\ 0 & 7 \\ \hline 1 & 7 \end{array} \end{bmatrix}
$$

$$
= \begin{bmatrix} 3 & 7 \\ 0 & 7 \\ 1 & 7 \end{bmatrix}.
$$

2. In the following product, the first matrix is partitioned into its columns and the second matrix is a diagonal matrix.

$$
\begin{bmatrix} \left. \begin{array}{c|c} 1 & -1 \\ 0 & 2 \end{array} \right. \end{bmatrix} \begin{bmatrix} 2 & 0 \\ 0 & 3 \end{bmatrix} = \begin{bmatrix} 2\begin{bmatrix} 1 \\ 0 \end{bmatrix} & 3\begin{bmatrix} -1 \\ 2 \end{bmatrix} \end{bmatrix} = \begin{bmatrix} 2 & -3 \\ 0 & 6 \end{bmatrix}.
$$

3. In the product that follows, the second matrix is partitioned into columns:

$$
\begin{bmatrix} 1 & -1 \\ 0 & 2 \end{bmatrix} \begin{bmatrix} \left. \begin{array}{c|c} 1 & -2 \\ 3 & 1 \end{array} \right. \end{bmatrix} = \begin{bmatrix} \begin{bmatrix} 1 & -1 \\ 0 & 2 \end{bmatrix}\begin{bmatrix} 1 \\ 3 \end{bmatrix} & \begin{bmatrix} 1 & -1 \\ 0 & 2 \end{bmatrix}\begin{bmatrix} -2 \\ 1 \end{bmatrix} \end{bmatrix} = \begin{bmatrix} \left. \begin{array}{c|c} -2 & -3 \\ 6 & 2 \end{array} \right. \end{bmatrix}
$$

$$
= \begin{bmatrix} -2 & -3 \\ 6 & 2 \end{bmatrix}.
$$

$\square$

The matrix in special case 2) above is called a **diagonal matrix** since the only nonzero entries lie on the diagonal of the matrix. We write $\mathrm{diag}(a_1, \ldots, a_n)$ for the $n \times n$ matrix

$$
\begin{bmatrix} a_1 & 0 & \cdots & 0 \\ 0 & a_2 & \cdots & 0 \\ \vdots & \vdots & \ddots & \vdots \\ 0 & 0 & \cdots & a_n \end{bmatrix}.
$$

Recall that in Section 1.3, the $n \times n$ identity matrix I_n was introduced using the Kronecker δ; that is, $I_n = \begin{bmatrix} \delta_{ij} \end{bmatrix}$. Using the diagonal matrix notation, $I_n = \text{diag}(1, 1, \ldots, 1)$. Recall also that in Theorem 1.3.8, we showed that I_n behaves like an "identity" with respect to matrix multiplication; that is, if A is an $m \times n$ matrix, then $I_m A = A$ and $A I_n = A$. Also in Section 1.3, the notion of a "nonsingular" matrix was defined. In Section 3.2, we defined the term "nonsingular" for linear transformations. We need to relate the two definitions.

NONSINGULARITY AGAIN

Let $\mathbf{V}$ be a finite-dimensional vector space with $\dim \mathbf{V} = n$ and $\mathbf{B}$ any basis for $\mathbf{V}$. For each $X \in \mathbf{B}, I(X) = X$, where $I : \mathbf{V} \to \mathbf{V}$ is the identity transformation. It is not hard to see that $[I : \mathbf{B}, \mathbf{B}] = I_n$.

In general, terminology applied to a linear transformation T is valid also for the matrix $[T : \mathbf{B}, \mathbf{C}]$ representing the transformation. For example, suppose $T : \mathbf{U} \to \mathbf{V}$ is a nonsingular transformation. Then there is a transformation $S : \mathbf{V} \to \mathbf{U}$ such that $ST = I$ and $TS = I$, where I is the identity transformation. Then for bases $\mathbf{B}$ and $\mathbf{C}$ of $\mathbf{U}$ and $\mathbf{V}$, respectively, $[T : \mathbf{B}, \mathbf{C}][S : \mathbf{C}, \mathbf{B}] = I$ and $[S : \mathbf{C}, \mathbf{B}][T : \mathbf{B}, \mathbf{C}] = I$. It follows that the matrix associated with a nonsingular linear transformation is a nonsingular matrix. (See Section 1.3 for the definition of nonsingular matrix and see Exercise 3.3.13.)

In Theorem 3.2.4, we saw that the inverse of a linear transformation is unique. This means that any transformation that behaves like the inverse of a given linear transformation, in fact is (or equals) that inverse. The same holds true for inverses of matrices.

> **Theorem 3.4.1.** *If A is a nonsingular $n \times n$ matrix and $AB_1 = B_1 A = I$ and $AB_2 = B_2 A = I$ for $n \times n$ matrices B_1 and B_2, then $B_1 = B_2$.*
>
> *(The inverse of a nonsingular matrix is unique.)*

Proof. Assume B_1 and B_2 are as above. Then:

$$(B_1 A)B_2 = IB_2 = B_2$$

and

$$(B_1 A)B_2 = B_1(AB_2) = B_1 I = B_1.$$

We see that $B_1 = B_2$, and so the inverse of A is unique. $\square$

We denote this unique inverse of the matrix A by A^{-1}. The implication of the theorem above is that any matrix that behaves like the inverse of a matrix, must in fact, be the inverse of the matrix. In Exercise 14, the reader is asked to prove that $(AB)^{-1} = B^{-1}A^{-1}$.

As a special case of the above remarks applied to the nonsingular transformation I, we can prove the following theorem. (See Exercise 3.3.14.)

> **Theorem 3.4.2.** *Let* **V** *be an n-dimensional vector space and let* **B** *and* **C** *be bases for* **V**. *If* $I : \mathbf{V} \to \mathbf{V}$ *is the identity transformation, then* $I_{\mathbf{B}}^{\mathbf{C}}$ *is nonsingular; in fact,* $(I_{\mathbf{B}}^{\mathbf{C}})^{-1} = I_{\mathbf{C}}^{\mathbf{B}}$.   *(The matrix of the identity transformation is nonsingular.)*

Proof. Using Theorem 3.3.1, we see that:

$$I_{\mathbf{B}}^{\mathbf{C}} I_{\mathbf{C}}^{\mathbf{B}} = (II)_{\mathbf{C}}^{\mathbf{C}} = I_{\mathbf{C}}^{\mathbf{C}} = I_n \text{ and}$$

$$I_{\mathbf{C}}^{\mathbf{B}} I_{\mathbf{B}}^{\mathbf{C}} = II_{\mathbf{B}}^{\mathbf{B}} = I_{\mathbf{B}}^{\mathbf{B}} = I_n$$

since the composite II equals I and $I_{\mathbf{B}}^{\mathbf{B}} = I_{\mathbf{C}}^{\mathbf{C}} = I_n$ by previous comments. Using the preceding theorem, we see that $(I_{\mathbf{B}}^{\mathbf{C}})^{-1} = I_{\mathbf{C}}^{\mathbf{B}}$.   □

The introduction of the notion of a nonsingular matrix gives rise to some problems: a) Which matrices are nonsingular? b) How can one determine whether a matrix is nonsingular? and c) Given that a matrix is nonsingular, how can one find its inverse? We will discover several methods for answering these questions later on, but for now let us use a crude method that might be described as "the method of undetermined entries." An example should suffice.

Example 3.4.2. Consider the 2×2 matrix $A = \begin{bmatrix} -1 & 3 \\ 2 & -4 \end{bmatrix}$ and suppose that we wish to determine whether A is nonsingular. We will assume that A is nonsingular and that its inverse is $B = \begin{bmatrix} a & b \\ c & d \end{bmatrix}$. Then:

$$AB = \begin{bmatrix} -1 & 3 \\ 2 & 4 \end{bmatrix} \begin{bmatrix} a & b \\ c & d \end{bmatrix} = \begin{bmatrix} -a + 3c & -b + 3d \\ 2a + 4c & 2b + 4d \end{bmatrix} = \begin{bmatrix} 1 & 0 \\ 0 & 1 \end{bmatrix}.$$

From this computation, we obtain the following two systems of two linear equations:

$$
\begin{array}{rrrr}
-a & + & 3c & = & 1 \\
2a & - & 4c & = & 0
\end{array}
\qquad
\begin{array}{rrrr}
-b & + & 3d & = & 0 \\
2b & - & 4d & = & 1
\end{array}
$$

Solving, we obtain $a = 2, b = 3/2, c = 1$, and $d = 1/2$, so that $B = \begin{bmatrix} 2 & 3/2 \\ 1 & 1/2 \end{bmatrix}$. Checking the product in the reverse order, we get:

$$BA = \begin{bmatrix} 2 & 3/2 \\ 1 & 1/2 \end{bmatrix} \begin{bmatrix} -1 & 3 \\ 2 & -4 \end{bmatrix} = \begin{bmatrix} 1 & 0 \\ 0 & 1 \end{bmatrix}.$$

It follows that B is the inverse of A.

□

Notice that to find the inverse of an $n \times n$ matrix using the above method we would have to solve n systems of n equations in n unknowns. There must be a better way!

Section 3.4 Exercises

In Exercises 1 - 6, verify that rules 1), 2), and 3) above do in fact work by performing the computations using both block multiplication and ordinary multiplication.

1. Check out rule 1) by computing the following product first by block multiplication and then by ordinary matrix multiplication.

$$\left[\begin{array}{cc|c} 1 & 0 & 2 \\ 2 & 1 & 0 \\ \hline -1 & 0 & 1 \end{array}\right] \left[\begin{array}{cc|cc} 0 & -1 & 0 & -1 \\ 1 & 2 & 1 & -3 \\ \hline 0 & 0 & 1 & 1 \end{array}\right]$$

2. Check out rule 1) by computing the following product first by block multiplication, then by ordinary matrix multiplication:

$$\left[\begin{array}{cc|c} 2 & 1 & 2 \\ 0 & 1 & 0 \\ \hline -2 & 2 & 2 \end{array}\right] \left[\begin{array}{cc|cc} 1 & -2 & 2 & -2 \\ 1 & 3 & 1 & -5 \\ 1 & 1 & 3 & 1 \end{array}\right]$$

3. Check out rule 2) by computing the following product first by block multiplication, then by ordinary matrix multiplication:

$$\left[\begin{array}{c|c} 1 & -1 \\ 1 & 2 \end{array}\right] \left[\begin{array}{cc} 2 & 0 \\ 0 & -1 \end{array}\right]$$

4. Check out rule 2) by computing the following product first by block multiplication, then by ordinary matrix multiplication:

$$\left[\begin{array}{c|c} 2 & -3 \\ 4 & 1 \end{array}\right] \left[\begin{array}{cc} 1 & 0 \\ 0 & -4 \end{array}\right]$$

5. Check out rule 3) by computing the following product first by block multiplication, then by ordinary matrix multiplication:

$$\left[\begin{array}{cc} 0 & -1 \\ 2 & -1 \end{array}\right] \left[\begin{array}{c|c} 0 & 1 \\ 0 & 1 \end{array}\right]$$

6. Check out rule 3) by computing the following product first by block multiplication, then by ordinary matrix multiplication:

$$\left[\begin{array}{cc} 1 & -2 \\ 3 & -2 \end{array}\right] \left[\begin{array}{c|c} 1 & 3 \\ -1 & 2 \end{array}\right]$$

7. Let $A = \text{diag}(a_1, \ldots, a_n)$ and $B = \text{diag}(b_1, \ldots, b_n)$. Show that $AB = \text{diag}(a_1 b_1, \ldots, a_n b_n)$.

8. Compute the inverse of $\left[\begin{array}{cc} 1 & 2 \\ 1 & 1 \end{array}\right]$.

9. Compute the inverse of $\begin{bmatrix} 1 & 1 \\ 0 & -2 \end{bmatrix}$.

10. Show that $\begin{bmatrix} -1 & 2 \\ 2 & -4 \end{bmatrix}$ is singular.

11. Determine whether $\begin{bmatrix} 0 & 1 \\ 1 & 1 \end{bmatrix}$ is singular or nonsingular.

12. Let $\mathbf{B} = \{(1, 2), (0, 1)\}$ and $\mathbf{C} = \{(-1, 0), (0, 2)\}$ be bases for $\mathbb{R}^2$. Find the inverse of the matrix $I_{\mathbf{B}}^{\mathbf{C}}$ and express it as the matrix of a linear transformation.

13. Let A and B be 3×3 matrices and assume $AB = I$. Show that $BA = I$.

14. Assume that A and B are nonsingular $n \times n$ matrices. Prove that AB is nonsingular and that $(AB)^{-1} = B^{-1}A^{-1}$.

3.5   Change of Basis

We return now to the idea of the matrix of a linear transformation that was defined in Section 3.2. In this section we propose to answer two questions: First, how does the matrix of a transformation relate to the operation performed on the vectors by the transformation, and second, what happens to the matrix of a transformation when the bases are changed? This investigation will lead us to important new relationships between matrices: equivalence and similarity. An idea that is explored in Chapters 5 and 6 emerges here: Given a linear transformation, perhaps a special basis can be found so that the matrix of the linear transformation, relative to this basis, is simple and the transformation can be more easily understood.

LINEAR TRANSFORMATIONS AND MATRIX MULTIPLICATION

Let $T : \mathbf{U} \to \mathbf{V}$ be a linear transformation and let $\mathbf{B} = \{X_1, \ldots, X_n\}$ and $\mathbf{C} = \{Y_1, \ldots, Y_m\}$ be bases for $\mathbf{U}$ and $\mathbf{V}$, respectively. Let $A = [a_{ij}] = T_{\mathbf{B}}^{\mathbf{C}}$. If $X \in \mathbf{U}$, then $X = x_1 X_1 + \ldots + x_n X_n$ for some scalars $x_1, \ldots, x_n$. The scalar x_i is the i-**th coordinate of** X **relative to the basis B**. The column vector consisting of these coordinates is denoted by $X^{\mathbf{B}}$. With the above notation,

$$X^{\mathbf{B}} = \begin{bmatrix} x_1 \\ x_2 \\ \vdots \\ x_n \end{bmatrix}.$$

Now computing $T(X)$ under the above assumptions, we get:

$$T(X) = T(x_1 X_1 + \ldots + x_n X_n) = \sum_{j=1}^{n} x_j T(X_j) = \sum_{j=1}^{n} x_j \sum_{i=1}^{m} a_{ij} Y_i = \sum_{i=1}^{m} \left(\sum_{j=1}^{n} a_{ij} x_j \right) Y_i.$$

Computing the product of the matrices A and $X^{\mathbf{B}}$, we get:

$$AX^{\mathbf{B}} = \begin{bmatrix} a_{11} & a_{12} & \cdots & a_{1n} \\ a_{21} & a_{22} & \cdots & a_{2n} \\ \vdots & \vdots & \ddots & \vdots \\ a_{m1} & a_{m2} & \cdots & a_{mn} \end{bmatrix} \begin{bmatrix} x_1 \\ \vdots \\ x_n \end{bmatrix} = \begin{bmatrix} a_{11}x_1 + \ldots a_{1n}x_n \\ \vdots \\ a_{m1}x_1 + \ldots + a_{mn}x_n \end{bmatrix}.$$

By comparing the coefficients of Y_i in the expansion of $T(X)$ with the i-th entry of $AX^{\mathbf{B}}$ we see that $T(X)^{\mathbf{C}} = AX^{\mathbf{B}}$, and so we have proved:

Theorem 3.5.1. *With the above notation,* $T(X)^{\mathbf{C}} = AX^{\mathbf{B}}$.

(The image of a vector under a linear transformation can be obtained by matrix multiplication.)

The above theorem almost says that $T(X) = AX$. This is true, provided that the vectors X are column vectors and the bases $\mathbf{B}$ and $\mathbf{C}$ are the standard bases. Consider the linear transformation $T : \mathbb{R}^3 \to \mathbb{R}^2$ defined by $T(x, y, z) = (2x + z, x + y + z)$. Let us calculate the matrix $A = T_{\mathbf{E}}^{\mathbf{E}'}$ where $\mathbf{E}$ and $\mathbf{E}'$ are the standard bases for $\mathbb{R}^3$ and $\mathbb{R}^2$ respectively. Now

$$T(1, 0, 0) = (2, 1) = 2(1, 0) + 1(0, 1)$$
$$T(0, 1, 0) = (0, 1) = 0(1, 0) + 1(0, 1)$$
$$T(0, 0, 1) = (1, 1) = 1(1, 0) + 1(0, 1)$$

and so

$$A = T_{\mathbf{E}}^{\mathbf{E}'} = \begin{bmatrix} 2 & 0 & 1 \\ 1 & 1 & 1 \end{bmatrix}.$$

Now

$$AX^{\mathbf{E}} = \begin{bmatrix} 2 & 0 & 1 \\ 1 & 1 & 1 \end{bmatrix} \begin{bmatrix} x \\ y \\ z \end{bmatrix} = \begin{bmatrix} 2x + z \\ x + y + z \end{bmatrix} \text{ and } T(X^{\mathbf{E}}) = \begin{bmatrix} 2x + z \\ x + y + z \end{bmatrix}.$$

Notice the unnaturalness in the switch of $(2x + z, x + y + z)$ to $\begin{bmatrix} 2x + z \\ x + y + z \end{bmatrix}$. This difficulty comes from two traditional practices - writing functions on the left and writing vectors as rows instead of columns. A change in either of these would simplify matters.

CHANGING THE BASIS

The following theorem describes what happens to the matrix of a linear transformation when the bases are changed.

> **Theorem 3.5.2.** *Let* $T : \mathbf{U} \to \mathbf{V}$ *be a linear transformation;* $\mathbf{B}, \mathbf{B}'$ *bases for* $\mathbf{U}$ *and* $\mathbf{C}, \mathbf{C}'$ *bases for* $\mathbf{V}$. *If* $P = [I_{\mathbf{U}} : \mathbf{B}', \mathbf{B}], Q = [I_{\mathbf{V}} : \mathbf{C}, \mathbf{C}']$, *and* $A = [T : \mathbf{B}, \mathbf{C}]$, *then* $[T : \mathbf{B}', \mathbf{C}'] = QAP$. *In addition,* P *and* Q *are nonsingular.*
>
> *(To change bases in the matrix of a transformation, left multiply by* $[I : \mathbf{B}', \mathbf{B}]$ *and right multiply by* $[I : \mathbf{C}, \mathbf{C}']$.)

Proof. By Theorem 3.3.1, we have $QAP = [I_{\mathbf{V}} : \mathbf{C}, \mathbf{C}'][T : \mathbf{B}, \mathbf{C}][I_{\mathbf{U}} : \mathbf{B}', \mathbf{B}] = [I_{\mathbf{V}}T : \mathbf{B}, \mathbf{C}'][I_{\mathbf{U}} : \mathbf{B}', \mathbf{B}] = [I_{\mathbf{V}}TI_{\mathbf{U}} : \mathbf{B}', \mathbf{C}'] = [T : \mathbf{B}', \mathbf{C}']$ since $T = I_{\mathbf{V}}TI_{\mathbf{U}}$. By Theorem 3.3.1, P and Q are nonsingular. $\square$

Corollary 3.5.3. *If* $T : \mathbf{U} \to \mathbf{U}$ *is a linear transformation,* $\mathbf{B}$ *and* $\mathbf{C}$ *are bases for* $\mathbf{U}$, *and* $A = T_{\mathbf{B}}^{\mathbf{B}}$, *then* $T_{\mathbf{C}}^{\mathbf{C}} = PAP^{-1}$, *where* $P = I_{\mathbf{B}}^{\mathbf{C}}$.

Proof. Since $P^{-1} = I_{\mathbf{C}}^{\mathbf{B}}$, we have by Theorem 3.5.2 that $PAP^{-1} = T_{\mathbf{C}}^{\mathbf{C}}$. $\square$

This "change of basis" theorem and its corollary are not easy to recall, but as with Theorem 3.2.1, there is a picture that makes it easier to remember. Let $T : \mathbf{U} \to \mathbf{V}$ be a linear transformation, where $\mathbf{U}$ and $\mathbf{V}$ are finite-dimensional vector spaces with bases $\mathbf{B}$ and $\mathbf{C}$ respectively. To change the basis $\mathbf{C}$, add the identity transformation $I_{\mathbf{V}} : \mathbf{V} \to \mathbf{V}$ on the right and to change $\mathbf{B}$, add the identity transformation $I_{\mathbf{U}} : \mathbf{U} \to \mathbf{U}$ on the left. Remember the picture:

$$
\begin{array}{ccccccc}
 & I_{\mathbf{U}} & & T & & I_{\mathbf{V}} & \\
\mathbf{U} & \to & \mathbf{U} & \to & \mathbf{V} & \to & \mathbf{V} \\
| & & | & & | & & | \\
\mathbf{B}' & & \mathbf{B} & & \mathbf{C} & & \mathbf{C}'
\end{array}
$$

and write the appropriate product of matrices using the order determined by the composite: $I_{\mathbf{V}}TI_{\mathbf{U}}$.

Given two matrices, they might both be matrices for a given linear transformation. This provides a relationship between matrices.

EQUIVALENCE AND SIMILARITY

We make two definitions concerning matrices that are related in a manner similar to the above theorem and corollary.

Definition 3.5.1. Two $m \times n$ matrices A and B are **equivalent** if and only if there is a nonsingular $m \times m$ matrix P and a nonsingular $n \times n$ matrix Q such that $A = PBQ$. Two $n \times n$ matrices A and B are **similar** if and only if there is a nonsingular matrix S such that $S^{-1}BS = A$.

In Chapters 5 and 6 we will study similarity of matrices. Recall that in Chapter 2, we introduced the notion of row-equivalence. In the next section, we will show the relationship between these two "equivalences."

We conclude this section with some examples illustrating the change of basis process.

Example 3.5.1. First, let's just change the basis with no other linear transformation involved. That is, we will take T to be the identify transformation I. Let $\mathbf{B} = \{(3,1), (2,-3)\}$ and $\mathbf{C} = \{(-1,4), (1,2)\}$ be bases for $\mathbb{R}^2$. It is more complicated than we would like to express the vectors in $\mathbf{B}$ in terms of those in $\mathbf{C}$, so let's work around that. Let $\mathbf{E_2}$ be the standard basis for $\mathbb{R}^2$, and notice that

$$I_{\mathbf{B}}^{\mathbf{E_2}} = \begin{bmatrix} 3 & 2 \\ 1 & -3 \end{bmatrix} \text{ and } I_{\mathbf{C}}^{\mathbf{E_2}} = \begin{bmatrix} -1 & 1 \\ 4 & 2 \end{bmatrix}.$$

Now

$$\begin{aligned} I_{\mathbf{B}}^{\mathbf{C}} &= I_{\mathbf{E_2}}^{\mathbf{C}} I_{\mathbf{B}}^{\mathbf{E_2}} \\ &= (I_{\mathbf{C}}^{\mathbf{E_2}})^{-1} I_{\mathbf{B}}^{\mathbf{E_2}} \\ &= -\frac{1}{6} \begin{bmatrix} 2 & -1 \\ -4 & -1 \end{bmatrix} \begin{bmatrix} 3 & 2 \\ 1 & -3 \end{bmatrix} \\ &= -\frac{1}{6} \begin{bmatrix} 5 & 7 \\ -13 & -5 \end{bmatrix}. \end{aligned}$$

Thus, if (for example) $v^{\mathbf{B}} = \begin{bmatrix} 4 \\ 1 \end{bmatrix}^{\mathbf{B}}$, then

$$v^{\mathbf{C}} = I_{\mathbf{B}}^{\mathbf{C}} v^{\mathbf{B}} = -\frac{1}{6} \begin{bmatrix} 5 & 7 \\ -13 & -5 \end{bmatrix} \begin{bmatrix} 4 \\ 1 \end{bmatrix}^{\mathbf{B}} = -\frac{1}{6} \begin{bmatrix} 27 \\ -57 \end{bmatrix} = -\frac{1}{2} \begin{bmatrix} 9 \\ -19 \end{bmatrix}^{\mathbf{C}}.$$

We can check that this is correct. In the standard basis,

$$v^{\mathbf{B}} = \begin{bmatrix} 4 \\ 1 \end{bmatrix}^{\mathbf{B}} = 4 \begin{bmatrix} 3 \\ 1 \end{bmatrix} + 1 \begin{bmatrix} 2 \\ -3 \end{bmatrix} = \begin{bmatrix} 14 \\ 1 \end{bmatrix}.$$

On the other hand,

$$v^{\mathbf{C}} = -\frac{1}{2} \begin{bmatrix} 9 \\ -19 \end{bmatrix}^{\mathbf{C}} = -\frac{1}{2} \left(9 \begin{bmatrix} -1 \\ 4 \end{bmatrix} - 19 \begin{bmatrix} 1 \\ 2 \end{bmatrix} \right) = -\frac{1}{2} \begin{bmatrix} -28 \\ -2 \end{bmatrix} = \begin{bmatrix} 14 \\ 1 \end{bmatrix},$$

the same thing!

Notice that we are not indicating the basis in a superscript when we are presenting vectors in the standard basis.

$\square$

Example 3.5.2. Let $T : \mathbb{R}^2 \to \mathbb{R}^2$ be defined by $T(x, y) = (x - y, 2x)$. Let $\mathbf{B} = \{(1, 0), (0, 1)\}$ be the standard basis for $\mathbb{R}^2$ and let $\mathbf{C} = \{(2, 1), (1, 0)\}$. Then $T(1, 0) = (1, 2)$ and $T(0, 1) = (-1, 0)$, so

$$T_{\mathbf{B}}^{\mathbf{B}} = \begin{bmatrix} 1 & -1 \\ 2 & 0 \end{bmatrix}.$$

We can now compute $I_{\mathbf{B}}^{\mathbf{C}}$:

$$I(1, 0) = (1, 0) = 0(2, 1) + 1(1, 0)$$
$$I(0, 1) = (0, 1) = 1(2, 1) + (-2)(1, 0)$$

and so $I_{\mathbf{B}}^{\mathbf{C}} = \begin{bmatrix} 0 & 1 \\ 1 & -2 \end{bmatrix}.$

Likewise, $I(2, 1) = (2, 1)$ and $I(1, 0) = (1, 0)$, so

$$I_{\mathbf{C}}^{\mathbf{B}} = \begin{bmatrix} 2 & 1 \\ 1 & 0 \end{bmatrix}.$$

(Recall that $I_{\mathbf{B}}^{\mathbf{C}} = (I_{\mathbf{C}}^{\mathbf{B}})^{-1}$.)

By Theorem 3.5.2 and its corollary, we then have the following:

$$T_{\mathbf{B}}^{\mathbf{C}} = I_{\mathbf{B}}^{\mathbf{C}} T_{\mathbf{B}}^{\mathbf{B}} = \begin{bmatrix} 0 & 1 \\ 1 & -2 \end{bmatrix} \begin{bmatrix} 1 & -1 \\ 2 & 0 \end{bmatrix} = \begin{bmatrix} 2 & 0 \\ -3 & -1 \end{bmatrix}$$

and

$$T_{\mathbf{C}}^{\mathbf{C}} = T_{\mathbf{B}}^{\mathbf{C}} I_{\mathbf{C}}^{\mathbf{B}} = \begin{bmatrix} 2 & 0 \\ -3 & -1 \end{bmatrix} \begin{bmatrix} 2 & 1 \\ 1 & 0 \end{bmatrix} = \begin{bmatrix} 4 & 2 \\ -7 & -3 \end{bmatrix}.$$

Also note

$$I_{\mathbf{C}}^{\mathbf{C}} I_{\mathbf{B}}^{\mathbf{C}} I_{\mathbf{C}}^{\mathbf{B}} = \begin{bmatrix} 0 & 1 \\ 1 & -2 \end{bmatrix} \begin{bmatrix} 2 & 1 \\ 1 & 0 \end{bmatrix} = \begin{bmatrix} 1 & 0 \\ 0 & 1 \end{bmatrix} = I_2,$$

but this is as expected.

$\square$

Example 3.5.3. Define $T : \mathbb{R}^2 \to \mathbb{R}^3$ by $T(x, y) = (4x - 2y, 2x + y, 3x - y)$, and let $\mathbf{B} = \{(3, 1), (2, 2)\}$ and $\mathbf{C} = \{(4, 1, 1), (2, 3, 1), (-1, 1, 2)\}$ be bases for $\mathbb{R}^2$ and $\mathbb{R}^3$, respectively. We will find $T_{\mathbf{B}}^{\mathbf{C}}$.

First, note that $T_{\mathbf{B}}^{\mathbf{C}} = I_{\mathbf{E}_3}^{\mathbf{C}} T_{\mathbf{E}_2}^{\mathbf{E}_3} I_{\mathbf{B}}^{\mathbf{E}_2}$, where $\mathbf{E}_2$ and $\mathbf{E}_3$ are the standard bases of $\mathbb{R}^2$ and $\mathbb{R}^3$, respectively. Since $T(1, 0) = (4, 2, 3)$ and $T(0, 1) = (-2, 1, -1)$, we find

$$I_{\mathbf{E}_3}^{\mathbf{C}} = (I_{\mathbf{C}}^{\mathbf{E}_3})^{-1} = \left(\begin{bmatrix} 4 & 2 & -1 \\ 1 & 3 & 1 \\ 1 & 1 & 2 \end{bmatrix}_{\mathbf{C}}^{\mathbf{E}_3} \right)^{-1} = \frac{1}{20} \begin{bmatrix} 5 & -5 & 5 \\ -1 & 9 & -5 \\ -2 & -2 & 10 \end{bmatrix}_{\mathbf{E}_3}^{\mathbf{C}} \quad \text{(courtesy of Python)},$$

$$T^{\mathbf{E_3}}_{\mathbf{E_2}} = \begin{bmatrix} 4 & -2 \\ 2 & 1 \\ 3 & -1 \end{bmatrix}^{\mathbf{E_3}}_{\mathbf{E_2}}, \text{ and}$$

$$I^{\mathbf{E_2}}_{\mathbf{B}} = \begin{bmatrix} 3 & 2 \\ 1 & 2 \end{bmatrix}^{\mathbf{E_2}}_{\mathbf{B}}, \text{ giving us}$$

$$T^{\mathbf{C}}_{\mathbf{B}} = I^{\mathbf{C}}_{\mathbf{E_3}} T^{\mathbf{E_3}}_{\mathbf{E_2}} I^{\mathbf{E_2}}_{\mathbf{B}} = \begin{bmatrix} 5 & -5 & 5 \\ -1 & 9 & -5 \\ -2 & -2 & 10 \end{bmatrix}^{\mathbf{C}}_{\mathbf{E_3}} \begin{bmatrix} 4 & -2 \\ 2 & 1 \\ 3 & -1 \end{bmatrix}^{\mathbf{E_3}}_{\mathbf{E_2}} \begin{bmatrix} 3 & 2 \\ 1 & 2 \end{bmatrix}^{\mathbf{E_2}}_{\mathbf{B}} = \frac{1}{20} \begin{bmatrix} 55 & 10 \\ 13 & 30 \\ 46 & 20 \end{bmatrix}^{\mathbf{C}}_{\mathbf{B}}.$$

As a partial check, note that $T(3,1) = \begin{bmatrix} 10 \\ 7 \\ 8 \end{bmatrix} = \frac{1}{20} \begin{bmatrix} 55 & 10 \\ 13 & 30 \\ 46 & 20 \end{bmatrix}^{\mathbf{C}}_{\mathbf{B}} \begin{bmatrix} 1 \\ 0 \end{bmatrix}^{\mathbf{B}}$. Verify that T

and multiplication by $T^{\mathbf{C}}_{\mathbf{B}}$ take $(2,2) = \begin{bmatrix} 0 \\ 1 \end{bmatrix}^{\mathbf{B}}$ to the same place. (You should get $(4,6,4)$ in the standard basis in both cases.) $\qquad\qquad\qquad\qquad\qquad\qquad\qquad\qquad\qquad\qquad\qquad\qquad\square$

Section 3.5 Exercises

1. Let $\mathbf{B}$ be the basis $\{(1,1),(1,-1)\}$ for $\mathbb{R}^2$ and let $X = (2,1) \in \mathbb{R}^2$. Calculate $X^{\mathbf{B}}$.

2. With $\mathbf{B}$ as in Exercise 1, calculate $X^{\mathbf{B}}$, where $X = (-1,3)$.

3. Let $\mathbf{B}$ be the basis $\{(1,0,0),(1,1,0),(1,1,1)\}$ for $\mathbb{R}^3$ and let $X = (0,1,2)$. Calculate $X^{\mathbf{B}}$.

4. With $\mathbf{B}$ as in Exercise 3, calculate $X^{\mathbf{B}}$, where $X = (2,3,-1)$.

5. Let $\mathbf{E}$ be the standard basis for $\mathbb{R}^2$. Determine $X^{\mathbf{E}}$ where $X = (-3,2)$.

6. Let $\mathbf{E}$ be the standard basis for $\mathbb{R}^2$. Show that for any $X = (x,y) \in \mathbb{R}^2$, $X^{\mathbf{E}} = \begin{bmatrix} x \\ y \end{bmatrix}$.

7. Let $D = \{(1,1),(0,1)\}$. With T and $\mathbf{B}$ as in Example 3.5.2, compute $[I : \mathbf{B},D], [T : \mathbf{B},D], [I : D,\mathbf{B}]$, and $[T : D,D]$. Show that

$$[T : D,D] = [T : \mathbf{B},D][I : D,\mathbf{B}] \text{ and that } [I : D,\mathbf{B}][I : \mathbf{B},D] = I_2.$$

8. With T and $\mathbf{B}$ and $\mathbf{C}$ as in Example 3.5.2 above, compute matrices P and Q such that $P[T : \mathbf{C},\mathbf{B}]Q = [T : \mathbf{B},\mathbf{C}]$. Express P and Q in the form $[I :?,?]$.

9. Define $T : \mathbb{R}^3 \to \mathbb{R}^3$ by $T(x,y,z) = (x - y, z, 2y)$. Let $\mathbf{B} = \{(0,1,0),(1,0,0),(0,0,1)\}$ and $\mathbf{C} = \{(1,0,0),(1,0,-1),(0,1,2)\}$. Compute $[T : \mathbf{C},\mathbf{B}]$ and then find matrices P and Q such that $P[T : \mathbf{C},\mathbf{B}]Q = [T : \mathbf{B},\mathbf{C}]$. Express P and Q in the form $[I :?,?]$.

10. Define $T : \mathbb{R}^3 \to \mathbb{R}^3$ by $T(x, y, z) = (x + 2y, x - y, 2y - z)$. Let $\mathbf{B} = \{(1, 0, 0), (1, 0, 1), (2, 1, -2)\}$ and $\mathbf{C} = \{(1, 0, 0), (0, 0, 1), (0, 1, 0)\}$. Compute $T_{\mathbf{C}}^{\mathbf{B}}$ and then find matrices P and Q such that $T_{\mathbf{C}}^{\mathbf{B}} Q = T_{\mathbf{B}}^{\mathbf{C}}$. Express P and Q in the form $I_?^?$.

11. With $T, \mathbf{B}$, and $\mathbf{C}$ as in Exercise 9, calculate $A = T_{\mathbf{B}}^{\mathbf{C}}$, $X^{\mathbf{B}}$, and $T(X)^{\mathbf{C}}$, where $X = (x, y, z)$ is an arbitrary vector in $\mathbb{R}^3$. Show that $AX^{\mathbf{B}} = T(X)^{\mathbf{C}}$.

12. Show that a nonsingular $n \times n$ matrix A is equivalent to I_n.

13. Show that if A is an $n \times n$ matrix and A is similar to I_n, then $A = I_n$.

3.6   Row and Column Operations and Change of Basis

In this section, we wish to investigate the relationship between operations on matrices for linear transformations and certain changes in basis. We will relate each elementary row and column operation to a corresponding change in one of the two bases.

In Section 1.3, we introduced elementary row operations. These operations arose naturally as "legal" operations that could be performed on the augmented matrix of a system of linear equations; that is, operations that would preserve the solution set. The three types of elementary row operations are denoted by $R_{ik}, R_i(a)$, and $R_{ik}(a)$. Recall that R_{ik} indicates that rows i and k are interchanged, $R_i(a)$ represents that row i is multiplied by a, and $R_{ik}(a)$ stands for the operation of adding a times row i to row k (remember that when the operation $R_{ik}(a)$ is performed, row i remains unchanged).

Associated with each elementary row operation there is an **elementary matrix** that is obtained by performing the given elementary row operation on the identity matrix. We use the same notation for the elementary row operation as the elementary matrix, so for $n \times n$ matrices:

R_{ik} is I_n with rows i and k switched
$R_i(a)$ is I_n with row i multiplied by a
$R_{ik}(a)$ is I_n with a times row i added to row k.

Now by Theorem 1.4.3, performing an elementary row operation on a matrix A and left multiplication of A by the corresponding elementary matrix E give the same result; that is, the result of the product EA is the same as the matrix A with the elementary row operation performed on it.

ROW OPERATIONS AND CHANGE OF BASIS

Let $T : \mathbf{U} \to \mathbf{V}$ be a linear transformation, let $\mathbf{B} = \{X_1, \ldots, X_n\}$ be a basis for $\mathbf{U}$ and let $\mathbf{C} = \{Y_1, \ldots, Y_m\}$ be a basis for $\mathbf{V}$. The matrix $A = [a_{ij}] = T_{\mathbf{B}}^{\mathbf{C}}$ is determined by:

$$
\begin{aligned}
T(X_1) &= a_{11}Y_1 + \ldots + a_{m1}Y_m \\
T(X_2) &= a_{12}Y_1 + \ldots + a_{m2}Y_m \\
&\;\;\vdots \qquad\quad \vdots \qquad\quad \vdots \\
T(X_n) &= a_{1n}Y_1 + \ldots + a_{mn}Y_m
\end{aligned}
\tag{3.6.1}
$$

For changes in $\mathbf{C}$ we list the corresponding change in $A = T_{\mathbf{B}}^{\mathbf{C}}$:

	Change in $\mathbf{C}$	Change in $T_{\mathbf{B}}^{\mathbf{C}}$
(1)	Interchange Y_i and Y_k	Row i and row k are interchanged.
(2)	Multiply Y_i by $1/a$, where a is a nonzero scalar.	Each element of row i is multiplied by a.
(3)	Replace Y_i by $Y_i - aY_k$, where a is some scalar.	Each element a_{kj} of row k replaced by $a_{kj} + aa_{ik}$; that is, a times the i-th row vector is added to the k-th row vector.

(1) and (2) are straightforward. To see that (3) holds, note that

$$T(X_j) = a_{1j}Y_1 + \ldots + a_{ij}Y_i + \ldots + a_{mj}Y_m$$
$$= a_{1j}Y_1 + \ldots + a_{ij}(Y_i - aY_j) + \ldots + (a_{kj} + aa_{ij})Y_k + \ldots + a_{mj}Y_m.$$

The reader should verify that each of the changes in $\mathbf{C}$ above yields a new basis $\mathbf{C}'$. Changing the basis $\mathbf{C}$ to the basis $\mathbf{C}'$ has the following effect (by the results of Section 3.4) on the matrix of $T : T_{\mathbf{B}}^{\mathbf{C}'} = I_{\mathbf{C}}^{\mathbf{C}'} T_{\mathbf{B}}^{\mathbf{C}}]$.

From Section 3.5 we know that to change $T_{\mathbf{B}}^{\mathbf{C}}$ to $T_{\mathbf{B}}^{\mathbf{C}'}$, we must multiply $T_{\mathbf{B}}^{\mathbf{C}}$ on the left by $I_{\mathbf{C}}^{\mathbf{C}'}$. To compute $I_{\mathbf{C}}^{\mathbf{C}'}$, we need only change the second basis $\mathbf{C}'$. But we have already determined how to do this for any transformation. It follows that $I_{\mathbf{C}}^{\mathbf{C}'}$ is just I_m with the corresponding operation performed on the rows of I_m. We see that elementary row operations and elementary matrices occur in the context of change of basis.

COLUMN OPERATIONS AND CHANGE OF BASIS

Let us now consider certain changes in the basis $\mathbf{B}$ using the relations 3.6.1. Because, in effect, the elements of the basis $\mathbf{B}$ correspond to the columns of the matrix of the linear transformation, changes in $\mathbf{B}$ correspond to changes in the columns of the matrix. Again we list the changes in $T_{\mathbf{B}}^{\mathbf{C}}$ corresponding to changes in $\mathbf{B}$.

	Change in $\mathbf{B}$	Change in $T_{\mathbf{B}}^{\mathbf{C}}$
(1)	Interchange X_k and X_j	Columns k and j are interchanged.
(2)	Multiply X_j by a	Column j is multiplied by a
(3)	Replace X_j by $X_j + aX_k$	a times column k is added to column j.

Now as before we can apply the results of Section 3.4 to obtain the relationship between $T_{\mathbf{B}}^{\mathbf{C}}$ and $T_{\mathbf{B}'}^{\mathbf{C}}$, where $\mathbf{B}'$ is the new basis obtained from B by making the changes listed under "change in $\mathbf{B}$" above. We get:

$$T_{\mathbf{B'}}^{\mathbf{C}} = T_{\mathbf{B}}^{\mathbf{C}} I_{\mathbf{B'}}^{\mathbf{B}}. \tag{3.6.2}$$

Just as before, the matrix $I_{\mathbf{B'}}^{\mathbf{B}}$ can be obtained from the $n \times n$ identity matrix I_n by performing the operations listed in the above table. This results in the elementary column matrices:

1. C_{kj} is I_n with columns k and j interchanged.

2. $C_j(a)$ is I_n with column j multiplied by a.

3. $C_{kj}(a)$ is I_n with a times column k added to column j.

The operations listed in the table above "change in $T_{\mathbf{B}}^{\mathbf{C}}$" are called **elementary column operations**, and as with elementary row operations, an elementary column operation can be performed by multiplying by the corresponding elementary column matrix. There is a dissimilarity, however: from Equation 3.6.2 we see that the column elementary matrix appears on the right, not the left. We have proved the following:

> **Theorem 3.6.1.** *Let A be an $m \times n$ matrix. Then any one of the elementary column operations can be performed on A by multiplying A on the right by the corresponding $n \times n$ elementary column matrix.*
>
> (A column operation is performed by right multiplying by the column elementary matrix.)

Theorem 3.6.1 makes it easy to compute products of elementary column matrices. The procedure is like that for the elementary row matrices, except that one works from the left to right. Consider the following product of 2×2 elementary column matrices:

$$
\begin{aligned}
C_{12} C_2(2) C_{12}(-1) C_{21}(2) &= \begin{bmatrix} 0 & 1 \\ 1 & 0 \end{bmatrix} C_2(2) C_{12}(-1) C_{21}(2) \\
&= \begin{bmatrix} 0 & 2 \\ 1 & 0 \end{bmatrix} C_{12}(-1) C_{21}(2) \\
&= \begin{bmatrix} 0 & 2 \\ 1 & -1 \end{bmatrix} C_{21}(2) \\
&= \begin{bmatrix} 4 & 2 \\ -1 & -1 \end{bmatrix}.
\end{aligned}
$$

ROW VS. COLUMN OPERATIONS

By considering the following "examples" of 3×3 and 4×4 matrices one can easily see the relationship between the row and column elementary matrices.

Example 3.6.1.

$$
R_{13} = \begin{bmatrix} 0 & 0 & 1 \\ 0 & 1 & 0 \\ 1 & 0 & 0 \end{bmatrix} = C_{13}, \; R_{23} = \begin{bmatrix} 1 & 0 & 0 & 0 \\ 0 & 0 & 1 & 0 \\ 0 & 1 & 0 & 0 \\ 0 & 0 & 0 & 1 \end{bmatrix} = C_{23}, \; R_2(a) = \begin{bmatrix} 1 & 0 & 0 \\ 0 & a & 0 \\ 0 & 0 & 1 \end{bmatrix} = C_2(a),
$$

$$R_{13}(a) = \begin{bmatrix} 1 & 0 & 0 \\ 0 & 1 & 0 \\ a & 0 & 1 \end{bmatrix} = C_{31}(a), \quad R_{23}(a) = \begin{bmatrix} 1 & 0 & 0 & 0 \\ 0 & 1 & 0 & 0 \\ 0 & a & 1 & 0 \\ 0 & 0 & 0 & 1 \end{bmatrix} = C_{32}(a).$$

$\square$

From this it is not hard to see the following theorem.

Theorem 3.6.2. *Every elementary row matrix is also an elementary column matrix. In fact:*

1. $R_{ij} = C_{ij}$

2. $R_i(a) = C_i(a)$

3. $R_{ij}(a) = C_{ji}(a).$

(Row elementary matrices are also column elementary matrices.)

Every elementary matrix is a nonsingular matrix. To see this one need only calculate the following products by considering the effect of multiplication by an elementary matrix:

$$\begin{array}{ll} R_{ik}R_{ik} = I & C_{kj}C_{kj} = I \\ R_i(1/a)R_i(a) = I & C_i(a)C_i(1/a) = I \\ R_{ik}(-a)R_{ik}(a) = I & C_{kj}(a)C_{kj}(-a) = I \end{array}$$

We see that the following theorem is true.

Theorem 3.6.3. *Every elementary matrix is nonsingular. In fact,*

1. $R_{ik}^{-1} = R_{ik}, C_{kj}^{-1} = C_{kj}$

2. $R_i(a)^{-1} = R_i(1/a), C_i(a)^{-1} = C_i(1/a)$

3. $R_{ik}(a)^{-1} = R_{ik}(-a), C_{kj}(a)^{-1} = C_{kj}(-a)$

(The inverse of an elementary matrix is an elementary matrix.)

We present now a sequence of results that show that every nonsingular matrix is a product of elementary matrices, and then find a practical method for finding the inverse of a nonsingular matrix.

Theorem 3.6.4. *Let $A_1, \ldots, A_m$ be nonsingular $n \times n$ matrices and let $P = A_1 \ldots A_m$ be their product. Then P is nonsingular and $P^{-1} = A_m^{-1} \ldots A_1^{-1}$.*

(The inverse of a product is the product of the inverses in the reverse order.)

Proof. A formal proof requires mathematical induction; we will use an informal approach. For $i = 1, \ldots, m$, $A_i A_i^{-1} = A_i^{-1} A_i = I$. Using this fact repeatedly, we see that

$$
\begin{aligned}
(A_m^{-1} \ldots A_1^{-1})P &= (A_m^{-1} \ldots A_1^{-1})(A_1 \ldots A_m) \\
&= (A_m^{-1} \ldots A_2^{-1})(A_1^{-1} A_1)(A_2 \ldots A_m) \\
&= (A_m^{-1} \ldots A_2^{-1})I(A_2 \ldots A_m) \\
&= (A_m^{-1} \ldots A_2^{-1})(A_2 \ldots A_m) \\
&\;\;\vdots \\
&= A_m^{-1} A_m \\
&= I.
\end{aligned}
$$

Likewise, $(A_1 \ldots A_m)(A_m^{-1} \ldots A_1^{-1}) = I$, and so P is nonsingular with $P^{-1} = (A_m^{-1} \ldots A_1^{-1})$. $\quad\square$

Combining Theorems 3.6.3 and 3.6.4, we get the following corollary.

Corollary 3.6.5. *A product $E_1 \ldots E_m$ of elementary matrices $E_1, \ldots, E_m$ is a nonsingular matrix.*   *(A product of elementary matrices is nonsingular.)*

We need two preliminary results for the next theorem. The proofs are left as exercises.

Lemma 3.6.6. *A product $P = E_1 \ldots E_m$ of $n \times n$ elementary row matrices $E_1, \ldots, E_m$ is a matrix of rank n.*   *(A product of elementary matrices has rank n.)*

Lemma 3.6.7. *If B is an $n \times n$ matrix in reduced echelon form and B has rank n, then $B = I_n$.*   *(The only $n \times n$ matrix in reduced echelon form with rank n is the identity matrix.)*

Note that the converse of the above Lemma is also true: If $B = I_n$, then B has rank n. If B has rank less than n, then B must have a row of zeros (remember that B is an $n \times n$ matrix in reduced echelon form). A nonsingular matrix cannot have a row of zeros. To see this, assume that A is an $n \times n$ matrix with a row of zeros. Then any product AC has a row of zeros and so $AC \neq I$ for any matrix C. This means that A cannot have an inverse and so A is singular. We will use this principle in Chapter 4.

Theorem 3.6.8. *Let A be an $n \times n$ nonsingular matrix. Then A is a product of elementary row matrices, and so also a product of elementary column matrices.*
(A nonsingular matrix is a product of elementary matrices.)

Proof. Let B be the reduced echelon form of A. We will see that B is the the identity matrix. Since B is obtained from A by elementary row operations, there is a product P of elementary row matrices with $PA = B$. Now assume that B has rank r and $r < n$. Then it follows, since B is $n \times n$, that the last row of B is all zeros. Now multiplying $PA = B$ by A^{-1} we get $(PA)A^{-1} = BA^{-1}$ or $P(AA^{-1}) = PI = P = BA^{-1}$. Now by Lemma 3.6.6, P has rank n, but the product BA^{-1} has a row of zeros (since B does), and so its rank must be less than n. This is a contradiction, and so our assumption that the rank r of B was less than n must be wrong. We conclude that the rank of B is n, and so using Lemma 3.6.7, we see that $B = I$. From this we see that $PA = I$ where P is a product of elementary row matrices, say $P = E_1 \ldots E_m$. Now solving $E_1 \ldots E_m A = I$ for A we get:

$$A = E_m^{-1} \ldots E_2^{-1} E_1^{-1} I = E_m^{-1} \ldots E_2^{-1} E_1^{-1}.$$

Since each E_i is an elementary row matrix, the inverse E_i^{-1} is also an elementary row matrix by Theorem 3.6.3. $\square$

We illustrate the proof of Theorem 3.6.8 with an example.

Example 3.6.2. Let $A = \begin{bmatrix} 0 & 1 & 0 \\ 2 & 2 & -1 \\ -1 & 0 & 1 \end{bmatrix}$. To express A as a product of elementary row matrices, we begin by reducing A to its reduced echelon form, keeping track of the row operations used:

$$\begin{bmatrix} 0 & 1 & 0 \\ 2 & 2 & -1 \\ -1 & 0 & 1 \end{bmatrix} \xrightarrow{R_{13}} \begin{bmatrix} -1 & 0 & 1 \\ 2 & 2 & -1 \\ 0 & 1 & 0 \end{bmatrix} \xrightarrow{R_1(-1)} \begin{bmatrix} 1 & 0 & -1 \\ 2 & 2 & -1 \\ 0 & 1 & 0 \end{bmatrix}$$

$$\xrightarrow{R_{12}(-2)} \begin{bmatrix} 1 & 0 & -1 \\ 0 & 2 & 1 \\ 0 & 1 & 0 \end{bmatrix} \xrightarrow{R_2(1/2)} \begin{bmatrix} 1 & 0 & -1 \\ 0 & 1 & 1/2 \\ 0 & 1 & 0 \end{bmatrix} \xrightarrow{R_{23}(-1)} \begin{bmatrix} 1 & 0 & -1 \\ 0 & 1 & 1/2 \\ 0 & 0 & -1/2 \end{bmatrix}$$

$$\xrightarrow{R_3(-2)} \begin{bmatrix} 1 & 0 & -1 \\ 0 & 1 & 1/2 \\ 0 & 0 & 1 \end{bmatrix} \xrightarrow{R_{32}(-1/2)} \begin{bmatrix} 1 & 0 & -1 \\ 0 & 1 & 0 \\ 0 & 0 & 1 \end{bmatrix} \xrightarrow{R_{31}(1)} \begin{bmatrix} 1 & 0 & 0 \\ 0 & 1 & 0 \\ 0 & 0 & 1 \end{bmatrix}$$

$$= \qquad B.$$

Since the matrix A is nonsingular, the reduced echelon form B of A is I_3 and so we get

$$R_{31}(1) R_2(1/2) R_3(-2) R_{23}(-1) R_2(1/2) R_{12}(-2) R_1(-1) R_{13} A = I. \qquad (3.6.3)$$

Multiplying both sides of the above identity successively by the inverses of the elementary matrices, we solve for A and obtain

$$A = R_{13}^{-1} R_1(-1)^{-1} R_{12}(-2)^{-1} R_2(1/2)^{-1} R_{23}(-1)^{-1} R_3(-2)^{-1} R_2(1/2)^{-1} R_{31}(1)^{-1} I$$
$$= R_{13} R_1(-1) R_{12}(2) R_2(2) R_{23}(1) R_3(-1/2) R_2(2) R_{31}(-1).$$

We see that we have expressed A as a product of elementary matrices.

$\square$

In the above example, the inverse of the matrix A appeared as the product of the elementary matrices in the identity 3.6.3. If A is an $n \times n$ matrix and $E_1, \ldots, E_m$ are elementary matrices with $E_1 \ldots E_m A = I_n$, then it appears that the product $P = E_1 \ldots E_m$ is the inverse of A. One must check, however, that $AP = I$. To see this, move the E's to the right side of the identity by successively multiplying on the left by inverses. We get

$$A = E_m^{-1} \ldots E_1^{-1} I = I E_m^{-1} \ldots E_1^{-1}$$

Now move the E's to the left side by successively multiplying on the right to obtain:

$$AE_1 \ldots E_m = I.$$

From this we can see that $PA = AP = I$, and so A is nonsingular and $P = A^{-1}$.

There is an easy method for obtaining the inverse of a nonsingular matrix, especially for those with a computer program for obtaining the reduced echelon form of a matrix. As above, assume that A is an $n \times n$ matrix with $E_1, \ldots, E_m$ elementary matrices with the product:

$$E_1 \ldots E_m A = I.$$

Adjoin the $n \times n$ identity matrix to obtain the "partitioned" matrix $[A|I_n]$. Now using block multiplication we see that $E_1 \ldots E_m[A|I_n] = [E_1 \ldots E_m A | E_1 \ldots E_m I_n] = [I_n | E_1 \ldots E_m]$, and $E_1 \ldots E_m$ is the inverse of A. We have shown:

Theorem 3.6.9. *If A is an $n \times n$ matrix and $[I_n|P]$ is the reduced echelon form of $[A|I_n]$, then $P = A^{-1}$.*   (If $[A|I]$ is reduced to reduced echelon form, the inverse of A appears on the right.)

Section 3.6 Exercises

1. Using row operations compute the following product of 2×2 elementary row matrices: $R_{12}R_2(-1)R_{12}(2)R_{21}(-3)$.

2. Using row operations compute the following product of 3×3 elementary row matrices: $R_{23}R_1(-3)R_{13}(2)R_{12}(-2)$.

3. Using column operations, compute the following product of 2×2 elementary column matrices: $C_{21}C_{21}(3)C_2(-2)C_{12}(-2)$.

4. Using column operations, compute the following product of 3×3 elementary column matrices: $C_{31}C_{32}(2)C_3(-1)C_{13}(-3)C_{23}$.

5. Using the "C-notation," express each of the following 2×2 elementary row matrices as column elementary matrices:

   (a) R_{12}

    (b) $R_2(1/2)$

    (c) $R_{21}(-1)$

    (d) $R_{12}(2)$

6. Using the "R-notation," express each of the 2×2 elementary column matrices as elementary row matrices:

    (a) C_{12}

    (b) $C_2(-1)$

    (c) $C_{12}(2)$

    (d) $C_{21}(-2)$

7. Using the "R-notation," express each of the following 3×3 elementary column matrices as elementary row matrices:

    (a) C_{13}

    (b) $C_1(-3)$

    (c) $C_{23}(-2)$

    (d) $C_{12}(3)$

8. Express the inverses of the 2×2 elementary row matrices in Exercise 5 as elementary row matrices.

9. Express the inverses of the 2×2 elementary column matrices in Exercise 6 as elementary column matrices.

10. Express the inverses of the 3×3 elementary column matrices in Exercise 7 as elementary column matrices.

11. Prove Lemma 3.6.6.

12. Prove Lemma 3.6.7.

13. Find the inverse of $\begin{bmatrix} 1 & 2 \\ -1 & 3 \end{bmatrix}$ by using the method of Theorem 3.6.9.

14. Express the matrix in Exercise 13 as a product of row elementary matrices.

15. Express the matrix $\begin{bmatrix} 0 & 0 & 2 \\ 1 & 0 & 1 \\ 2 & -1 & 3 \end{bmatrix}$ as a product of elementary matrices and find its inverse.

16. Let A be a nonsingular $n \times n$ matrix. Show that A^{-1} is nonsingular and that $(A^{-1})^{-1} = A$.

Chapter 4

DETERMINANTS

4.1 Definitions and Elementary Properties

Associated with every square (that is, $n \times n$ for some n) matrix there is a scalar quantity called the determinant of the matrix. The determinant of a matrix is rich in interesting properties and we will see that using the determinant, the inverse of a nonsingular matrix can be calculated and certain systems of linear equations can be solved. On the surface, it appears that the determinant is of importance as a computational tool, but we will see that it is of more importance as a theoretical tool and that computationally it is not worthwhile for work on large matrices. Systems of linear equations can be solved and inverses can be calculated more efficiently using row operations and the technique of reduction to echelon form. The determinant will be essential in calculating eigenvalues in Chapter 5.

From the complicated definition of the determinant given below it is unclear how anyone could have ever thought to study this quantity. Consider a general system of two linear equations in the unknowns x and y:

$$\begin{aligned} ax + by &= h \\ cx + dy &= k \end{aligned} \qquad (4.1.1)$$

Solving for x and y (if possible), we get:

$$x = \frac{dh - bk}{ad - bc}$$

$$y = \frac{ak - ch}{ad - bc}$$

The quantity $ad - bc$ in the denominator is, according to the definition below, the **determinant** of the coefficient matrix of the system 4.1.1.

4.1.1 THE DEFINITION

We define the **determinant**, $|A|$, of an $n \times n$ matrix $A = [a_{ij}]$ inductively as follows:

If $A = [a_{ij}]$ is 1×1, we define $|A| = a_{11}$. Assume that the determinant of any $(n-1) \times (n-1)$ or smaller matrix has been defined and let $A = [a_{ij}]$ be $n \times n$. For any $i, j, 1 \leq i, j \leq n$, let M_{ij} be the matrix obtained by omitting row i and column j from the matrix A; M_{ij} is called the **minor** of a_{ij}. Then M_{ij} is $(n-1) \times (n-1)$, so $|M_{ij}|$ is defined. The **cofactor** of a_{ij} is defined to be $(-1)^{i+j}|M_{ij}|$. The determinant of the $n \times n$ matrix $A = [a_{ij}]$ is now defined by:

$$|A| = \sum_{j=1}^{n} a_{1j}A_{1j},$$

where A_{1j} is the cofactor of a_{1j}.

To illustrate the definition we compute the determinants of general 2×2 and 3×3 matrices:

$$\begin{vmatrix} a_{11} & a_{12} \\ a_{21} & a_{22} \end{vmatrix} = a_{11}(-1)^{1+1}|[a_{22}]| + a_{12}(-1)^{1+2}|[a_{21}]| = a_{11}a_{22} - a_{12}a_{21}$$

$$\begin{vmatrix} a_{11} & a_{12} & a_{13} \\ a_{21} & a_{22} & a_{23} \\ a_{31} & a_{32} & a_{33} \end{vmatrix} = a_{11}\begin{vmatrix} a_{22} & a_{23} \\ a_{32} & a_{33} \end{vmatrix} - a_{12}\begin{vmatrix} a_{21} & a_{23} \\ a_{31} & a_{33} \end{vmatrix} + a_{13}\begin{vmatrix} a_{21} & a_{22} \\ a_{31} & a_{32} \end{vmatrix}$$

$$= a_{11}a_{22}a_{33} - a_{11}a_{23}a_{32} - a_{12}a_{21}a_{33} + a_{12}a_{23}a_{31} + a_{13}a_{21}a_{32} - a_{13}a_{22}a_{31}.$$

Note that in these two cases the determinants are the sum, with the appropriate sign $+$ or $-$, of all possible products of entries of the matrix, where exactly one entry in each product comes from each row and each column. This rule holds in general; that is, $|[a_{ij}]|$ is the sum of all products $\pm a_{1j_1} \ldots a_{nj_n}$, where $j_1, j_2, \ldots, j_n$ is a listing of the integers $1, 2, \ldots, n$ in some particular order. Knowing this, the following two theorems are believable, but the proofs are complicated and we postpone them until Appendix 6.

PROPERTIES OF THE DETERMINANT

The following three theorems give properties of the determinant. The definition of the determinants states that the determinant of a matrix is the sum of the products of each entry in row one times their cofactors. The first theorem states that any row or column can be used instead of row one and that the resulting sum of the products will be the same.

Theorem 4.1.1. *Let $A = [a_{ij}]$ be an $n \times n$ matrix. Then*

$$|A| = \sum_{j=1}^{n} a_{ij}A_{ij} \quad \textit{(expansion along row i), and}$$

$$|A| = \sum_{i=1}^{n} a_{ij}A_{ij} \quad \textit{(expansion down column j).}$$

(Expansion along any row or column gives the determinant.)

For example, expanding the determinant of a general 3×3 matrix first along row 2 and then along column 1, we get:

$$
\begin{vmatrix} a_{11} & a_{12} & a_{13} \\ a_{21} & a_{22} & a_{23} \\ a_{31} & a_{32} & a_{33} \end{vmatrix} = a_{21}(-1)^3 \begin{vmatrix} a_{12} & a_{13} \\ a_{32} & a_{33} \end{vmatrix} + a_{22}(-1)^4 \begin{vmatrix} a_{11} & a_{13} \\ a_{31} & a_{33} \end{vmatrix} + a_{23}(-1)^5 \begin{vmatrix} a_{11} & a_{12} \\ a_{31} & a_{32} \end{vmatrix}
$$

$$
= a_{11}(-1)^2 \begin{vmatrix} a_{22} & a_{23} \\ a_{32} & a_{33} \end{vmatrix} + a_{21}(-1)^3 \begin{vmatrix} a_{12} & a_{13} \\ a_{32} & a_{33} \end{vmatrix} + a_{31}(-1)^4 \begin{vmatrix} a_{12} & a_{13} \\ a_{22} & a_{23} \end{vmatrix}.
$$

You should verify that the two expansions above are in fact equal.

The above theorem allows one to take advantage of a row or column containing several zeros to simplify the computation of the determinant of a matrix.

Example 4.1.1.

$$
\begin{vmatrix} 1 & 2 & -1 & 0 \\ 4 & 3 & 1 & 2 \\ -1 & 0 & 1 & 0 \\ 3 & 0 & 2 & 0 \end{vmatrix} = 2(-1)^6 \begin{vmatrix} 1 & 2 & -1 \\ -1 & 0 & 1 \\ 3 & 0 & 2 \end{vmatrix}
$$

$$
= 2(2(-1)^3) \begin{vmatrix} -1 & 1 \\ 3 & 2 \end{vmatrix}
$$

$$
= -4(-2 - 3)
$$

$$
= 20,
$$

where the first expression used column four and the second expression used column two of the 3×3 determinant.

$\square$

Theorem 4.1.2. *Let A be an $n \times n$ matrix and let B be the matrix obtained from A by interchanging two of the rows (or columns) of A. Then $|B| = -|A|$.*

(Switching rows or columns changes the sign of the determinant.)

Using elementary row and column matrices, the above result can be stated as:

$$
|R_{ik}A| = -|A| \text{ and } |AC_{kj}| = -|A|.
$$

The theorem is seen to be true for 2×2 matrices since

$$
\begin{vmatrix} a & b \\ c & d \end{vmatrix} = ad - bc,
$$

but

$$
\begin{vmatrix} c & d \\ a & b \end{vmatrix} = cb - ad = -(ad - bc) \text{ and } \begin{vmatrix} b & a \\ d & c \end{vmatrix} = bc - ad = -(ad - bc).
$$

Knowing the theorems above, it is easy to establish many properties of determinants. We must recall a definition that we made in Section 1.2: for an $m \times n$ matrix $A = [a_{ij}]$, the

transpose of A is the $n \times m$ matrix $A^t = [b_{ij}]$, where $b_{ij} = a_{ji}$ for all i, j. In effect, the transpose of a matrix is simply the matrix obtained by switching the rows and columns of the original matrix.

Example 4.1.2. If

$$A = \begin{bmatrix} 1 & 2 & 1 \\ 3 & 1 & 2 \end{bmatrix},$$

then

$$A^t = \begin{bmatrix} 1 & 3 \\ 2 & 1 \\ 1 & 2 \end{bmatrix}.$$

□

Theorem 4.1.3. *Let $A = [a_{ij}]$ be an $n \times n$ matrix. Then*

(a) $|A| \quad = \quad |A^t|, \quad$ *where* $\quad A^t \quad$ *is* $\quad$ *the* $\quad$ *transpose* $\quad$ *of* $\quad A$.
   (A matrix and its transpose have the same determinant.)

(b) *If all the entries in one row or column of A are zero, then $|A| = 0$.*
   (A row of zeros makes the determinant zero.)

(c) *If B is obtained from A by multiplying each entry of one row or one column by a scalar c, then $|B| = c|A|$; that is, $|R_i(c)A| = |AC_j(c)| = c|A|$.*
   (If a row is multiplied by a scalar, the determinant is multiplied by that scalar.)

(d) $|cA| = c^n|A|$.

(e) *If A is partitioned into column vectors $A = [X_1 \ldots X_n]$ and B is an $n \times n$ matrix that differs from A only in column k, say $B = [X_1 \ldots Y_k \ldots X_n]$, then $|A| + |B| = |[X_1 \ldots X_k \ldots X_n]| + |[X_1 \ldots Y_k \ldots X_n]| = |[X_1 \ldots X_k + Y_k \ldots X_n]|$. The corresponding result is true for rows.* $\quad$ *(The determinant preserves addition in rows or columns.)*

(f) *If A has two rows that are the same, then $|A| = 0$. If one row of A is a multiple of another row of A, then $|A| = 0$. A similar result holds true for columns.*
   (If two rows are the same, the determinant is 0.)

(g) *If a multiple of one row of A is added to another row of A, then the determinant of A is unchanged; that is, $|R_{ik}(a)A| = |A|$. The corresponding result is true for columns.* $\quad$ *(Adding a multiple of a row or a column to another doesn't change the determinant.)*

Proof. (a) The expansion of $|A|$ by row 1 is identical to the expansion of $|A^t|$ by column 1, except that the minors of one expression are the transposes of the minors in the other expression. This allows an easy induction proof. See Exercise 6.

(b) Expand along the row or column of zeros.

(c) Expand along the row or column in question and factor out the scalar c.

(d) Apply (c) to each row of cA.

(e) Expand along column k and apply the distributive law.

(f) Interchanging the identical rows we see that $|A| = -|A|$, so $|A| = 0$. For the second part, apply (c) and then the first part.

(g) This is a little harder. Assume $k < j$.

$$
\begin{aligned}
|R_{ij}(a)A| &= |[X_1 \dots X_k + cX_j \dots X_n]| \\
&= |[X_1 \dots X_k \dots X_n]| + |[X_1 \dots cX_j \dots X_j \dots X_n]| \\
&= |[X_1 \dots X_k \dots X_n]| + c|[X_1 \dots X_j \dots X_j \dots X_n]| \\
&= |[X_1 \dots X_k \dots X_n]|
\end{aligned}
$$

since $|[X_1 \dots X_j \dots X_j \dots X_n]| = 0$ (because two columns are the same).

Similar proofs work for either rows or columns.

$\square$

Parts (c) and (e) of the theorem above state that the determinant, when considered as a function of a single row or column, has the properties of a linear transformation.

Example 4.1.3. We can use row and column operations to simplify the computation of a determinant:

$$
\begin{vmatrix}
2 & 1 & 0 & 1 \\
1 & -1 & 2 & 0 \\
0 & 3 & 1 & 2 \\
0 & 1 & 4 & 0
\end{vmatrix}
=
\begin{vmatrix}
0 & 3 & -4 & 1 \\
1 & -1 & 2 & 0 \\
0 & 3 & 1 & 2 \\
0 & 1 & 4 & 0
\end{vmatrix}
\quad \text{performing } R_{21}[-2]
$$

$$
= -1
\begin{vmatrix}
3 & -4 & 1 \\
3 & 1 & 2 \\
1 & 4 & 0
\end{vmatrix}
\quad \text{expanding along column 1}
$$

$$
= -
\begin{vmatrix}
3 & -4 & 1 \\
0 & 5 & 1 \\
1 & 4 & 0
\end{vmatrix}
\quad \text{performing } R_{12}[-1]
$$

$$
= -\left(3
\begin{vmatrix}
5 & 1 \\
4 & 0
\end{vmatrix}
+ 1
\begin{vmatrix}
-4 & 1 \\
5 & 1
\end{vmatrix}
\right)
\quad \text{expanding down column 1}
$$

$$
= -(-12 - 9)
$$

$$
= 21.
$$

$\square$

Section 4.1 Exercises

In Exercises 1 - 6 evaluate the determinant.

1. $\begin{vmatrix} 1 & -3 \\ 2 & 1 \end{vmatrix}$

2. $\begin{vmatrix} -1 & 5 \\ -2 & 3 \end{vmatrix}$

3. $\begin{vmatrix} 2 & 3 & -1 \\ 1 & -3 & 2 \\ 4 & 2 & 6 \end{vmatrix}$

4. $\begin{vmatrix} 2 & -3 & 2 \\ 1 & 3 & 2 \\ -4 & 8 & 6 \end{vmatrix}$

5. $\begin{vmatrix} 1 & 3 & 1 & 2 \\ 0 & 2 & -1 & 0 \\ -2 & 4 & -2 & 1 \\ 2 & 6 & -5 & 2 \end{vmatrix}$

6. $\begin{vmatrix} 0 & 1 & 0 & 4 \\ -1 & 0 & 2 & 3 \\ -2 & 0 & 3 & -3 \\ 1 & 0 & -3 & -2 \end{vmatrix}$

7. Calculate the following determinant:

$$\begin{vmatrix} 1 & 1 & 3 & 4 & 2 \\ 1 & 2 & 6 & 4 & 5 \\ 1 & 2 & 1 & 2 & 3 \\ 1 & 2 & 1 & 2 & 4 \\ 3 & 4 & 0 & 0 & 1 \end{vmatrix}.$$

8. Expand the following determinant along row 1. Then expand along row 3, and finally down column 3.

$$\begin{vmatrix} 1 & -1 & 0 \\ 2 & 1 & -1 \\ 4 & 3 & 1 \end{vmatrix}$$

9. Compute the following determinant and evaluate the cofactor of the entry 6 in row 4 column 3.

$$\begin{vmatrix} 1 & 2 & 1 & 0 \\ 0 & 0 & 2 & 0 \\ 1 & -1 & 1 & 2 \\ 0 & 4 & 6 & 1 \end{vmatrix}$$

10. Let $A = \begin{bmatrix} 1 & 2 \\ -1 & 3 \end{bmatrix}$. Compute $|A|$ and compute all of the cofactors $A_{11}, A_{21}, A_{12}, A_{22}$.

    Now let $B = \begin{bmatrix} A_{11} & A_{21} \\ A_{12} & A_{22} \end{bmatrix}$. Compute AB and BA.

11. Prove Theorems 4.1.1 and 4.1.2 for 2×2 matrices.

12. Let $A = \begin{bmatrix} 1 & 2 \\ 2 & 1 \end{bmatrix}$ and $B = \begin{bmatrix} 2 & 2 \\ 1 & -1 \end{bmatrix}$. Compute $|A|, |B|, |AB|$, and $|BA|$.

13. Complete the proof of Theorem 4.1.3 part (a) by using induction on the order n of the matrix A.

14. Prove: If $D = \text{diag}(a_1, \ldots, a_n)$ is an $n \times n$ diagonal matrix, then $|D| = a_1 a_2 \ldots a_n$. [Hint: Use mathematical induction.]

15. Let A and B be $m \times n$ matrices. Prove: $(A + B)^t = A^t + B^t$

16. Let A be an $m \times n$ matrix and B be an $n \times r$ matrix. Prove: $(AB)^t = B^t A^t$.

4.2 Determinant of a Product, Adjoints, etc.

In this section, we show that the determinant of a product of two matrices is the product of the determinants, and then find a method, using determinants, for calculating the inverse of a matrix.

The first theorem lists the determinants of the identity matrix and the elementary matrices.

Theorem 4.2.1. *1.* $|I_n| = 1$

2. $|R_{ij}| = -1 = |C_{ij}|$

3. $|R_i(c)| = c = |C_i(c)|$

4. $|R_{ik}(c)| = 1 = |C_{ik}(c)|$

(The determinants of the elementary matrices can be calculated using the properties of determinants.)

Proof. 1. The proof uses mathematical induction. For 1×1 matrices, we see that $|I_1| = 1$. Assume inductively that $|I_{n-1}| = 1$. Expand $|I_n|$ along the first row. Then $|I_n| = (1)(|I_{n-1}|) = 1$, and so by induction, $|I_n| = 1$ for all n.

2. Apply Theorem 4.1.2 to I.

3. Apply Theorem 4.1.3, part (c), to I.

4. Apply Theorem 4.1.3, part (g), to I.

$\square$

Since the determinants of the elementary matrices follow from properties of the determinant and since multiplication by an elementary matrix performs the corresponding row operation of the given matrix, it is not hard to see that the determinant function preserves products, that is, the determinant of a product of two matrices is the product of the determinants of the matrices in the product. As a first step we consider the product of an elementary matrix times an arbitrary matrix.

THE DETERMINANT OF A PRODUCT

Theorem 4.2.2. *If E is an elementary row matrix and A is any matrix, then $|EA| = |E||A|$.*

Proof. Multiplying A on the left by an elementary matrix performs the corresponding row operation on A. Knowing this, apply Theorem 4.1.2 and the appropriate parts of Theorem 4.1.3 to each of the three cases of elementary matrices to finish the proof. □

Theorem 4.2.3. *If A is an $n \times n$ matrix, then A is nonsingular if and only if $|A| \neq 0$.*

(A matrix is nonsingular iff it has nonzero determinant.)

Proof. If A is nonsingular, then A can be written as a product of elementary matrices by Theorem 3.6.8, say $A = E_1 \ldots E_k$. Then $|A| = |E_1 \ldots E_k| = |E_1||E_2 \ldots E_k| = \ldots = |E_1||E_2| \ldots |E_k|$. Since the determinant of every elementary matrix is nonzero, $|A| \neq 0$.

Assume that A is singular, and let B be the row reduced echelon form for A; say $PA = B$, where P is nonsingular. Then $A = P^{-1}B = E_1 \ldots E_k B$, where each E_i is an elementary matrix. By Theorem 4.2.2, $|A| = |E_1| \ldots |E_k||B|$. Since A is singular, B is not the identity matrix and must have at least one row of zeros (the bottom row) . It follows that $|B| = 0$, and so $|A| = 0$. □

Theorem 4.2.4. *Let A and B be $n \times n$ matrices. Then $|AB| = |A||B|$.*

(The determinant of a product is the product of the determinants.)

Proof. Assume first that A is nonsingular. Then $A = E_1 \ldots E_n$, so $|AB| = |E_1 \ldots E_k B| = |E_1| \ldots |E_k||B| = |E_1 \ldots E_k||B| = |A||B|$.

Conversely, assume that A is singular. Then AB is singular (See Exercise 10), so $|AB| = 0$ and $|A| = 0$. Again, we see that, $|AB| = |A||B|$. □

We already know that

$$|A| = \sum_{i=1}^{n} a_{ij} A_{ij}.$$

The following lemma completes this investigation and is essential in what follows.

Lemma 4.2.5. *Let $A = [a_{ij}]$ be an $n \times n$ matrix. Then:*

$$\sum_{i=1}^{n} a_{ij} A_{ik} = \begin{cases} |A| & \text{if } j = k \\ 0 & \text{if } j \neq k \end{cases}$$

Proof. The $j = k$ part is clear from Theorem 4.1.1. Assume $j \neq k$. Let $B = [b_{ij}]$ be the matrix obtained from A by replacing the k-th column of A by the j-th column of A. Then B has two columns that are the same, so $|B| = 0$. Expanding $|B|$ down its k-th column we get

$$|B| = \sum_{i=1}^{n} b_{ik} B_{ik}.$$

But the k-th column of B is the same as the j-th column of A, so $b_{ik} = a_{ij}$. Also B differs from A only in column k and so $B_{ik} = A_{ik}$ for $i = 1, \ldots, n$. We get

$$0 = |B| = \sum_{i=1}^{n} b_{ik} B_{ik} = \sum_{i=1}^{n} a_{ij} A_{ik}.$$

$\square$

4.2.1 THE ADJOINT AND THE INVERSE OF A MATRIX

The theorem above makes it possible to calculate the product of a matrix and its transposed matrix of cofactors. This transposed matrix of cofactors is called the **adjoint** of the matrix.

Definition 4.2.1. Let $A = [a_{ij}]$ be an $n \times n$ matrix, and let A_{ij} be the cofactor of a_{ij}. The **adjoint** of A is the matrix $[A_{ij}]^t$ and is denoted by $\text{Adj}A$. (The adjoint is the transposed matrix of cofactors.)

It is a bit of a chore calculating the adjoint of a matrix. It is usually safest to calculate the matrix of cofactors, and then in a separate step, take the transpose.

Example 4.2.1. Let

$$A = \begin{bmatrix} 1 & 2 & -1 \\ 0 & 1 & 1 \\ 1 & 3 & -2 \end{bmatrix}.$$

Then

$$\text{Adj}A = [A_{ij}]^t = \begin{bmatrix} -5 & 1 & -1 \\ 1 & -1 & -1 \\ 3 & -1 & 1 \end{bmatrix}^t = \begin{bmatrix} -5 & 1 & 3 \\ 1 & -1 & -1 \\ -1 & -1 & 1 \end{bmatrix}.$$

$\square$

Theorem 4.2.6. *Let $A = [a_{ij}]$ be an $n \times n$ matrix. Then $(AdjA)(A) = (|A|)(I_n)$. If $|A| \neq 0$, A is nonsingular and $A^{-1} = \dfrac{1}{|A|}(AdjA)$.*

(The inverse of a nonsingular matrix is the adjoint divided by the determinant.)

Proof. Just compute it. Multiplying row k of $\text{Adj}A$ and column j of A we get

$$\sum_{i=1}^{n} A_{ik} a_{ij} = \delta_{jk} |A|,$$

and so

$$(\text{Adj}A)A = |A|I_n.$$

If we multiply both sides by A^{-1} and divide both sides by the scalar $|A|$, then we get

$$\frac{1}{|A|}(\text{Adj}A)AA^{-1} = \frac{|A|}{|A|}I_nA^{-1}$$

and so

$$A^{-1} = \frac{1}{|A|}(\text{Adj}A).$$

□

Example 4.2.2. For an easy illustration, let

$$A = \begin{bmatrix} 1 & 2 \\ -1 & 0 \end{bmatrix}.$$

Then $|A| = 2, A_{11} = 0, A_{12} = 1, A_{21} = -2$, and $A_{22} = 1$, and so

$$\text{Adj}A = \begin{bmatrix} A_{11} & A_{12} \\ A_{21} & A_{22} \end{bmatrix}^t = \begin{bmatrix} A_{11} & A_{21} \\ A_{12} & A_{22} \end{bmatrix} = \begin{bmatrix} 0 & -2 \\ 1 & 1 \end{bmatrix}$$

and

$$A^{-1} = \frac{1}{|A|}(\text{Adj}A) = \frac{1}{2}\begin{bmatrix} 0 & -2 \\ 1 & 1 \end{bmatrix} = \begin{bmatrix} 0 & -1 \\ 1/2 & 1/2 \end{bmatrix}.$$

Computing, we get

$$\begin{bmatrix} 1 & 2 \\ -1 & 0 \end{bmatrix}\begin{bmatrix} 0 & -1 \\ 1/2 & 1/2 \end{bmatrix} = \begin{bmatrix} 1 & 0 \\ 0 & 1 \end{bmatrix}.$$

□

Section 4.2 Exercises

1. Let $A = \begin{bmatrix} 2 & 1 \\ -1 & 3 \end{bmatrix}$. Compute $\text{Adj}A, A\text{Adj}(A)$, and A^{-1}.

2. Let $A = \begin{bmatrix} 3 & 2 \\ -2 & 1 \end{bmatrix}$. Compute $\text{Adj}A, A\text{Adj}(A)$, and A^{-1}.

3. Let $A = \begin{bmatrix} -1 & 0 & 2 \\ 2 & 1 & -1 \\ 1 & 0 & 3 \end{bmatrix}$. Compute $\text{Adj}A$ and A^{-1}.

4. Let $A = \begin{bmatrix} 1 & 0 & -2 \\ 2 & 1 & -1 \\ 1 & 0 & 3 \end{bmatrix}$. Compute $\text{Adj}A$ and $A\text{Adj}(A)$.

5. Let $A = \begin{bmatrix} 1 & 0 & -2 \\ 0 & 1 & 0 \\ 1 & 1 & 3 \end{bmatrix}$. Compute $\text{Adj}A$ and $A\text{Adj}(A)$.

6. Let A and B be $n \times n$ matrices with $|A| = 2$ and $|B| = 3$. Find $|AB|$.

7. Prove that if A is $n \times n$ and nonsingular, $|\text{Adj}A| = |A|^{n-1}$. singular?

8. Let P be a matrix in which each row and each column contains exactly one 1, and all other entries are zero. Prove that $|P| = \pm 1$. (Such a matrix is called a **permutation matrix**.)

9. Let A and B be nonsingular $n \times n$ matrices. Discover and prove a nontrivial theorem concerning $\text{Adj}(AB)$.

10. Complete the proof of Theorem 4.2.4 by proving the following: If A and B are $n \times n$ matrices and A is singular, then AB is singular. (Do not, of course, use Theorem 4.2.4 in your proof.) [Hint: Let PA be the reduced echelon form of A with P nonsingular. Then PA has a row of zeros - Why?]

11. Prove: If A and B are $n \times n$ matrices, then $|AB| = |BA|$.

12. Let B be an $n \times n$ matrix that is in echelon form. Prove the following:

    (a) If $\text{rank}(B) = n$, then $|B| = 1$.
    (b) If $\text{rank}(B) < n$, then $|B| = 0$.

13. Theorems 4.2.1 and 4.2.4 can be used to find a method of calculating the determinant of a matrix using the reduction of a matrix to echelon form. Use the following principle. Let B be a matrix in echelon form that is row equivalent to the $n \times n$ matrix A, say with $B = E_1 \ldots E_n A$ where $E_1, \ldots, E_n$ are elementary matrices. Then $|B| = |E_1| \ldots |E_n||A|$. Now $|E_1|, \ldots, |E_n|$ are given by Theorem 4.2.1 and $|B|$ can be found using Exercise 12 Describe the method.

4.3 Cramer's Rule

In the previous section we saw that determinants can be used to calculate the inverse of a matrix when one exists. This technique can be of use in solving systems of linear equations when the coefficient matrix is nonsingular. In this section, we consider $n \times n$ systems of equations $AX = H$ where A is a nonsingular matrix. We have discussed nonsingular matrices is several different places: Section 1.3 contained the definition; in Section 3.4 we saw that the inverse of a matrix is unique, and that the identity transformation has a nonsingular matrix; in Section 3.6 we proved that a nonsingular matrix is the product of elementary matrices. It is time to collect these results together. The first theorem summarizes equivalent characterizations of nonsingular matrices.

> **Theorem 4.3.1.** *Let A be an $n \times n$ matrix. Then the following statements are equivalent:*
>
> 1. *A is nonsingular (that is, A has an inverse)*
>
> 2. *$|A| \neq 0$*
>
> 3. *A is a product of elementary matrices*
>
> 4. *A has rank n (that is, the rows of A are linearly independent)*
>
> *(Rank n, nonzero determinant, equal to a product of elementary matrices, and nonsingularity are equivalent.)*

Proof. The equivalence of (a) and (b) is proved in Theorem 4.2.3; that is, (a) $\Leftrightarrow$ b).

Assume that A is nonsingular. Then by Theorem 3.6.8, $A = E_1 \ldots E_m$ for some elementary matrices $E_1, \ldots, E_m$, and so we see that (a) $\implies$ (c). Since A is a product of elementary matrices, A can be obtained from the identity matrix I_n by a sequence of row operations. This means that A is row equivalent to I_n. By Theorem 2.8.1, row equivalent matrices have the same rank, and since I_n has rank n, A must have rank n. We see that (c) $\implies$ (d).

Next, assume that A has rank n. If B is the reduced echelon form of A, then B has rank n and since B is $n \times n$, it follows that $B = I_n$. But the reduced echelon form of A is obtained by a sequence of elementary row operations, and so $B = E_1 \ldots E_m A = I_n$ for some elementary matrices $E_1, \ldots, E_m$. Since each E_i is nonsingular by Theorem 3.6.3, we may solve for A and obtain $A = E_m^{-1} \ldots E_1^{-1}$. It follows from Theorems 3.6.3 and 3.6.4 that A is nonsingular. This proves that (d) $\implies$ (a), and so the theorem is proved. $\square$

In beginning algebra we learn that to solve the equation $ax = b$, with $a \neq 0$, we can divide both sides by a. If the coefficient matrix A is nonsingular, the same trick works for the matrix equation $AX = H$.

> **Theorem 4.3.2.** *If A is nonsingular, then $AX = H$ has the unique solution $X = A^{-1}H$.*
> *(To solve a matrix equation, divide by the coefficient matrix, if possible.)*

Proof. To see that $X = A^{-1}H$ is a solution, we just calculate:

$$A(A^{-1}H) = (AA^{-1})H = IH = H.$$

So we see that $X = A^{-1}H$ is a solution.

If X_1 and X_2 are both solutions, then $AX_1 = H = AX_2$. Multiplying both sides on the left by A^{-1}, we obtain

$$A^{-1}(AX_1) = A^{-1}(AX_2)$$
$$(A^{-1}A)X_1 = (A^{-1}A)X_2$$
$$IX_1 = IX_2$$
$$X_1 = X_2.$$

It follows that the solution is unique. $\qquad\square$

A homogeneous system of equations $AX = 0$ always has the solution $X = 0$. Applying the above theorem to this situation, we see that if A is $n \times n$ and nonsingular, then the solution $X = 0$ is unique. From this it follows that if $AX = 0$ has a nonzero solution, then A must be singular.

Now consider a system of equations $AX = H$, where

$$A = \begin{bmatrix} a_{11} & a_{12} & \cdots & a_{1n} \\ a_{21} & a_{22} & \cdots & a_{2n} \\ \vdots & \vdots & \ddots & \vdots \\ a_{m1} & a_{m2} & \cdots & a_{mn} \end{bmatrix}, X = \begin{bmatrix} x_1 \\ \vdots \\ x_n \end{bmatrix}, \text{ and } H = \begin{bmatrix} h_1 \\ \vdots \\ h_n \end{bmatrix}.$$

We can apply the above theorem and the adjoint method of calculating the inverse of a matrix to obtain the solution of a system of linear equations. By Theorem 4.2.6, $A^{-1} = (1/|A|)\text{Adj}A = (1/|A|)[A_{ij}]^t$, where A_{ij} is the cofactor of a_{ij}, and so

$$X = \begin{bmatrix} x_1 \\ \vdots \\ x_n \end{bmatrix} = \frac{1}{|A|} \begin{bmatrix} A_{11} & A_{12} & \cdots & A_{1n} \\ A_{21} & A_{22} & \cdots & A_{2n} \\ \vdots & \vdots & \ddots & \vdots \\ A_{m1} & A_{m2} & \cdots & A_{mn} \end{bmatrix} \begin{bmatrix} h_1 \\ \vdots \\ h_n \end{bmatrix}.$$

To compute x_i, multiply row i of $\text{Adj}A$ by H and divide by $|A|$. We get

$$x_i = (1/|A|)(h_1 A_{1i} + h_2 A_{2i} + \ldots + h_n A_{ni}).$$

This sum looks like the determinant of a matrix expanded down the i-th column! In fact, let A_i be the matrix obtained from A by replacing column i of A by the column vector H. Then A_i is given by

$$A_i = \begin{bmatrix} a_{11} & \ldots & h_1 & \ldots & a_{1n} \\ a_{21} & \ldots & h_2 & \ldots & a_{2n} \\ \vdots & \ldots & \vdots & \ldots & \vdots \\ a_{n1} & \ldots & h_n & \ldots & a_{nn} \end{bmatrix},$$

where the h's appear in column i. It is easy to see that the expansion of $|A_i|$ down column i gives

$$|A_i| = h_1 A_{1i} + h_2 A_{2i} + \ldots + h_n A_{ni}.$$

This proves **Cramer's Rule**.[1]

---

[1]Named for Swiss mathematician Gabriel Cramer (1704-1752)

Theorem 4.3.3 (Cramer's Rule). *If A is a nonsingular $n \times n$ matrix, the system of equations $AX = H$ has the unique solution*

$$x_1 = |A_1|/|A|, x_2 = |A_2|/|A|, \ldots, x_n = |A_n|/|A|,$$

where A_i is the matrix obtained from A by replacing column i of A by the column vector H.
(The variables in a system of linear equations with a nonsingular coefficient matrix can be expressed as a quotient of two determinants.)

Example 4.3.1.   1. Let us apply Cramer's Rule to a general 2×2 system of equations:

$$ax + by = h$$
$$cx + dy = k$$

By Cramer's Rule, we obtain the solution for the variables x and y by forming quotients of two determinants. The denominator of each quotient is the determinant of the coefficient matrix. The numerator of the solution for x is the matrix obtained by replacing the first column of the coefficient matrix with the constants h and k, and, similarly, y has the determinant of the coefficient matrix with the second column replaced by h and k on top. So, by Cramer's Rule, we see that the solution for x and y is given by

$$x = \frac{\begin{vmatrix} h & b \\ k & d \end{vmatrix}}{\begin{vmatrix} a & b \\ c & d \end{vmatrix}}$$

$$y = \frac{\begin{vmatrix} a & h \\ c & k \end{vmatrix}}{\begin{vmatrix} a & b \\ c & d \end{vmatrix}}$$

2. Consider the following system of equations:

$$x_1 - 2x_2 = 2$$
$$3x_1 - 4x_2 = 1.$$

By Cramer's Rule, the solution is

$$x_1 = \frac{\begin{vmatrix} 2 & -2 \\ 1 & -4 \end{vmatrix}}{\begin{vmatrix} 1 & -2 \\ 3 & -4 \end{vmatrix}} = \frac{-6}{2} = -3$$

$$x_2 = \frac{\begin{vmatrix} 1 & 2 \\ 3 & 1 \end{vmatrix}}{\begin{vmatrix} 1 & -2 \\ 3 & -4 \end{vmatrix}} = \frac{-5}{2}$$

3. Consider the system of equations

$$2x - 3y = 4$$
$$-4x + 6y = 5.$$

The coefficient matrix is $\begin{bmatrix} 2 & -3 \\ -4 & 6 \end{bmatrix}$, and calculating the determinant, we see that $\begin{vmatrix} 2 & -3 \\ -4 & 6 \end{vmatrix} = 12 - 12 = 0$. In this case, Cramer's Rule does not apply since the coefficient matrix is singular.

4. Let us solve the following system of equations:

$$\begin{array}{rcrcrcr} 2x & + & 3y & - & z & = & 2 \\ x & - & y & + & 2z & = & 6 \\ 3x & + & y & - & z & = & -2 \end{array}$$

for the value of z. Using Cramer's Rule and applying row operations to simplify the computations, we get

$$z = \frac{\begin{vmatrix} 2 & 3 & 2 \\ 1 & -1 & 6 \\ 3 & 1 & -2 \end{vmatrix}}{\begin{vmatrix} 2 & 3 & -1 \\ 1 & -1 & 2 \\ 3 & 1 & -1 \end{vmatrix}} = \frac{\begin{vmatrix} 2 & 3 & 2 \\ 1 & -1 & 6 \\ 4 & 0 & 4 \end{vmatrix}}{\begin{vmatrix} 2 & 3 & -1 \\ 1 & -1 & 2 \\ 4 & 0 & 1 \end{vmatrix}} = \frac{\begin{vmatrix} 5 & 0 & 20 \\ 1 & -1 & 6 \\ 4 & 0 & 4 \end{vmatrix}}{\begin{vmatrix} 5 & 0 & 5 \\ 1 & -1 & 2 \\ 4 & 0 & 1 \end{vmatrix}} = \frac{-(20 - 80)}{-(5 - 20)} = \frac{60}{15} = 4.$$

$\square$

We will see in Section 4.5 that Cramer's Rule is neither effective nor efficient as a computational tool. The problem lies in the computation of the determinant. If the definition of the determinant is used for the computation of a determinant of a 10×10 matrix, then it is necessary to calculate $10!$ products. The problem is that $10! = 3,628,800$, and 10 is not a very large number. The true value of the determinant and of Cramer's Rule lies in theoretical considerations.

SYSTEMS OF EQUATIONS - A SUMMARY

Let us collect a list of all of the basic results that we have established regarding solution of systems of linear equations. We will suppose that we have a system of equations expressed in the form of a matrix equation $AX = H$, where A is an $m \times n$ matrix and X and H are $n \times 1$ and $m \times 1$ column vectors, respectively. The matrix A is the coefficient matrix and $[A|H]$ is the augmented matrix.

1. The system $AX = H$ may have no solutions (inconsistent), a unique solution, or many solutions. If A and $[A|H]$ have the same rank, (this can be easily determined from the reduced echelon form of $[A|H]$), the system is consistent.

2. To find the general solution in a consistent system, reduce the augmented matrix $[A|H]$ to reduced echelon form. If the constants in the reduced echelon form are r and $j_1, \ldots, j_r$, then the variables corresponding to $j_1, \ldots, j_r$ are called the **basic variables**. The other variables are called the **free variables**: these variables may be arbitrarily chosen and their values determine the value of the basic variables. The number of free variables is $n - r$.

3. The system $AX = 0$ is said to be **homogeneous** and is called the associated homogeneous equation of the system $AX = H$. The zero vector is always a solution of $AX = 0$, and the set of all solutions forms a subspace (of the space of all $n \times 1$ column vectors) of dimension $n - r$. To find a basis, take solutions in which the free variables take on the values in the standard basis vectors: $(1, 0, 0, \ldots, 0); (0, 1, 0, \ldots, 0)$; etc.

4. If X_p is any particular solution of $AX = H$ and X_h is a solution of $AX = 0$, then $X = X_p + X_h$ is also a solution of $AX = H$. We see that the solution set of $AX = H$ is the set $\{X_p + X_h | X_h$ is a solution of $AX = 0\}$.

5. If A is a square matrix $(m = n)$ then the system $AX = H$ has a unique solution if and only if A is nonsingular. In this case, the solution may be expressed in the form $X = A^{-1}H$ or, using Cramer's Rule, as above.

6. We have not listed here all that there is to know about solving systems of equations. Section 1.7 contains an outline of methods which are useful in solving large systems using computers, and in Section 4.5 further issues involving systems and their solutions using computers are investigated. A topic of great importance that we have not covered is roundoff error. We will illustrate with an example.

   Some matrices are "ill-conditioned" in that a small change in a entry in the matrix can produce a large change in the solution of an associated system of equations. Let us consider the following system:

$$
\begin{aligned}
x + 0.999y &= 1 \\
x + y &= 2.
\end{aligned}
$$

   We can solve: pivot on x in the first equation and get

$$
\begin{aligned}
x + 0.999y &= 1 \\
0.001y &= 1.
\end{aligned}
$$

   We see that $y = 1000$ and $x = -998$.

Now suppose that an error of 0.001 was somehow introduced in the coefficient of y in the first equation giving the system of equations:

$$\begin{aligned} x &+ 0.998y &= 1 \\ x &+ y &= 2. \end{aligned}$$

Pivoting on x in the first equation, we get

$$\begin{aligned} x &+ 0.998y &= 1 \\ & 0.002y &= 1. \end{aligned}$$

We see that $y = 500$ and $x = -498$. A change of 500 in the values of x and y is produced by a change of 0.001 in one of the coefficients. Why does this happen? Notice that the two lines given by the equations in the original system are very nearly parallel. Because of this, their point of intersection lies relatively far from the origin and a small change in the slope of one of the lines moves the point of intersection a great distance. Notice also that the determinant of the coefficient matrix is 0.001 - close to zero.

Section 4.3 Exercises

Solve the systems of equations in Exercises 1 - 4 using Cramer's Rule.

1. $\begin{aligned} x_1 &+ x_2 &= -1 \\ 2x_1 &+ 3x_2 &= -2 \end{aligned}$

3. $\begin{aligned} 2x_1 &+ 5x_2 &= -3 \\ 3x_1 &+ 2x_2 &= -1 \end{aligned}$

2. $\begin{aligned} 2x_1 &+ 3x_2 &+ x_3 &= 1 \\ x_1 &+ 2x_2 &+ 3x_3 &= 2 \\ 3x_1 &+ x_2 &+ 2x_3 &= 2 \end{aligned}$

4. $\begin{aligned} 3x_1 &+ 2x_2 &+ x_3 &= 2 \\ x_1 &+ x_2 &+ 5x_3 &= 3 \\ 2x_1 &+ 2x_2 &+ 2x_3 &= 2 \end{aligned}$

5. Solve the systems of equations in Exercise 1 using the method of Theorem 4.3.2.

6. Solve the systems of equations in Exercise 2 using the method of Theorem 4.3.2.

7. Solve the systems of equations in Exercise 3 using the method of Theorem 4.3.2.

8. Solve the systems of equations in Exercise 4 using the method of Theorem 4.3.2.

9. Consider the system of equations

$$\begin{aligned} x &+ y &- z &= 1 \\ 2x &- y &+ 3z &= 2. \end{aligned}$$

Cramer's Rule does not apply since the system has three unknowns but only two equations. Cramer's Rule can be applied if one chooses a value for one of the variables. Find solutions of the above system that satisfy each of the following conditions.

    (a) $z = 1$

    (b) $y = 2$

    (c) $x = 0$

10. Let N be an $n \times n$ matrix over $\mathbb{R}$ and assume $N^2 = 0$. Prove that $I_n + N$ is nonsingular.

11. Prove: If A is a real $n \times n$ matrix with n odd and $A^t = -A$, then A is singular.

12. Prove: If A and B are $n \times n$ matrices and $AX = 0$ has a nonzero solution, then $ABX = 0$ has a nonzero solution.

4.4   Some Geometrical Aspects of Three-dimensional Vectors (optional)

In this section, we present some of the elementary properties of vectors in the 3-dimensional vector space over the real numbers. Some of these topics are included in calculus courses. The determinant is used to define the triple scalar product of three vectors and we will see that this determinant is related to the volume of a solid figure generated by the three vectors. The cross product and the dot product are defined and these are related to the triple scalar product. Vector operations are used to find equations of planes and lines in 3-dimensional space. The applications found at the end of the section are often covered in a beginning physics course.

    All vectors in this section are in $\mathbb{R}^3$ and for this section only, we adopt the following somewhat inconsistent but more traditional notation: $\mathbf{i}, \mathbf{j}$, and $\mathbf{k}$ are used to denote the standard basis for $\mathbb{R}^3$:

$$\mathbf{i} = (1,0,0), \mathbf{j} = (0,1,0), \text{ and } \mathbf{k} = (0,0,1).$$

    In this section, if A is a vector in $\mathbb{R}^3$, we use A_x, A_y, and A_z for the coordinates of the vector A relative to the basis $\mathbf{i}, \mathbf{j}$, and $\mathbf{k}$, respectively, so that $A = A_x\mathbf{i} + A_y\mathbf{j} + A_z\mathbf{k}$. Suppose that we wish to find an equation of the line through some point and in the direction of a given vector.

EQUATION OF A LINE

Let $V = a\mathbf{i} + b\mathbf{j} + c\mathbf{k}$ and $R_0 = x_0\mathbf{i} + y_0\mathbf{j} + z_0\mathbf{k}$. We want to find an equation of the line L that is parallel to the vector V and passes through the endpoint of R_0, as in Figure 4.1.

    Let $R = x\mathbf{i} + y\mathbf{j} + z\mathbf{k}$ be the vector whose endpoint lies on an arbitrary point (x, y, z) of the line. The vector $R - R_0$ is in the direction of the line – consequently, in the direction of V – so we get the vector equation $R - R_0 = tV$, where t is some scalar. Equating x, y, and z components of the above vector equation, we get the following "parametric equations" for the line:

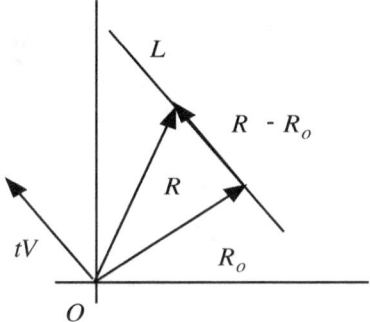

Figure 4.1: The line determined by two vectors

$$x = x_0 + ta$$
$$y = y_0 + tb$$
$$z = z_0 + tc.$$

THE DOT PRODUCT

A useful quantity associated with a pair of vectors is the dot product (or inner product) of the two vectors. A more thorough treatment of the inner product is found in Section 5.1. Here we discuss only 3-dimensional vectors over the real field.

Let A and B be two 3-dimensional vectors with $A = A_x\mathbf{i} + A_y\mathbf{j} + A_z\mathbf{k}$ and $B = B_x\mathbf{i} + B_y\mathbf{j} + B_z\mathbf{k}$. The **dot product** of A and B is the scalar $A \cdot B$ defined by

$$A \cdot B = A_x B_x + A_y B_y + A_z B_z.$$

The dot product of two vectors has many interesting properties and is related to the length of a vector and to the angle between vectors. The **length of a vector** A given by $A_x\mathbf{i} + A_y\mathbf{j} + A_z\mathbf{k}$ is the distance from the origin $(0,0,0)$ to the endpoint of A, (A_x, A_y, A_z), and is denoted by $||A||$. By the distance formula,

$$||A||^2 = A_x^2 + A_y^2 + A_z^2 = A \cdot A.$$

A vector of length 1 is called a **unit vector**.

Two vectors A and B in $\mathbb{R}^3$ determine a triangle with sides of lengths $||A||$, $||B||$, and $||A - B||$. Applying the law of cosines to this triangle, we can obtain the following result:

Need to draw this figure

Figure 4.2: A triangle formed by two vectors

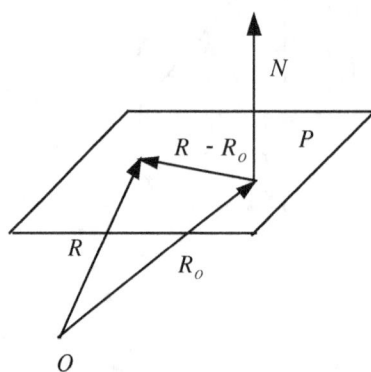

Figure 4.3: Vector Plane

Theorem 4.4.1. *Let A and B be vectors in* $\mathbb{R}^3$*. Then:*

$$A \cdot B = ||A||||B|| \cos\theta,$$

where θ is the smallest angle between A and B.

(The dot product is the product of the magnitudes times the cosine of the angle between the vectors.)

Proof. See Exercise 1 and Figure 4.2.   □

EQUATION OF A PLANE

If A and B are nonzero vectors in $\mathbb{R}^3$, then $||A|| \neq 0 \neq ||B||$ and so by the preceding theorem we get

$$\cos\theta = \frac{A \cdot B}{||A||||B||}.$$

For $0 < \theta < 180°$, $\cos\theta = 0$ if and only if $\theta = 90°$, and so A is perpendicular to B if and only if $A \cdot B = 0$. We can use this fact to find an equation for a plane P perpendicular to a vector $N = a\mathbf{i} + b\mathbf{j} + c\mathbf{k}$ and passing through a point (x_0, y_0, z_0). See Figure 4.3.

As before, let $R_0 = x_0\mathbf{i} + y_0\mathbf{j} + z_0\mathbf{k}$ and let $R = x\mathbf{i} + y\mathbf{j} + z\mathbf{k}$ be an arbitrary point on the plane. Then $R - R_0$ is a vector parallel to the plane P, so $R - R_0$ is perpendicular to N; that is, $(R - R_0) \cdot N = 0$. From this we get an equation of the plane:

$$(x - x_0, y - y_0, z - z_0) \cdot (a, b, c) = a(x - x_0) + b(y - y_0) + c(z - z_0) = 0.$$

Example 4.4.1. The plane through $(1, 2, -1)$ perpendicular to the vector $2\mathbf{i} + 3\mathbf{j} - 2\mathbf{k}$ is given by the equation $2(x - 1) + 3(y - 2) - 2(z + 1) = 0$ or $2x + 3y - 2z = 10$.

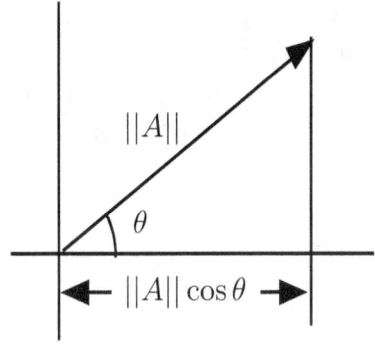

Figure 4.4: Projection of a vector

PROJECTION OF A VECTOR

Theorem 4.4.1 can be used to find the length of the projection of one vector in the direction of another. If U is a unit vector and A any vector, then the length of the projection of the vector A on the line through U is given by $||A|| \cos \theta$ where θ is the angle between A and U. See Figure 4.4.

But note that $||A|| \cos \theta = ||A|| ||U|| \cos \theta = A \cdot U$, and so the magnitude of the projection is $A \cdot U$.

In Exercise 10, it is shown that if U, V, W are mutually perpendicular unit vectors, then U, V, W form a basis for $\mathbb{R}^3$. If $A \in \mathbb{R}^3$, then the projections of A onto U, V, and W determine the expression of A as a linear combination of U, V, and W:

$$A = (A \cdot U)U + (A \cdot V)V + (A \cdot W)W.$$

THE CROSS PRODUCT AND THE TRIPLE SCALAR PRODUCT

The cross product and the triple scalar product provide useful combinations of 3-dimensional vectors, and we will see in Theorem 4.4.3 that the two are related. The **cross product** of two vectors A and B is defined by

$$A \times B = (A_y B_z - A_z B_y)\mathbf{i} + (A_z B_x - A_x B_z)\mathbf{j} + (A_x B_y - A_y B_x)\mathbf{k}.$$

This definition is almost impossible to remember as stated, but notice that if one formally expands the following determinant along the first row, one has exactly $A \times B$:

$$A \times B = \begin{vmatrix} \mathbf{i} & \mathbf{j} & \mathbf{k} \\ A_x & A_y & A_z \\ B_x & B_y & B_z \end{vmatrix}.$$

Note that this is not the determinant of a 3×3 matrix since $\mathbf{i}, \mathbf{j},$ and $\mathbf{k}$ are vectors. Because of this, the properties of determinants that were proved in Section 4.1 do not necessarily apply to this sort of "determinant" and must not be used in proving properties of $A \times B$. While different methods must be used in proving them, many of the expected properties of $A \times B$ do hold.

Theorem 4.4.2. *If $A, B,$ and C are any three vectors and r is a scalar, then*

    1. $A \times B = -(B \times A)$ *(Anti-commutativity)*

    2. $(A + B) \times C = (A \times C) + (B \times C)$ *(Distributive Property.)*

    3. $A \times (B + C) = (A \times B) + (A \times C)$ *(Distributive Property.)*

    4. $A \times (rB) = r(A \times B) = (rA) \times B.$ *(Associative Property.)*

Proof. See Exercise 2. $\square$

The triple scalar product of three vectors is useful in both studying and applying the cross product of vectors. Let $A, B,$ and C be vectors. The **triple scalar product** of $A, B,$ and C is the scalar $[A, B, C]$ defined by

$$[A, B, C] = \begin{vmatrix} A_x & A_y & A_z \\ B_x & B_y & B_z \\ C_x & C_y & C_z \end{vmatrix}.$$

We have now two "products" that have no apparent property other than being hard to compute. We will see that the cross product of two vectors that do not lie in the same direction is a vector that is perpendicular to the plane of the two. This, for example, gives us a method of constructing a normal to the plane containing two vectors. The triple scalar product will be of use in understanding the orientation of two vectors and their cross product.

Theorem 4.4.3. *For any three three-dimensional vectors $A, B,$ and C, $(A \times B) \cdot C =$ $[A, B, C]$.*   *(Cross, dot, and triple scalar products property)*

Proof. By definition,

$$A \times B \cdot C = (A_y B_z - A_z B_y)C_x + (A_z B_x - A_x B_z)C_y + (A_x B_y - A_y B_x)C_z$$

$$= \begin{vmatrix} A_x & A_y & A_z \\ B_x & B_y & B_z \\ C_x & C_y & C_z \end{vmatrix} \text{ (expanding along the third row).}$$

                                                                      $\square$

Corollary 4.4.4. *$A \times B$ is perpendicular to both A and B.*

Proof. $(A \times B) \cdot A = [A, B, A] = 0$ since $[A, B, A]$ is the determinant of a matrix with two rows the same. It follows that $A \times B$ is perpendicular to B. □

This corollary tells us that if $A \times B \neq 0$ then it is a vector in the direction of one of the two unit vectors that are perpendicular to the plane containing the vectors A and B. We also have the following corollary.

Corollary 4.4.5. *If $A \times B \neq 0$, then $[A, B, A \times B] > 0$.*

Proof. $[A, B, A \times B] = (A \times B) \cdot (A \times B) = ||A \times B||^2 > 0$. □

We next determine the length of $A \times B$.

Theorem 4.4.6. *If A and B are any two vectors, then $||A \times B|| = ||A||||B|| \sin \theta$, where θ is the angle between A and B.*

(The magnitude of the cross product is the product of the magnitudes times the sine of the angle.)

Proof. If either A or B is the zero vector, the result is trivial, so we will assume A and B are nonzero vectors. First assume that A and B are unit vectors. We will prove $||A \times B||^2 = \sin^2 \theta$. Compute:

$$
\begin{aligned}
||A \times B||^2 &= (A_y B_z - A_z B_y)^2 + (A_z B_x - A_x B_z)^2 + (A_x B_y - A_y B_x)^2 \\
&= (A_y B_z)^2 + (A_z B_y)^2 + (A_z B_x)^2 + (A_x B_z)^2 + (A_x B_y)^2 + (A_y B_x)^2 \\
&\quad - 2(A_y B_z A_z B_y + A_x B_z A_z B_x + A_x B_y A_y B_x) \\
&= A_x^2(B_x^2 + B_y^2 + B_z^2) + A_y^2(B_x^2 + B_y^2 + B_z^2) + A_z^2(B_x^2 + B_y^2 + B_z^2) \\
&\quad - [(A_x B_x)^2 + (A_y B_y)^2 + (A_z B_z)^2 + 2(A_y B_z A_z B_y + A_x B_z A_z B_x + A_x B_y A_y B_x)] \\
&= A_x^2 + A_y^2 + A_z^2 - [(A_x B_x + A_y B_y + A_z B_z)^2] \\
&= 1 - \cos^2 \theta \\
&= \sin^2 \theta
\end{aligned}
$$

(since $||B||^2 = 1 = B_x^2 + B_y^2 + B_z^2$ and $||A||^2 = 1 = A_x^2 + A_y^2 + A_z^2$).

Thus, the theorem is proved in the case where A and B are unit vectors.

Now let A and B be any two vectors with $||A|| = a$ and $||B|| = b$. Let $U = (1/a)A$ and $V = (1/b)B$. Then U and V are unit vectors and we have

$$
\begin{aligned}
||A \times B|| &= ||(aU) \times (bV)|| \\
&= ||ab(U \times V)|| \\
&= ab||U \times V|| \\
&= ab \sin \theta \\
&= ||A||||B|| \sin \theta.
\end{aligned}
$$

□

Figure 4.5: A parallelepiped determined by A, B, and C

Theorem 4.4.7. *Let A, B, and C be three vectors in $\mathbb{R}^3$. Then the absolute value of $[A, B, C]$ is the volume of the parallelepiped determined by A, B, and C.*

(The triple scalar product is $+$ or $-$ the volume of the solid determined by the vectors.)

Proof. The area a of the parallelogram determined by A and B is $h_1\|A\|$ (See Figure 4.5); that is, $a = \|A\|\|B\|\sin\theta$. The volume v of the parallelepiped is therefore $v = h_2 a$. Now $h_2 = |U \cdot C|$, where $U = (A \times B)/\|A \times B\|$, since $A \times B$ is perpendicular to the plane of A and B. It follows that

$$
\begin{aligned}
v &= ah_2 \\
&= \|A\|\|B\|(\sin\theta)|U \cdot C| \\
&= \|A \times B\|\left|\frac{A \times B}{\|A \times B\|} \cdot C\right| \\
&= (A \times B) \cdot C \\
&= [A, B, C].
\end{aligned}
$$

$\square$

Because of the previous theorem, it is easy to see that the following theorem is true.

Theorem 4.4.8. *Three vectors A, B, and C lie in a plane if and only if $[A, B, C] = 0$.*

(Three vectors are dependent iff their triple scalar product is 0.)

ORIENTATION

Because of the previous results, we know almost everything about $A \times B$. The length of $A \times B$ is $\|A\|\|B\|\sin\theta$, where θ is the angle between A and B, and $A \times B$ is perpendicular to both A and B. We have thus eliminated all but two possibilities for $A \times B$. (See Figures 4.6 and 4.7.)

There are two possible ways for a vector to be perpendicular to the plane of two other vectors. One configuration is called a "right-handed system," the other is a "left-handed system." The term "right-handed" comes from the fact that a screw with a right-hand thread advances in the direction of $A \times B$ (in the right-hand system) when rotated in the direction that carries A into B. This comment is not very useful since almost no one knows which direction a right-hand screw travels. There is an easier rule. If the fingers of your right hand point in the direction of rotation that carries A into B and your thumb is stretched out, then the thumb points in the direction of $A \times B$ in the right hand system. (See Figure 4.8.) The left hand works in a similar manner for a left-hand system.

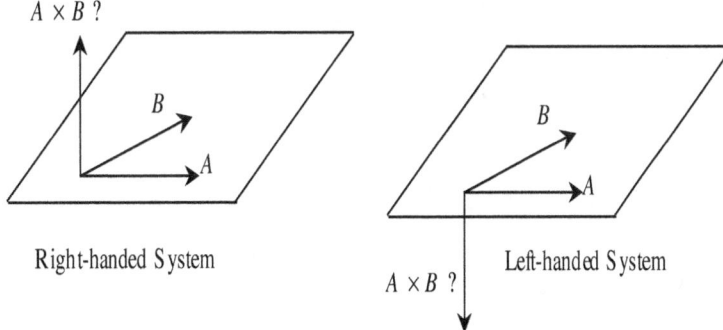

Figure 4.6: Right-handed system   Figure 4.7: Left-handed system

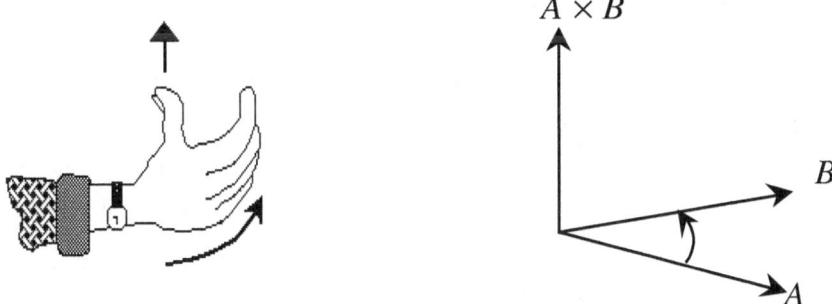

Figure 4.8: Right-hand rule   Figure 4.9: Vectors forming a right-handed system

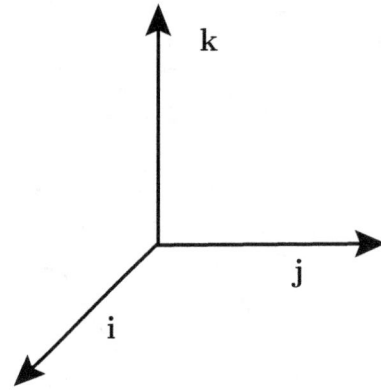

Figure 4.10: $\mathbf{i}, \mathbf{j}, \mathbf{k}$

The standard basis vectors $\mathbf{i}, \mathbf{j}, \mathbf{k}$, as usually pictured (see Figure 4.10), form a right-handed system and the triple scalar product $[\mathbf{i}, \mathbf{j}, \mathbf{k}] = 1 > 0$. Since $\mathbf{i} \times \mathbf{j} = \mathbf{k}$, one would guess that in general for two vectors A and B, the vector A, B and $A \times B$ form a right-handed system. We will argue this as follows:

If for two nonzero vectors A and B we have that $A, B, A \times B$ form a left-handed system, then through a sequence of rotations and stretchings or shrinkings of these vectors, the system could be transformed into the system $\mathbf{i}, \mathbf{j}, -\mathbf{k}$. This could be done without ever making the three vectors lie in a plane. Now the triple scalar product is a continuous function of its three vectors (or their nine components) and $[\mathbf{i}, \mathbf{j}, -\mathbf{k}] = -1 < 0$, so by the intermediate value theorem, it must have been the case that $[A, B, A \times B] < 0$, which is a contradiction. While we have not given a formal proof, the following theorem seems plausible:

Theorem 4.4.9. *If A and B are nonzero, nonparallel vectors, then $A, B,$ and $A \times B$ form a right-handed system.*   *(The cross product is perpendicular in the right-handed direction.)*

In physics, much use is made of the vector algebra that has been presented in this section. In particular, both the cross product and the dot product arise in the area of mechanics. The following examples give an indication of these elementary applications.

SOME APPLICATIONS FROM PHYSICS

1. Consider a body rotating about a line through the origin with angular speed (in radians per unit of time) ω. The **angular velocity** is defined to be the vector Ω of length ω and direction perpendicular to the plane of rotation in the right-handed sense. That is, if the fingers of the right hand point in the direction of rotation, then the thumb points in the direction of Ω as in Figure 4.11.

   Let R be the position vector of a point on the rotating body. Notice that the radius of the path followed by the point is $||R|| \sin \theta$. Its speed will then be $\omega ||R|| \sin \theta = ||\Omega \times R||$ (where as before, θ is the angle between R and the axis of rotation, or, equivalently,

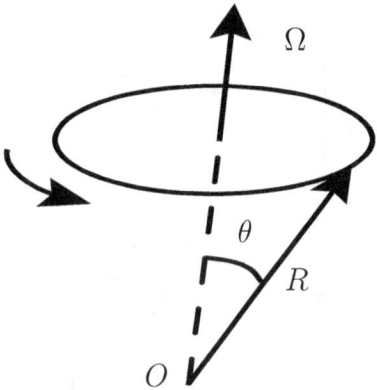

Figure 4.11: Angular velocity

θ is the angle between R and Ω.) Considering the direction of $\Omega \times R$, we see that $V = \Omega \times R$ is exactly the velocity vector of the point with position vector R.

2. A body is rotating about the line through $(0,0,0)$ and $(1,1,1)$ with an angular speed of 2 revolutions per second. The problem is to find the velocity of a point on the body that is passing through the point $(2, 1, -1)$.

   The angular speed of the body is 4π rad/sec and a unit vector along the axis of rotation is $\dfrac{1}{\sqrt{3}}(1, 1, 1) = \dfrac{1}{\sqrt{3}}(\mathbf{i} + \mathbf{j} + \mathbf{k})$. From this we see that the angular velocity is given by $\dfrac{4\pi}{\sqrt{3}}(\mathbf{i}+\mathbf{j}+\mathbf{k})$ (assuming that the body, as seen from the origin, is rotating in a clockwise direction), and the position vector R of the point is given by $R = 2\mathbf{i}+\mathbf{j} - \mathbf{k}$. From this we find that the velocity of the point is given by

   $$V = \Omega \times R = \frac{4\pi}{\sqrt{3}}(-2\mathbf{i} + 3\mathbf{j} - \mathbf{k})$$

   and the speed of the point is $s = ||\Omega \times R|| \approx 27.1464$ units/sec.

3. In physics, work is defined as the product of the displacement caused by a certain force and the component of that force in the direction of the displacement. Using the dot product, this is very easily expressed. Let F be the force vector and D the displacement vector. Then the work is given by $W = F \cdot D$.

   For example, if a force of 4 pounds is exerted on an object at an angle of $30°$ to the path of motion (see Figure 4.12) and because of this force, the object moves 6 feet, then the work done by the force is given by

   $$\begin{aligned} W &= F \cdot D \\ &= 4(\cos 30°\mathbf{i} + \sin 30°\mathbf{j}) \cdot 6\mathbf{i} \\ &= 24 \cos 30° \\ &= 12\sqrt{3} \text{ foot-pounds.} \end{aligned}$$

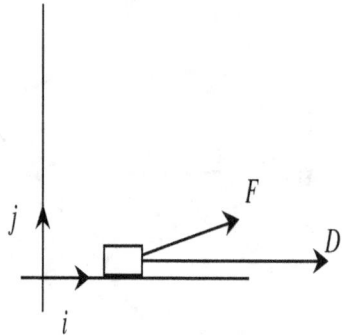

Figure 4.12: Work done on an object

Section 4.4 Exercises

1. Prove Theorem 4.4.1.

2. Prove Theorem 4.4.2.

3. Write an equation of the line through the point $(1, 2, 1)$ and parallel to the line given by the parametric equations $x = t, y = 2t + 1, z = -t - 1$.

4. Write an equation of the plane that contains the line segments from $(0, 0, 0)$ to $(1, 1, 0)$ and $(1, 1, 0)$ to $(1, -1, 1)$.

5. Write an equation of the plane passing through $(1, 2, -1), (0, 1, 1)$, and $(-1, 1, 2)$.

6. Find an equation of a line that passes through the point $(1, -2, 3)$ that is parallel to (does not intersect) the plane that has equation $2x + 3y - z = 6$.

7. A is body rotating at 6 rev/min about the line joining $(1, 2, 1)$ and $(-1, 3, 2)$. What is the velocity and speed of the point on the body that is passing through $(0, 0, 0)$? Assume that the angular velocity vector points in the direction from $(1, 2, 1)$ to $(-1, 3, 2)$.

8. A body moves 2 feet along a line from $(0, 0, 0)$ to $(1, 1, 1)$ because of a force of 4 pounds in the direction of the vector $\mathbf{k}$; that is, along the z-axis. What force in the direction of motion would accomplish this displacement?

9. Let $A, B, C \in \mathbb{R}^3$ and let r be some scalar. Prove the following:

    (a) $A \cdot B = B \cdot A$

    (b) $(rA) \cdot B = r(A \cdot B)$

    (c) $A \cdot (rB) = r(A \cdot B)$

    (d) $A \cdot (B + C) = A \cdot B + A \cdot C$

    (e) $(B + C) \cdot A = B \cdot A + C \cdot A$

10. Let U, V, and W be mutually perpendicular nonzero vectors in $\mathbb{R}^3$. Show that U, V, and W are linearly independent. (Note: U, V, and W mutually perpendicular implies that $U \cdot V = U \cdot W = V \cdot W = 0$.)

11. Let U, V, and W be mutually perpendicular unit vectors in $\mathbb{R}^3$ and let $A \in \mathbb{R}^3$ be arbitrary. Use Exercises 9 and 10 to show that $A = (A \cdot U)U + (A \cdot V)V + (A \cdot W)W$.

12. Let $A \in \mathbb{R}^3$ be a nonzero vector. Show that $U = A/\|A\|$ is a unit vector.

13. For $A, B, C \in \mathbb{R}^3$, prove the following identities:

    (a) $A \cdot (B \times C) = (A \times B) \cdot C$

    (b) $A \times (B \times C) = B(A \cdot C) - C(A \cdot B)$

    (c) $(A \times B) \times C = B(A \cdot C) - A(B \cdot C)$

4.5 Computational Complexity (optional)

Computer scientists study the efficiency of various methods of solving problems. This general area of study is referred to as the "analysis of algorithms," and one attempts to determine the "computational complexity" of the algorithm or method. A major part of the idea is to try to understand how long it will take for a computer to arrive at the solution of a given problem. Of course, the time needed to solve a problem must depend on some notion of the "size" of the problem. In general, the size of a problem is measured by the amount of data that the problem involves. Now it might be said that computers are simple, albeit very fast, devices. Basically, computers perform arithmetic operations (add, subtract, multiply, and divide) and certain other simple operations such as comparing numbers to determine the largest. The speed of a computer is a measure of how fast these simple operations can be performed and the length of time it takes a computer to solve a problem depends on how many of the operations it must perform.

We will consider the relative efficiency of solving a system of equations by Cramer's Rule and by the Gauss-Jordan reduction method. For a large system (say 30 equations in 30 unknowns), it would no doubt be necessary to use a computer, and so, in order to investigate the efficiency of the various methods, we must consider the speed with which the computer performs the various operations. Of course, the speed and efficiency of computers has increased dramatically over the last twenty or thirty years and these advances promise to continue.

While computers have become much faster in recent years, there has also been a change in the relationship among the times required for the various operations. In order to understand the historical development of this area, we begin by discussing the addition and multiplication times for an outdated computer, the IBM 1620, which dates back to about 1955.

Let p be a number with D_p digits and q a number with D_q digits. The IBM 1620 had formulas for the time required for the operations it could perform. To add the numbers p and q it took $160 + 80D_p$ microseconds. To multiply p and q it took $560 + 40D_q + 168D_q D_p$ microseconds. Thus, if p and q each had 8 digits, a sum would take 800 microseconds and a product would take 11,632 microseconds, (recall that a microsecond is one-millionth of a second or 10^{-6} seconds). Because a product took so much more time to perform on the computer than a sum, it was by far more important to count the number of multiplications in a program than the number of sums.

In the years since the development of the IBM 1620, computers have changed dramatically. New computers are smaller and faster, have bigger memories, and cost much less. The development of better algorithms for computing products and the development of better hardware have resulted in a dramatic decrease in the difference between addition and multiplication times. The speeds of various types of computers that were in production in about 1980, are listed in the following table. The times listed are in microseconds.

Computer/Type	Addition Time	Multiplication Time
North Star/Microcomputer	3830.00	6210.00
HP 3000/30/Minicomputer	43.90	61.00
PDP 11/70/MainFrame	0.90	3.30
Cyber 205/Supercomputer	0.005	0.005

We see that the IBM 1620 could perform about 86 multiplications per second while the Cyber 205 could perform 200,000,000 multiplications in a second - an increase of 2.3 million times.

Today, with the reduction of the difference between addition and multiplication times, efficiency of computers is measured in FLOPS - Floating-point Operations Per Second - and the standard unit is the megaflop (MFLOP) which represents one million floating point operations per second. By 1989, the following speeds were reported for various types of computers:

Computer	MFLOPS
Cray Y-MP	195.
IBM 3090	16.
Sun 4/260	1.1
DEC VAX 8550	0.99
IBM PC	0.012
Apple Macintosh	0.0038

The development of ever more powerful computers continues: In November of 1990 Intel announced a machine rated at 32 gigaflops (32 billion flops or 32,000 megaflops) and in July of 1991, an article quoted a projection made by Cray Research that a teraflop machine (one trillion floating point operations per second) would be available by 1997. In a talk in early 1993, an Intel executive talked about a teraflop machine which they had under development and hoped to have available in 1996. In December of 1996, a computer at Sandia National Laboratories, using 9,072 Intel Pentium Pro processors, reportedly achieved 1.8 trillion floating point operations per second. And in 2013, the Tinahe-2 topped 30 *peta*flops: over 30 *quadrillion* flops, or 30,000 teraflops.

Let us now return to our problem, analyzing the methods of Gauss-Jordan Reduction and Cramer's Rule. We will need to count or estimate the numbers of operations required. Since Cramer's Rule requires the computation of determinants, the following theorem is crucial.

> **Theorem 4.5.1.** *The calculation of an $n \times n$ determinant entails the computation of $n!$ products of n numbers each, assuming that the definition of determinant is used.*

Proof. Since $\begin{vmatrix} a & b \\ c & d \end{vmatrix} = ad - bc$, the result is seen to be true for $n = 2$.

Assume the theorem is true for $n = k - 1$ and let $A = [a_{ij}]$ be $k \times k$. Then $|A| = a_{11}A_{11} + \ldots + a_{1k}A_{1k}$, where each of the cofactors A_{1i} is a $(k-1) \times (k-1)$ determinant and consequently involves $(k-1)!$ products of $k-1$ numbers. Now, distributing the a_{1i} over the products in A_{1i} we see that $a_{1i}A_{1i}$ involves the computation of $(k-1)!$ products of k terms each. The computation of $|A|$ therefore involves the computation of $k(k-1)! = k!$ such products. It follows that the proposition is true for $n = k$ and so by mathematical induction the theorem is true for all positive integers n. $\square$

We now estimate the computing time necessary to solve a system of 30 linear equations in 30 unknowns using (a) Cramer's Rule and (b) the Gauss-Jordan reduction. Actually, we will only approximate the time required to do the multiplication.

(a) **Cramer's Rule:** To solve a system of 30 equations in 30 unknowns using Cramer's Rule, we must evaluate the 31 determinants each of which is 30×30. Each 30×30 determinant involves 30! products of 30 factors each (and so 29 multiplications each) and since there are 31 of them, $31 \cdot 29 \cdot 30!$ products are necessary. Now $31! \approx 8.22 \times 10^{33}$ so the number of required multiplications is $29 \cdot 31 \cdot 30! = 29 \cdot 31! \approx 29(8.22 \times 10^{33}) \approx 2.3838 \times 10^{35}$. So for the IBM 1620, the multiplication time is

$$(2.3838 \times 10^{35})(11,632 \times 10^{-6}) \approx 27728.362 \times 10^{29}$$
$$\approx 2.77 \times 10^{33} \text{ seconds}$$
$$\approx 8.78 \times 10^{25} \text{ years.}$$

This is clearly too long!

Let us consider the solution with a modern machine operating at 200 megaflops. Here the multiplication time would be .005 microseconds. For this machine the time required is:

$$(2.3838 \times 10^{35})(0.005 \times 10^{-6})sec \approx 1.1919 \times 10^{27} \text{ sec}$$
$$\approx 3.78 \times 10^{19} \text{ years.}$$

This is still a long time to wait! For a teraflop machine, the time would be reduced by a factor of $0.0002(= 200 \times 10^6 / 10^{12})$. It would then require "only" about 8×10^{15} years.

Finally, on the Tianhe-2, we have 33.86 petaflops, which corresponds to 2.9533×10^{-17} seconds per operation, giving us

$$(2.3838 \times 10^{35})(2.9533 \times 10^{-17}) \approx 7.040 \times 10^{18} \text{ seconds}$$
$$\approx 2.232 \times 10^{11} \text{ years.}$$

That's 223 billion years.

(b) **Gauss-Jordan Reduction:** Let us count the approximate number of multiplications required to solve a 30×30 system of linear equations using the method of Gauss-Jordan Reduction. This will entail the reduction of a 30×31 matrix to reduced echelon form, and so we estimate how many multiplications are necessary to accomplish this. It will be necessary to perform one pivot operation for each of the 30 rows. For each row i, we must multiply each entry in the row by the reciprocal of the first nonzero entry (in order to make the first nonzero entry a one) - so 31 multiplications are required. Now a multiple of row i must be added to each of the other rows, so $29 \cdot 31$ multiplications are necessary, bringing the total for row i to $31 + 29 \cdot 31 = 30 \cdot 31$ multiplications. This process must be done for each row and so a total of $30 \cdot 30 \cdot 31 = 27,900$ multiplications must be performed. The time required for these multiplications on the IBM 1620 is therefore

$$27,900 \cdot 1.1632 \times 10^{-2} \text{ seconds} \approx 324.5328 \text{ seconds.}$$

For a machine operating at 200 megaflops, the time required to perform the multiplications is just

$$27,900 \cdot (0.005 \times 10^{-6}) \approx 1.395 \times 10^{-4} \text{ seconds} = 0.0001395 \text{ seconds.}$$

From this we see that for practical purposes, the time required for Cramer's Rule is infinite, while the time required for Gauss-Jordan reduction is zero!

The same reasoning applies to systems of n linear equations in n unknowns where n is an arbitrary positive integer. We see that for Cramer's Rule, the multiplication time is proportional to $(n + 1)!$ and for Gauss-Jordan Reduction, time is proportional to $n \times n \times (n + 1) = n^3 + n^2$. By Stirling's formula, $n! > e^n$ for large n. In the jargon of computational complexity, one says that Cramer's Rule "runs in exponential time" while the method of Gauss-Jordan Reduction "runs in polynomial time."[2]

The above analysis of Cramer's Rule versus Gauss-Jordan Reduction is a little naive. Additions are not counted and furthermore no one would consider solving a system of equations using Cramer's Rule. A better study might be the comparison of Gaussian Elimination and the method of Gauss-Jordan Reduction. What is the difference between these methods?

---

[2]For further information on the analysis of algorithms, see Robert Sedgewick, *Algorithms*, *Addison-Wesley, Reading, MA, 1983* or *A.V. Aho, J.E. Hopcraft, and J.D. Ullman*, The Design and Analysis of Computer Algorithms, Addison-Wesley, Reading, MA, 1974.

Both methods call for the reduction of the augmented matrix of the system of equations to a matrix in echelon form. Under Gauss-Jordan, the matrix is reduced to reduced row echelon form, while Gaussian Elimination as it is usually explained, calls for "forward elimination" followed by "back substitution." Forward elimination begins at the upper left of the matrix and, at each step, is the process of making the first entry in a row a 1, and then making 0s below it. The result of forward elimination is a matrix in echelon (not reduced echelon) form. Most authors describe back substitution as follows: After reducing to echelon form, write the system of equations that results. Solve the last equation for the variable corresponding to the column with the first nonzero entry (a 1), and then substitute the result into the equation above. Continue until all of the equations have been solved.

There is no reason to forsake the augmented matrix in the back substitution process. We need some teriminology to help us explain. Pivoting is the process by which one divides a row by a certain entry in order to make a 1 in a certain row and column, and then performing the necessary row operations to make zeros above and below that 1. The process of making zeros below the 1 will be called "downward elimination" while the process of making zeros above the entry will be called "upward elimination." Gaussian Elimination might then be described as follows: Proceed down the rows making 1s as the first entry in each row and then performing downward elimination on the 1. When the last row is reached, proceed upward through the rows performing upward elimination on each of the 1s.

As observed in Section 1.5, Gaussian Elimination is more efficient since the row operations are performed on shorter rows. We will count the additions and multiplications needed under both the Gauss and Gauss-Jordan methods as applied to a system of n equations in n unknowns, but a preliminary observation will be helpful. In an $m \times n$ matrix, an operation applied to the entries in columns j through n requires $n - j + 1$ operations. If we are dividing the entries by the entry in column j, only $n - j$ operations will be required for we know that a 1 will appear in column j so there is no need to calculate it. A similar statement holds for making a zero in some row below a 1.

Let us now count the operations needed in both the Gaussian Elimination and Gauss-Jordan Reduction methods. We will need to make heavy use to the summation formulas below.

$$\sum_{i=1}^{n} i = \frac{n(n+1)}{2} \qquad \sum_{i=1}^{n} i^2 = \frac{n(n+1)(2n+1)}{6}.$$

We will assume that the matrix is $n \times (n+1)$ and that there is a unique solution so that operations must be performed on each of the rows.

(c) **Gauss-Jordan Reduction:** For row i, we will need $n + 1 - i$ multiplications to make a 1 in column i and $(n-1)(n+1-i)$ multiplications to make 0s. Thus, the number of multiplications for row i is $n(n+1-i)$. The number of additions required to make 0s is the same as the number of multiplications needed to make 0s, that is, $(n-1)(n+1-i)$. Adding for rows 1 to n we get

$$\sum_{i=1}^{n} n(n+1-i) = n \sum_{i=1}^{n} (n+1-i) = n\frac{n(n+1)}{2} = \frac{n^3 + n^2}{2}$$

multiplications and

$$\sum_{i=1}^{n} (n-1)(n+1-i) = (n-1) \sum_{i=1}^{n} (n+1-i) = (n-1)\frac{n(n+1)}{2} = \frac{n^3 - n}{2}$$

additions. The total number of operations required for Gauss-Jordan Reduction is $\frac{2n^3 + n^2 - n}{2}$.

(d) **Gaussian Elimination:** First consider downward elimination. For row i, we will need $n + 1 - i$ multiplications to make a 1 and $(n - i)(n + 1 - i)$ multiplications to make 0s below. The total number of multiplications for row i is $(n + 1 - i)(n + 1 - i)$. The number of additions is, as before, related to the multiplications needed to make 0s: $(n - i)(n + 1 - i)$. Summarizing and summing we get:

$$\sum_{i=1}^{n} (n+1-i)^2 = \sum_{i=1}^{n} i^2 = \frac{n(n+1)(2n+1)}{6} = \frac{2n^3 + 3n^2 + n}{6}$$

multiplications in downward elimination, and, letting M represent the number of multiplications,

$$\sum_{i=1}^{n} (n-i)(n+1-i) = M - \sum_{i=1}^{n} (n+1-i)$$

$$= M - \sum_{i=1}^{n} i$$

$$= \frac{n(n+1)(2n+1)}{6} - \frac{n(n+1)}{2}$$

$$= \frac{n^3}{3} + \frac{n^2}{2} - \frac{5n}{6}$$

additions in downward elimination.

For upward elimination, we start in row n and make zeros above the entry in column n (remember that we are assuming that the system has a unique solution, that is, the coefficient matrix has rank n). Proceeding upward, for row i we need only make zeros above the entry in column i, so $i - 1$ multiplications and additions are required to place the correct values in column $n + 1$. Summing we get:

$$\sum_{i=n}^{2} (i-1) = \sum_{i=1}^{n-1} i = \frac{n(n-1)}{2}$$

multiplications and additions in upward elimination.

Adding all four of these terms and simplifying we find that the total number of operations required for Gaussian Elimination is $\frac{2n^3}{3} + \frac{3n^2}{2} - \frac{7n}{6}$.

In comparing the expressions for the number of operations required by the above two methods, it is appropriate to consider only the dominant term – in this case the term involving n^3 – as an approximation of the total value. Making this estimate we see that Gauss-Jordan Reduction requires approximately n^3 operations while Gaussian Elimination needs only $\dfrac{2n^3}{3}$. It appears that Gauss-Jordan Reduction requires 50% more effort than Gaussian Elimination.

Section 4.5 Exercises

1. As above, estimate multiplication times for a 30×30 system using both Cramer's Rule and Gauss-Jordan Reduction for machines with the following speeds:

   (a) a machine with a multiplication time of 3 microseconds.

   (b) a machine which can perform 2 megaflops

   (c) a machine which can perform 10 gigaflops

2. Proceeding as above, estimate multiplication time for both methods for a system of 10 linear equations and 10 unknowns using a machine which can perform 1.8 megaflops.

3. Estimate both the number of additions and the number of multiplications required to reduce an $m \times n$ matrix to echelon form (as required in Gauss-Jordan Reduction). Calculate the time required to solve a system of 50 equations and 50 unknowns using a machine which can perform 2 megaflops.

4. Estimate the number of multiplications required to calculate the inverse of a nonsingular $n \times n$ matrix using each of the following methods:

   (a) The adjoint method of Section 4.2 and Theorem 4.2.6.

   (b) The method of Theorem 3.6.9.

5. Find an algorithm for computing the determinant of a matrix that runs in polynomial time and make an estimate of the time required. (HINT: Reduce the matrix to echelon form and keep track of what happens to the determinant. - See Exercise 13.)

6. Let A and B be $n \times n$ matrices. Assume that B is an arbitrary matrix. We want to count the number of multiplications needed to calculate the product AB under the various assumptions. (Note that A is **upper-triangular** when the entries below the diagonal are 0; that is, $a_{ij} = 0$ if $i > j$.)

   (a) A is arbitrary

   (b) A is diagonal

   (c) A is upper-triangular

Chapter 5

SIMILARITY, EIGENVALUES and DIAGONALIZATION

5.1 Inner Product, Orthogonality, and the Gram-Schmidt Process

In Chapters 1-4, we developed a lot of machinery: In Chapter 1, we investigated systems of linear equations and we studied the reduced echelon form of a matrix. In Chapter 2, we presented the idea of a vector space, and we used linear independence, basis and dimension to characterize certain vector spaces. We found that solution spaces of homogeneous systems of linear equations could be described in terms of a basis. In Chapter 3 we introduced the idea of a linear transformation and we saw that certain linear transformations are represented by matrices. In Chapter 4, we saw how the determinant can be used to determine nonsingularity of matrices and to find inverses of matrices. Why have we expended all of this effort?

Linear transformations and their associated matrices arise in many different endeavors. Mathematicians encounter quadratic forms (see Section 5.7), physicists make use of the inertia tensor (see Section 5.8), and biologists, statisticians, and economists find matrices useful in applications involving probabilities (Section 1.9). In many of these situations the reduction of a matrix to "diagonal form" is not only useful, but essential. Often in applications, the matrix that arises is relative to a given coordinate system. Different coordinate systems produce different matrices. Perhaps some coordinate system will produce a diagonal matrix.

This chapter brings some reward for all of our hard work. We will study similarity and we will discover that certain matrices are similar to diagonal matrices. The methods used will make use of the theory that we have developed. We will have occasion to solve systems of equations (using the methods of Chapter 1), determinants (Chapter 4) will be used, an understanding of linear independence (Chapter 2) will be essential, and knowledge of linear transformations and their associated matrices (Chapter 3) will provide motivation for our work. In other words, Chapter 5 will provide an excellent review of the first four chapters.

If T is a linear transformation from a finite-dimensional vector space into itself and $\mathbf{B}$ is a basis for the vector space, then there is an associated matrix $[T : \mathbf{B}, \mathbf{B}]$. If the basis $\mathbf{B}$ is changed, then the matrix changes and the new matrix is, by definition, similar to the previous one. One wonders whether the basis might be changed in some way so that the new matrix is somehow "nice" or easy to understand. We will investigate this question and arrive at a satisfying conclusion. We will see that the theory that we develop has applications in other areas of mathematics as well as other areas of science.

Throughout the first four chapters of this text, we used the assumption that scalars were always elements of some field, but we did not need to make further assumptions regarding the properties of the collection of scalars. In Chapters Five and Six, we will be more restrictive and make the assumption that the scalar field in question is either $\mathbb{R}$, the field of real numbers, or $\mathbb{C}$, the field of complex numbers. This additional assumption is made in order to take advantage of some of the geometrical aspects of real and complex vectors such as length, direction, etc. In addition, we will want to have available some of the special properties of polynomials over the real and complex numbers. Appendix C contains information concerning fields. Appendix D reviews facts about polynomials.

Recall that the field $\mathbb{C}$ of complex numbers consists of all numbers of the form $a + bi$ where a and b are real numbers and i is the so-called imaginary unit satisfying $i^2 = -1$. We identify a and $a + 0i$ so that we can assume $\mathbb{R} \subseteq \mathbb{C}$. For a complex number $z = a + bi$, the **conjugate** of z is denoted by $\overline{z}$ and is defined by $\overline{z} = a - bi$. Several useful properties of the conjugate of a complex number are listed in the following theorem.

Theorem 5.1.1. *Let $z = a + bi$ and $w = c + di$ be complex numbers. Then*

(a) *The magnitude or absolute value of z is given by $|z| = (z\overline{z})^{1/2} = (a^2 + b^2)^{1/2}$.*

(b) $\overline{z + w} = \overline{z} + \overline{w}$ *(The conjugate of a sum is the sum of the conjugates.)*

(c) $\overline{zw} = \overline{z}\ \overline{w}$ *(The conjugate of a product is the product of the conjugates.)*

(d) z *is a real number (that is, $b = 0$) if and only if $z = \overline{z}$.* *(Real numbers equal their conjugates.)*

Proof. See Exercise 13. □

We will make use of the conjugate in defining a generalization of the notion of the dot product encountered earlier.

THE INNER PRODUCT

In Section 4.4 we defined the dot product of two vectors in $\mathbb{R}^3$ and we saw how the dot product was related to the idea of the length of a vector and to the angle between two vectors. We now extend this notion to arbitrary vectors in $\mathbb{R}^n$ or $\mathbb{C}^n$. Let $X = (x_1, \ldots, x_n)$ and $Y = (y_1, \ldots, y_n)$ be vectors in either $\mathbb{R}^n$ or $\mathbb{C}^n$. The **inner product** (also called the **dot product**) of X and Y is the scalar $X \cdot Y$ given by

$$X \cdot Y = \overline{x_1}y_1 + \overline{x_2}y_2 + \ldots + \overline{x_n}y_n.$$

If $X, Y \in \mathbb{R}^n$, then x_i is a real number for $i = 1, \ldots, n$ and so $\overline{x_i} = x_i$. Thus, for real vectors X and Y,

$$X \cdot Y = x_1 y_1 + \ldots + x_n y_n$$

and so this definition of inner product coincides with the definition of inner product or dot product that the reader may have encountered in an introductory physics or calculus course. There is nothing special in our assumption that X and Y were, in effect, row vectors; we could make the same definition for column vectors, that is, $n \times 1$ matrices. Later in this chapter we will find it more convenient to use column vectors and we will use this definition and the following theory in that environment.

Example 5.1.1. $(2, -1, 3) \cdot (4, 1, -2) = 4 \cdot 2 + (-1) \cdot 1 + 3 \cdot (-2) = 1$ and $(-i, 1-i) \cdot (1+i, 2+i) = i \cdot (1+i) + (1+i)(2+i) = 4i$.

□

The **length** of X is the quantity $||X||$ given by $||X|| = \sqrt{X \cdot X}$. Notice that for any complex number z, $z\overline{z}$ is a real number, so for any real or complex vector X, $||X||$ is a real number. For vectors in $\mathbb{R}^2$ and $\mathbb{R}^3$, this definition of length agrees with the usual definition. If $||X|| = 1$, we say that X is a **unit vector** or of **unit length**.

Theorem 5.1.2. *Let X, Y and Z be real or complex n-dimensional vectors and let c be a real or complex scalar. Then:*

(a) $X \cdot Y = \overline{Y \cdot X}$ *so* $X \cdot Y = Y \cdot X$ *if* $X, Y \in \mathbb{R}^n$.   *(Commutative Property)*

(b) $(cX) \cdot Y = \overline{c}(X \cdot Y)$ *(Associative property)*

(c) $X \cdot (cY) = c(X \cdot Y)$ *(Associative property)*

(d) $X \cdot (Y + Z) = X \cdot Y + X \cdot Z$ *and* $(X + Y) \cdot Z = X \cdot Z + Y \cdot Z$.   *(Distributive Property)*

(e) $||X + Y|| \leq ||X|| + ||Y||$ *(Triangle Inequality)*

(f) $|X \cdot Y| \leq ||X|| ||Y||$ *(Schwarz Inequality)*

Proof. We omit the proofs of parts (e) and (f) (although we will revisit them in Section 5.1.1. Proofs of parts (a), (b), and (c) are left as Exercise 14. We will prove part (d).

Let $X = (x_1, \ldots, x_n), Y = (y_1, \ldots, y_n), Z = (z_1, \ldots, z_n)$. Then

$$\begin{aligned}
X \cdot (Y + Z) &= (x_1, \ldots, x_n) \cdot (y_1 + z_1, \ldots, y_n + z_n) \\
&= \overline{x_1}(y_1 + z_1) + \ldots + \overline{x_n}(y_n + z_n) \\
&= (\overline{x_1}y_1 + \ldots + \overline{x_n}y_n) + (\overline{x_1}z_1 + \ldots + \overline{x_n}z_n) \\
&= X \cdot Y + X \cdot Z.
\end{aligned}$$

□

Notice that the inner product of two vectors is a "sum of products" much like one computes in taking the product of two matrices. With some terminology and conventions, one can make the inner product a product of two matrices. Recall that in Section 3.3, we adopted the convention of identifying a 1×1 matrix $[a]$ with its single scalar entry a, and in Section 4.1, the transpose A^t of a matrix A was defined. A further definition is needed: If $A = [a_{ij}]$ is a matrix over the complex numbers, the **conjugate** $\overline{A}$ of A is defined by $\overline{A} = [\overline{a}_{ij}]$. Now let $X = (x_1, \ldots, x_n)$ and $Y = (y_1, \ldots, y_n)$ be vectors over the complex numbers. We may regard X and Y as $1 \times n$ matrices and apply the above terminology:

$$\overline{X}Y^t = [\overline{x}_1, \ldots, \overline{x}_n] = \overline{x}_1 y_1 + \ldots + \overline{x}_n y_n = X \cdot Y.$$

We see that the inner product $X \cdot Y$ may be regarded as a matrix product $\overline{X}Y^t$. If the vectors X and Y are $n \times 1$ column vectors, the inner product may be obtained as the matrix product $\overline{X}^t Y$. We make use of the inner product in defining what is meant for two n-dimensional vectors to be "orthogonal."

ORTHOGONALITY

In Exercise 15, it is shown that for $X, Y \in \mathbb{R}^3$, $X \cdot Y = ||X|| \, ||Y|| \cos \theta$, where θ is the angle between X and Y. It follows that X and Y are orthogonal (perpendicular) if and only if $X \cdot Y = 0$ (for $X, Y \neq 0$). How might these ideas be generalized to vector spaces of higher dimension? Our problem lies in the fact that in other vector spaces, the angle between vectors has no clear meaning[1] We take an opposite approach. By the Schwartz Inequality, Theorem 5.1.2 (f), $|X \cdot Y| \leq ||X|| \, ||Y||$, so that $\dfrac{|X \cdot Y|}{||X|| \, ||Y||} \leq 1$. From this we see that we may define the angle between X and Y to be the angle θ where $\cos \theta = \dfrac{|X \cdot Y|}{||X|| \, ||Y||}$. We therefore make the following definition.

Definition 5.1.1. Vectors X and Y (in $\mathbb{R}^n$ or $\mathbb{C}^n$) are **orthogonal** if and only if $X \cdot Y = 0$. A set of vectors $\{X_1, \ldots, X_k\}$ is **orthogonal** if and only if $X_i \cdot X_j = 0$ for $i \neq j$. The set of vectors $\{X_1, \ldots, X_k\}$ is **orthonormal** if and only if $\{X_1, \ldots, X_k\}$ is orthogonal and each X_i is a unit vector.

The standard basis vectors, $E_1, \ldots, E_n$, in $\mathbb{R}^n$ form an orthonormal set since one can easily see that $||E_i|| = 1$ and $E_i \cdot E_j = 0$ for $i \neq j$. There are other orthonormal sets: In $\mathbb{R}^2$ consider the vectors $X_1 = (1, 1)$ and $X_2 = (1, -1)$. Clearly, $X_1 \cdot X_2 = 1 - 1 = 0$ so that X_1 and X_2 are orthogonal, but X_1 and X_2 are not unit vectors. Dividing each vector by its length produces unit vectors that remain orthogonal. We get:

$$\frac{1}{||X_1||} X_1 = \frac{1}{\sqrt{2}}(1, 1) = (1/\sqrt{2}, 1/\sqrt{2})$$

$$\frac{1}{||X_2||} X_2 = \frac{1}{\sqrt{2}}(1, -1) = (1/\sqrt{2}, -1/\sqrt{2})$$

---

[1]Two noncollinear vectors with their tails in the same place determine a unique plane, so we can define the angle between them as the angle in that plane. However, for other vector spaces, there may be no geometric intuition underlying the notion of "angle."

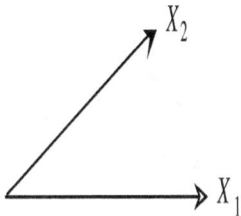

Figure 5.1: Two linearly independent vectors

In Exercise 18, it is shown that if X is a nonzero vector, then $(1/\|X\|)X$ is a unit vector. We call this process of dividing a vector by its length **normalizing**. If we normalize each vector in an orthogonal set, the result is an orthonormal set.

It is easy to show that any orthogonal set of nonzero vectors in linearly independent.

Theorem 5.1.3. *If $\{X_1, \ldots, X_k\}$ is an orthogonal set of nonzero vectors, then $\{X_1, \ldots, X_k\}$ is linearly independent.* *(Orthogonal nonzero vectors are independent.)*

Proof. Assume $c_1 X_1 + \ldots + c_k X_k = 0$. We must show that each $c_i = 0$. Now

$$
\begin{aligned}
0 &= X_i \cdot 0 \\
&= X_i \cdot (c_1 X_1 + \ldots + c_k X_k) \\
&= c_1 (X_i \cdot X_1) + \ldots + c_i (X_i \cdot X_i) + \ldots + c_k (X_i \cdot X_k) \\
&= c_i (X_i \cdot X_i).
\end{aligned}
$$

Now we have that $c_i (X_i \cdot X_i) = 0$, but $X_i \neq 0$ and so by Exercise 16, $X_i \cdot X_i \neq 0$. It follows that $c_i = 0$ and so $X_1, \ldots, X_k$ are linearly independent. $\qquad \square$

Since an orthogonal set of nonzero vectors is linearly independent, one wonders whether every vector space has a basis which is an orthogonal, or orthonormal, set.

FINDING AN ORTHONORMAL BASIS

It is always possible to find an orthonormal basis for any subspace of $\mathbb{R}^n$ or C^n. The theorem below tells how. The procedure is called the **Gram-Schmidt Process**. The formulas look messy, but the idea is rather simple. Suppose that X_1 and X_2 are two independent vectors in $\mathbb{R}^n$. Then they are not collinear.

The first step is to divide X_1 by its length (normalize) in order to obtain a unit vector Z_1 in the same direction, as pictured in Figure 5.2.

Next, the vector X_2 must be "straightened up" so that it is perpendicular to X_1. To do this, we subtract the component of X_2 in the direction of X_1. This component has length

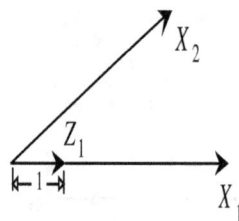

Figure 5.2: Finding Z_1

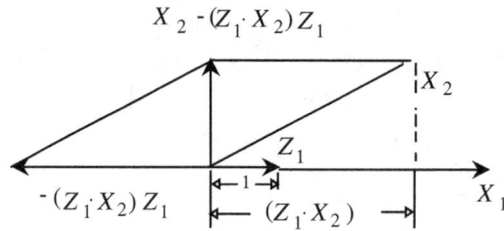

Figure 5.3: Two linearly independent vectors

$Z_1 \cdot X_2$ (See Exercise 15, and note that $||Z_1|| = 1$) and so $(Z_1 \cdot X_2)Z_1$ is the appropriate vector to subtract.

Finally, we have found a vector perpendicular to Z_1, and so all that remains is to divide by the length of the vector in order to obtain a unit vector. The Gram-Schmidt process extends this method to an arbitrary collection of linearly independent vectors.

Theorem 5.1.4. *Let* **W** *be a subspace of* $\mathbb{R}^n$ *or* C^n *and let* $X_1, \ldots, X_m$ *be any basis for* **W**. *Let*

$$
\begin{aligned}
Y_1 &= X_1 & Z_1 &= Y_1/||Y_1|| \\
Y_2 &= X_2 - (Z_1 \cdot X_2)Z_1 & Z_2 &= Y_2/||Y_2|| \\
Y_3 &= X_3 - (Z_1 \cdot X_3)Z_1 - (Z_2 \cdot X_3)Z_2 & Z_3 &= Y_3/||Y_3|| \\
&\;\;\vdots & &\;\;\vdots \\
Y_m &= X_m - (Z_1 \cdot X_m)Z_1 - \ldots - (Z_{m-1} \cdot X_m)Z_{m-1} & Z_m &= Y_m/||Y_m||.
\end{aligned}
$$

Then $Z_1, \ldots, Z_m$ *is an orthonormal basis for* **W**.

(Certain linear combinations of a basis will form an orthonormal basis.)

Proof. By Exercise 18, each Z_i is a unit vector, so we need only prove that $Z_1, \ldots, Z_m$ are orthogonal. Let's assume that the vectors are in $\mathbb{R}^n$ so that we may avoid the conjugates.

Since $Z_i \cdot Z_j = Z_j \cdot Z_i$ we need only show that Z_k is orthogonal to $Z_1, \ldots, Z_{k-1}$ for $k = 2, \ldots, m$. We prove this by induction on the integer k.

For $k = 2$:

$$
\begin{aligned}
Z_1 \cdot Z_2 &= Z_1 \cdot (1/\|Y_2\|)(X_2 - (Z_i \cdot Z_2)Z_1) \\
&= (1/\|Y_2\|)(Z_1 \cdot X_2 - (Z_1 \cdot X_2)(Z_1 \cdot Z_1)) \\
&= (1/\|Y_2\|)(0) \\
&= 0
\end{aligned}
$$

since $Z_1 \cdot Z_1 = \|Z_1\|^2 = 1$.

Assume that Z_k is orthogonal to $Z_1, \ldots, Z_{k-1}$. We must show Z_{k+1} is orthogonal to $Z_1, \ldots, Z_k$. Let i be an integer, $1 \le i \le k$. Then

$$
\begin{aligned}
Z_i \cdot Z_{k+1} &= Z_i \cdot \left(\frac{1}{\|Y_{k+1}\|}(X_{k+1} - (Z_1 \cdot X_{k+1})Z_1 - \ldots - (Z_k \cdot X_{k+1})Z_k) \right) \\
&= \frac{1}{\|Y_k + 1\|}(Z_i \cdot X_{k+1} - (Z_1 \cdot X_{k+1})Z_i \cdot Z_1 - \cdots - (Z_i \cdot X_{k+1})(Z_i \cdot Z_i)) \\
&\quad - \cdots - (Z_k \cdot X_{k+1})Z_i \cdot Z_k) \\
&= \frac{1}{\|Y_k + 1\|}(Z_i \cdot X_{k+1} - (Z_i \cdot X_{k+1})(Z_i \cdot Z_i)) \\
&= 0
\end{aligned}
$$

since $Z_i \cdot Z_i = 1$ and $Z_i \cdot Z_j = 0$ for $i \ne j \le k + 1$.

Finally, then, we see that $Z_1, \ldots, Z_m$ is an orthonormal basis for **W**. $\square$

Example 5.1.2. 1. Consider the subspace of $\mathbb{R}^3$ spanned by the vectors $X_1 = (1, -1, 2)$ and $X_2 = (2, 0, 1)$. Note that X_1 and X_2 are linearly independent, but $X_1 \cdot X_2 = 4$ so that X_1 and X_2 are not orthogonal. We apply the Gram-Schmidt Process:

$$
\begin{aligned}
Z_1 &= \frac{1}{\|X_1\|}X_1 \\
&= \frac{1}{\sqrt{6}}(1, -1, 2) \\
Y_2 &= X_2 - (Z_1 \cdot X_2)Z_1 \\
&= (2, 0, 1) - \frac{1}{\sqrt{6}}(4)\frac{1}{\sqrt{6}}(1, -1, 2) \\
&= (2, 0, 1) - \frac{4}{6}(1, -1, 2) \\
&= (2, 0, 1) - (2/3, -2/3, 4/3) \\
&= (4/3, 2/3, -1/3), \text{ and} \\
Z_2 &= \frac{1}{\sqrt{21/9}}(4/3, 2/3, -1/3) \\
&= \frac{3}{\sqrt{21}}(4, 2, -1).
\end{aligned}
$$

Notice that in calculating Z_2 we could have more easily obtained the same vector if we had divided $(4, 2, -1) = 3Y_2$ by its length rather than performing the same operation on Y_2. See Exercise 14.

2. The standard basis for $\mathbb{R}^3$ is an orthonormal basis, but we will need in Section 5.5 to find an orthonormal basis in which the first vector in the basis is a scalar multiple of a given vector. This can always be done without using the Gram-Schmidt process, and it can usually be done by inspection.

Suppose we consider the vector $X_1 = (1, -2, 1)$ and suppose we try to find an orthonormal basis in which a scalar multiple of X_1 is the first vector. Recall from Chapter 2 that we can find vectors X_2 and X_3 so that X_1, X_2, and X_3 form a linearly independent set. If we apply the Gram-Schmidt process to this set we obtain an orthonormal set with the first vector a multiple of X_1. But the Gram-Schmidt process is messy. By inspection, we can find a vector orthogonal to X_1. Take $X_2 = (2, 1, 0)$ (this is the old switch-the-values-and-change-one-sign trick). Now find a vector X_3 that is orthogonal to both X_1 and X_2, say $X_3 = (1, -2, -5)$. Notice that in X_3 the first two entries assure that X_3 is orthogonal to X_2 no matter what the third entry is. We then choose the third entry so that X_3 is orthogonal to X_1. We now have three orthogonal vectors and to obtain the orthonormal basis it is only necessary to normalize. Dividing by the lengths we get

$$Z_1 = \frac{1}{\sqrt{6}}(1, -2, 1), Z_2 = \frac{1}{\sqrt{5}}(2, 1, 0), Z_3 = \frac{1}{\sqrt{30}}(1, -2, -5).$$

Notice that Z_1 is a scalar multiple of X_1.

$\square$

5.1.1   GENERAL INNER PRODUCT SPACES (optional)

The vector spaces $\mathbb{R}^n$ and $\mathbb{C}^n$ have an inner product as defined above, but there are other spaces on which such a product can be defined. In general, we have the following definition.

Definition 5.1.2. Let $\mathbf{V}$ be a vector space over $F = \mathbb{R}$ or $F = \mathbb{C}$. An **inner product** on $\mathbf{V}$ is a function $<, >$ from $\mathbf{V} \times \mathbf{V} \to F$ with the following properties:

1. For all $X, Y \in \mathbf{V}, < X, Y > = \overline{< Y, X >}$.   Antisymmetry

2. For all $X, Y \in \mathbf{V}$ and $r \in F, < rX, Y > = \bar{r} < X, Y >$.[2]

   Conjugate linearity in the first coordinate/linearity in the second coordinate

3. For all $X, Y, Z \in \mathbf{V}, < X+Y, Z > = < X, Z > + < Y, Z >$ (Conjugate) linearity in the first coordinate

4. For all $X \in \mathbf{V}, < X, X > \geq 0$, with equality if and only if $X = 0$.   Positive definiteness

---

[2]In keeping with the flavor of this text, we adopt the convention more commonly used by physicists rather than that used by mathematicians. The conditions $< rX, Y > = \bar{r} < X, Y >$ and $< X, rY > = \bar{r} < X, Y >$ lead to the the same properties.

A vector space with an inner product as defined above is called an **inner product space**. We often write $X \cdot Y$ for $< X, Y >$. In addition, we define the **norm** of $X \in \mathbf{V}$ by $||X||^2 =< X, X >$.

There are several elementary properties of inner product spaces that are worth noting.

Theorem 5.1.5. *Let* $\mathbf{V}$ *be an inner product space over* $F = \mathbb{R}$ *or* $F = \mathbb{C}$*. Let* $X, Y, Z \in$ $\mathbf{V}$*, and let* $r \in F$*.*

1. $< X, X >\in \mathbb{R}$.

2. $< X, rY >= r < X, Y >$

3. $< X, Y + Z >=< X, Y > + < X, Z >$

Proof. The proofs are left as exercises. $\square$

Example 5.1.3. We are already familiar with the inner product space $\mathbb{R}^n$ with the usual dot product as an inner product. It is straightforward to verify that the required conditions all hold (see Exercise 22).

$\square$

Example 5.1.4. Consider $C[0, 1]$, the space of all continuous functions on $[0, 1]$, and define the inner product of $f, g \in C[0, 1]$ by

$$< f, g >= \int_0^1 f(x)g(x)dx.$$

Certainly $< f, g >= \overline{< g, f >} =< g, f >$ since $\int_0^1 f(x)g(x)dx = \int_0^1 g(x)f(x)dx$. Also, $<$ $rf, g >= \int_0^1 (rf)(x)g(x)dx = \int_0^1 rf(x)g(x)dx = r \int_0^1 f(x)g(x)dx = r < f, g >= \bar{r} < f, g >$ for any real number r. In addition, for any $f, g, h \in C[0, 1]$, we have

$$\begin{aligned}
< f + g, h > &= \int_0^1 (f + g)(x)h(x)dx \\
&= \int_0^1 (f(x) + g(x))h(x)dx \\
&= \int_0^1 (f(x)h(x) + g(x)h(x))dx \\
&= \int_0^1 f(x)h(x)dx + \int_0^1 g(x)h(x)dx \\
&=< f, h > + < g, h > .
\end{aligned}$$

Finally, $< f, f >= \int_0^1 (f(x))^2 dx \geq 0$ for any $f \in C[0, 1]$, and if $< f, f >= 0$, then, since $(f(x))^2 \geq 0$ for $x \in [0, 1]$, we must have $f(x) = 0$ for $x \in [0, 1]$.

A similar result holds for $C[a, b]$ for any interval $[a, b]$.

□

We return now to the Cauchy-Schwarz inequality and the triangle inequality.

Theorem 5.1.6. *Let* **V** *be an inner product space over* $F = \mathbb{R}$ *or* $F = \mathbb{C}$, *and let* $X, Y \in$ **V**. *Then* $|X \cdot Y| \leq ||X||||Y||$.

Proof. If $X = 0$ or $Y = 0$, then the inequality holds since both sides are zero. Thus, we may assume that X and Y are nonzero. Consider the **projection** P of Y onto X: $P = \dfrac{Y \cdot X}{||X||^2}X$. Notice that

$$\begin{aligned}
(Y - P) \cdot X &= Y \cdot X - P \cdot X \\
&= Y \cdot X - \left(\frac{Y \cdot X}{||X||^2}X\right) \cdot X \\
&= Y \cdot X - Y \cdot X \\
&= 0.
\end{aligned}$$

This also implies that $(Y - P) \cdot P = 0$ since P is parallel to X (see Exercise 23)). Now $Y = P + (Y - P)$, so

$$\begin{aligned}
||Y||^2 &= Y \cdot Y \\
&= (P + (Y - P)) \cdot (P + (Y - P)) \\
&= P \cdot P + (Y - P) \cdot P + P \cdot (Y - P) + (Y - P) \cdot (Y - P) \\
&= \left(\frac{Y \cdot X}{||X||^2}X\right) \cdot \left(\frac{Y \cdot X}{||X||^2}X\right) + ||Y - P||^2 \\
&= \left(\frac{|Y \cdot X|}{||X||^2}\right)^2 X \cdot X + ||Y - P||^2 \\
&\geq \left(\frac{|Y \cdot X|}{||X||^2}\right)^2 ||X||^2 \\
&= \frac{(|Y \cdot X|)^2}{||X||^2}.
\end{aligned}$$

Thus $||X||^2||Y||^2 \geq (|Y \cdot X|)^2$, so $|X \cdot Y| \leq ||X||^2||Y||^2$, as desired.   □

This leads to the famous Triangle Inequality:

Theorem 5.1.7. *Let* **V** *be an inner product space over* $F = \mathbb{R}$ *or* $F = \mathbb{C}$, *and let* $X, Y \in$ **V**. *Then* $||X + Y|| \leq ||X|| + ||Y||$.

Proof.

$$
\begin{aligned}
||X + Y||^2 &= (X + Y) \cdot (X + Y) \\
&= X \cdot X + X \cdot Y + Y \cdot X + Y \cdot Y \\
&= ||X||^2 + X \cdot Y + \overline{X \cdot Y} + ||Y||^2 \\
&= ||X||^2 + 2\Re(X \cdot Y) + ||Y||^2 \\
&\leq ||X||^2 + 2|X \cdot Y| + ||Y||^2 \\
&\leq ||X||^2 + 2||X||||Y|| + ||Y||^2 \\
&= (||X|| + ||Y||)^2.
\end{aligned}
$$

Thus, $||X + Y|| \leq ||X|| + ||Y||$. Note the use of the Cauchy-Schwarz Inequality. $\qquad\square$

The Cauchy-Schwarz Inequality also gives us a way to measure the "angle" between vectors in an arbitrary inner product space even when such vectors do not lend themselves to geometric interpretations. If V is an inner product space, then the Cauchy-Schwarz Inequality tells us that for all $X, Y \in V$, $|X \cdot Y| \leq ||X||||Y||$ and thus (as before)

$$
\frac{|X \cdot Y|}{||X||||Y||} \leq 1.
$$

By analogy with $\mathbb{R}^n$, we equate the left-hand side of this inequality with the cosine of some angle θ and call θ the angle between the vectors.

Definition 5.1.3. Let V be an inner product space, and let $X, Y \in V$ be nonzero. The **angle** between X and Y is given by

$$
\cos \theta = \frac{|X \cdot Y|}{||X||||Y||}.
$$

We say that two vectors are **orthogonal** if $\theta = 0$, or, equivalently, if their inner product equals 0.

Example 5.1.5. Consider $S = \{f(x) \in C[-\pi, \pi] | f(x) = \cos(nx) \text{ or } f(x) = \sin(nx) \text{ for some } n \in \mathbb{Z}_{\geq 0}\}$. For an integers $m \neq n$, $\cos(mx)$ and $\cos(nx)$ are orthogonal and $\sin(mx)$ and $\sin(nx)$ are orthogonal. In addition, $\cos(mx)$ and $\sin(nx)$ are orthogonal for any integers m and n. (See the exercises.) By Exercise 24, it follows that S is an orthonormal basis for a subspace of $C[-\pi, \pi]$.

The set S is used to create **Fourier series**, infinite series representing periodic functions (with period 2π, in this case) on $(-\infty, \infty)$ akin to Taylor polynomials for other functions.

$\qquad\square$

Example 5.1.6. Find the angle between $f(x) = 5x^2$ and $g(x) = 12x$ in the inner product space $C[0, 1]$.

We need $< f, f >, < f, g >$, and $< g, g >$ in order to find the angle θ between f and g.

$$
< f, f > = \int_0^1 (5x^2)^2 dx = \int_0^1 25x^4 dx = 5
$$

$$< g, g >= \int_0^1 (12x)^2 dx = \int_0^1 144x^2 dx = 48$$

$$< f, g >= \int_0^1 (5x^2)(12x) dx = \int_0^1 60x^3 dx = 15$$

Thus $\cos \theta = \dfrac{15}{5 \cdot 48} = \dfrac{1}{16}$. This gives $\theta \approx 1.51$ radians or about $86.4°$.

On the other hand, $< x, 3x - 2 >= \int_0^1 x(3x - 2)dx = \int_0^1 (3x^2 - 2x)dx = 0$, so x and $3x - 2$ are orthogonal.

$\square$

Section 5.1 Exercises

In Exercises 1-4, let X = (1, 3, -1), Y = (3, 2, 5) and Z = (-2, 3, -1).

1. Compute $X \cdot Y$ and $X \cdot Z$.

2. Compute $Y \cdot Y$ and $Z \cdot X$.

3. Compute $||X||, ||Y||$, and $||Z||$.

4. Let $U = \dfrac{1}{||X||} X$. Compute $||U||$.

5. Let $X_1 = (1, 0, 3), X_2 = (2, 1, -1)$. Find $X_1 \cdot X_2$ and $||X_2||$.

6. Use the Gram-Schmidt Process to obtain an orthonormal basis for the subspace of $\mathbb{R}^3$ spanned by the vectors $(1, -1, 0)$ and $(2, -1, -2)$.

7. Show that $X_1 = (1 + i, i, 1), X_2 = (i, 1 - i, 0)$, and $X_3 = (1 - i, 1, 3i)$ are orthogonal.

8. Use The Gram-Schmidt Process to obtain an orthonormal basis for the subspace of $\mathbb{R}^4$ spanned by the vectors $X_1 = (1, 0, 1, 0), X_2 = (1, 3, 1, 0), X_3 = (3, 2, -1, 0)$.

   In Exercises 9 - 12, proceed as in Example 5.1.2 above.

9. Find an orthonormal basis for $\mathbb{R}^2$ in which the first vector in the basis is a scalar multiple of the vector $(1, 2)$.

10. Find an orthonormal basis for $\mathbb{R}^2$ in which the first vector in the basis is a scalar multiple of the vector $(-2, 3)$.

11. Find an orthonormal basis for $\mathbb{R}^3$ in which the first vector in the basis is a scalar multiple of the vector $(0, -3, 2)$.

12. Find an orthonormal basis for $\mathbb{R}^3$ in which the first vector in the basis is a scalar multiple of the vector $(-1, 2, -2)$.

13. Prove Theorem 5.1.1.

14. Prove parts (a), (b), and (c) of Theorem 5.1.2.

15. For vectors $X, Y \in \mathbb{R}^3$, prove that $X \cdot Y = ||X|| \, ||Y|| \cos \theta$, where θ is the angle between the vectors X and Y. (Hint: Consider the triangle formed by the vectors X and Y. The third side of the triangle has length $||Y - X||$. Apply the law of cosines.)

16. Show that if X is a nonzero real or complex vector, then $X \cdot X \neq 0$ so that $||X|| > 0$.

17. Show that if X is a nonzero real or complex vector, and c is a scalar, then $||cX|| = |c| \, ||X||$.

18. Show that if X is a nonzero real or complex vector, then $U = (1/||X||)X$ is a unit vector.

19. Let X be a nonzero real or complex vector and let c be a positive real scalar. Show that $\dfrac{1}{||X||}X = \dfrac{1}{||cX||}cX$.

20. Let A and B be respectively $m \times n$ and $n \times r$ matrices over the complex numbers. Show that $\overline{AB} = \overline{A} \, \overline{B}$.

21. Prove Theorem 5.1.5.

22. Show that $\mathbb{R}^n$ is an inner product space with respect to the usual dot product.

23. Prove that if $X \cdot Y = 0$ and Z is parallel to Y, then $X \cdot Z = 0$.

24. Let V be an inner product space. Show that if $S \subseteq V$, where $0 \notin S$, such that for any $X, Y \in S$, $< X, Y >= 0$, then S is linearly independent in V.

25. Show that for any integers $m \neq n$, $\cos(mx)$ and $\cos(nx)$ are orthogonal on $[-\pi, \pi]$ and $\sin(mx)$ and $\sin(nx)$ are orthogonal on $[-\pi, \pi]$.

26. Show that for any integers $m, n \in \mathbb{Z}$, $\cos(mx)$ and $\sin(nx)$ are orthogonal on $[-\pi, \pi]$.

27. Let $f(x) = x$ and $g(x) = \cos x$ in $C[-\pi, \pi]$. Find the angle between f and g.

28. Let $f(x) = x$ and $g(x) = \sin x$ in $C[-\pi, \pi]$. Find the angle between f and g.

29. Let $f(x) = 4x^2 + 3$ and $g(x) = 7x - 1$ in $C[0, 1]$. Find the angle between f and g.

30. Show that if X is a non-zero vector in an inner product space V, then the angle between X and itself is 0.

Let V be an inner product space, and let W be a subspace of V. We define the **orthogonal complement** $W^\perp$ of W by $W^\perp = \{v \in V \mid < v, w >= 0 \text{ for all } w \in W\}$.

31. Prove that $W^\perp$ is a subspace of V.

32. Let $\{w_1, \ldots, w_k\}$ be an orthonormal basis for W, and define $T : V \to V$ by $T(v) = v - \dfrac{v \cdot w_1}{w_1 \cdot w_1} w_1 - \ldots - \dfrac{v \cdot w_k}{w_k \cdot w_k} w_k$ for each $v \in V$ (by analogy with the Gram-Schmidt process). Prove the following:

   (a) T is a linear transformation.

   (b) $\mathrm{Im}(T) = W^{\perp}$.

   (c) $\ker(T) = W$.

   (d) If $\dim(V) = n$, then $\dim W + \dim W^{\perp} = n$. [Hint: use T.]

   (e) $(W^{\perp})^{\perp} = W$.

33. Prove that $R(A)^{\perp} = N(A)$.

5.2   Eigenvalues and Eigenvectors

We begin now the study of an interesting problem. Given a linear transformation, $T : \mathbf{U} \to \mathbf{U}$, we wonder whether a basis $\mathbf{B}$ for the vector space $\mathbf{U}$ can be found so that $T_{\mathbf{B}}^{\mathbf{B}}$ is a "nice" matrix. The term "nice" might be interpreted to mean "diagonal" or "nearly diagonal". Recall that by the results regarding change of basis that if $\mathbf{C}$ is another basis for $\mathbf{U}$, then

$$T_{\mathbf{C}}^{\mathbf{C}} = I_{\mathbf{B}}^{\mathbf{C}} T_{\mathbf{B}}^{\mathbf{B}} I_{\mathbf{C}}^{\mathbf{B}} = (I_{\mathbf{C}}^{\mathbf{B}})^{-1} T_{\mathbf{B}}^{\mathbf{B}} I_{\mathbf{C}}^{\mathbf{B}}.$$

This relationship between $T_{\mathbf{C}}^{\mathbf{C}}$ and $T_{\mathbf{B}}^{\mathbf{B}}$ gave rise to the definition of similarity of matrices: two $n \times n$ matrices A and B are **similar** if and only if there is a nonsingular $n \times n$ matrix S with $B = S^{-1}AS$. By the results of Section 3.5, any nonsingular matrix can be written as a product of either row or column elementary matrices and these elementary matrices are associated with certain elementary row or column operations. Using the results (in Section 3.5) concerning the relationship between the elementary row and column operations and certain changes in basis, any nonsingular matrix S can be represented as a matrix $I_{\mathbf{C}}^{\mathbf{B}}$ where $\mathbf{B}$ is a given basis and $\mathbf{C}$ is calculated from $\mathbf{B}$ by performing the necessary operations on the elements of $\mathbf{B}$. Because of this we will study matrices and similarity of matrices without mention of linear transformations and bases for vector spaces. To motivate our basic definitions, however, we will consider now a property of a linear transformation.

Let $T : \mathbf{U} \to \mathbf{U}$ be a linear transformation of the vector space $\mathbf{U}$ into itself. A vector X with the property that $T(X) = \lambda X$ for some scalar, λ, is called an **invariant vector of T**. Geometrically, this condition means that the action of T shrinks or stretches the vector X, but does not change its direction. Invariant vectors are interesting, in part, for the following reason: suppose $\mathbf{U}$ has a basis $\mathbf{B} = \{X_1, \ldots, X_n\}$ consisting of invariant vectors, say $T(X_i) = \lambda_i X_i$. Then computing the matrix for T relative to this basis $\mathbf{B}$, we get $T_{\mathbf{B}}^{\mathbf{B}} = \mathrm{diag}(\lambda_1, \ldots, \lambda_n)$. Because of this interesting property, we make the following definition regarding an $n \times n$ matrix.

EIGENVALUES AND EIGENVECTORS

Definition 5.2.1. Let A be an $n \times n$ matrix. A scalar λ is an **eigenvalue** of A if there is a nonzero (column) vector X such that $AX = \lambda X$. An **eigenvector** of A (associated with the eigenvalue λ) is a nonzero (column) vector X with $AX = \lambda X$ for some scalar λ.

[Note: λ is the lower-case Greek letter lambda.]

For a scalar λ, the equation $AX = \lambda X$ has a nonzero solution if and only if $(A - \lambda I)X = 0$ has a nonzero solution, where I is the $n \times n$ identity matrix. If the eigenvalue λ could be found, we could find the associated eigenvector X by solving the homogeneous system of linear equations $(A - \lambda I)X = 0$ using the methods of Sections 1.6 and 2.9. Now $(A - \lambda I)X = 0$ is a matrix equation with $n \times n$ coefficient matrix and so, using facts about determinants from Section 4.2, we see that $(A - \lambda I)X = 0$ has a nonzero solution if and only if $|A - \lambda I| = 0$. It follows that λ is an eigenvalue if and only if $|A - \lambda I| = 0$. Considering $|A - \lambda I|$ as a function of λ, we see that it is a polynomial in the variable λ. For example, if

$$A = \begin{bmatrix} 1 & 3 \\ 2 & -1 \end{bmatrix},$$

then

$$|A - \lambda I| = \begin{vmatrix} 1 - \lambda & 3 \\ 2 & -1 - \lambda \end{vmatrix} = (1 - \lambda)(-1 - \lambda) - 6 = \lambda^2 - 7.$$

THE CHARACTERISTIC POLYNOMIAL - FINDING EIGENVALUES

Definition 5.2.2. Let A be an $n \times n$ matrix. The polynomial $p_A(\lambda) = |A - \lambda I|$ is called the **characteristic polynomial** of A. The equation $p_A(\lambda) = |A - \lambda I| = 0$ is called the **characteristic equation** of A.

As we have observed:

> **Theorem 5.2.1.** *The eigenvalues of A are the roots of the characteristic polynomial $p_A(\lambda)$ of A.* *(The eigenvalues are the roots of the characteristic polynomial.)*

Example 5.2.1. Let us use the above theorem to find eigenvalues and corresponding eigenvectors for a given matrix. Let

$$A = \begin{bmatrix} 3 & 1 & 2 \\ -3 & 2 & 3 \\ 2 & 1 & 3 \end{bmatrix}.$$

Then expanding across the first row gives

$$p_A(\lambda) = |A - \lambda I|$$

$$= \begin{vmatrix} 3 - \lambda & 1 & 2 \\ -3 & 2 - \lambda & 3 \\ 2 & 1 & 3 - \lambda \end{vmatrix}$$

$$= (3 - \lambda)[((2 - \lambda)(3 - \lambda) - 3)] - (-3(3 - \lambda) - 2) + 2(3 - 2(2 - \lambda))$$

$$= -\lambda^3 + 8\lambda^2 - 17\lambda + 10$$

$$= (\lambda - 1)(-\lambda^2 + 7\lambda - 10)$$

$$= -(\lambda - 1)(\lambda - 2)(\lambda - 5),$$

and so the eigenvalues of A are $\lambda_1 = 1, \lambda_2 = 2$, and $\lambda_3 = 5$. The reader may wish to consult Appendix D for facts about polynomials and hints about finding solutions of polynomial equations.

To find eigenvectors corresponding to these eigenvalues, one finds solutions of the systems of homogeneous equations $(A - \lambda I)X = 0$ where λ is one of the eigenvalues. We calculate the coefficient matrices for each eigenvalue and then use row operations to simplify the solution. Since we need only find one solution for each system, we perform only enough row operations to enable us to find a solution by inspection:

$$A - 1I = \begin{bmatrix} 2 & 1 & 2 \\ -3 & 1 & 3 \\ 2 & 1 & 2 \end{bmatrix} \begin{matrix} R_{13}(-1) \\ \rightarrow \\ R_{12}(-1) \end{matrix} \begin{bmatrix} 2 & 1 & 2 \\ -5 & 0 & 1 \\ 0 & 0 & 0 \end{bmatrix}, X_1 = \begin{bmatrix} 1 \\ -12 \\ 5 \end{bmatrix}$$

$$A - 2I = \begin{bmatrix} 1 & 1 & 2 \\ -3 & 0 & 3 \\ 2 & 1 & 1 \end{bmatrix} \begin{matrix} R_{13}(-1) \\ \rightarrow \end{matrix} \begin{bmatrix} 1 & 1 & 2 \\ -3 & 0 & 3 \\ 1 & 0 & -1 \end{bmatrix}, X_2 = \begin{bmatrix} 1 \\ -3 \\ 1 \end{bmatrix}$$

$$A - 5I = \begin{bmatrix} -2 & 1 & 2 \\ -3 & -3 & 3 \\ 2 & 1 & -2 \end{bmatrix} \begin{matrix} R_{13}(1) \\ \rightarrow \end{matrix} \begin{bmatrix} -2 & 1 & 2 \\ -3 & -3 & 3 \\ 0 & 2 & 0 \end{bmatrix}, X_3 = \begin{bmatrix} 1 \\ 0 \\ 1 \end{bmatrix}$$

We see that X_1, X_2, and X_3 are eigenvectors of A corresponding to the eigenvalues $\lambda_1 = 1, \lambda_2 = 2$, and $\lambda_3 = 5$.

$\square$

Theorem 5.2.2. Let A be an $n \times n$ matrix over $\mathbb{C}$ (or $\mathbb{R}$) with eigenvalues $\lambda_1, \ldots, \lambda_n$ (not necessarily distinct). Then $|A| = \lambda_1 \lambda + 2 \cdots \lambda_n$.

Proof. Notice that $p_A(0) = |A - 0I| = |A|$. Also, $p_A(\lambda)$ factors completely over $\mathbb{C}$, so we have $p_A(\lambda) = (\lambda_1 - \lambda)(\lambda_2 - \lambda) \cdots (\lambda_n - \lambda)$, so $p_A(0) = \lambda_1 \lambda_2 \cdots \lambda_n$. This gives the result. $\square$

The following properties of the characteristic polynomial are easy to establish.

Theorem 5.2.3. *Let A be an $n \times n$ matrix. Then*

(a) $p_A(\lambda) = p_{A^t}(\lambda)$, so A and A^t have the same eigenvalues.

(A matrix and its transpose have the same eigenvalues.)

(b) If A is similar to B, $p_A(\lambda) = p_B(\lambda)$, so A and B have the same eigenvalues.

(Similar matrices have the same eigenvalues.)

Proof. (a) $p_A(\lambda) = |A - \lambda I| = |(A - \lambda I)^t| = |A^t - (\lambda I)^t| = |A^t - \lambda I| = p_{A^t}(\lambda)$.

(b) If A is similar to B, then $B = S^{-1}AS$ for some nonsingular matrix S. Then we have

$$
\begin{aligned}
p_B(\lambda) &= |B - \lambda I| \\
&= |S^{-1}AS - \lambda I| \\
&= |S^{-1}AS - \lambda S^{-1}IS| \\
&= |S^{-1}(A - \lambda I)S| \\
&= |S^{-1}||A - \lambda I||S| \\
&= |S^{-1}||S||A - \lambda I| \\
&= |S^{-1}S||A - \lambda I| \\
&= |I||A - \lambda I| \\
&= |A - \lambda I| \\
&= p_A(\lambda).
\end{aligned}
$$

$\square$

As mentioned in our introductory remarks, if one can find a basis $\mathbf{B}$ of invariant vectors for a linear transformation T, then $[T : \mathbf{B}, \mathbf{B}]$ is a diagonal matrix. This is not always possible, but the following theorem provides important information regarding eigenvectors corresponding to distinct eigenvalues.

Theorem 5.2.4. *If $\lambda_1, \ldots, \lambda_k$ are distinct eigenvalues of the $n \times n$ matrix A and $X_1, \ldots, X_k$ are corresponding eigenvectors, then $X_1, \ldots, X_k$ are linearly independent.*

(Eigenvectors corresponding to distinct eigenvalues are independent.)

Proof. Assume not. Then $X_1, \ldots, X_m$ are linearly independent for some m, but $X_1, \ldots, X_m$, X_{m+1} are linearly dependent. Then $a_1 X_1 + \ldots + a_m X_m + a_{m+1} X_{m+1} = 0$ for some scalars $a_1, \ldots, a_{m+1}$, not all of which are zero. Multiply by A and get

$$
A a_1 X_1 + \ldots + A a_{m+1} X_{m+1} = a_1 \lambda_1 X_1 + \ldots + a_{m+1} \lambda_{m+1} X_{m+1} = 0.
$$

Now multiply by λ_{m+1} and subtract to get

$$0 = a_1(\lambda_1 - \lambda_{m+1})X_1 + \ldots + a_m(\lambda_m - \lambda_{m+1})X_m + a_{m+1}(\lambda_{m+1} - \lambda_{m+1})X_{m+1}$$
$$= a_1(\lambda_1 - \lambda_{m+1})X_1 + \ldots + a_m(\lambda_m - \lambda_m + 1)X_m = 0.$$

But $X_1, \ldots, X_m$ are linearly independent and $\lambda_i - \lambda_{m+1} \neq 0$ (because the λ_is are all different), so we get $a_1 = \ldots = a_m = 0$. It follows that $a_1X_1 + \ldots + a_mX_m + a_{m+1}X_{m+1} = a_{m+1}X_{m+1} = 0$ and since $X_{m+1} \neq 0$ (it is an eigenvector), we get $a_{m+1} = 0$. It follows that $X_1, \ldots, X_k$ must be linearly independent. $\qquad \square$

Two related classes of matrices arise often in applications and have interesting properties. We will see in Section 5.5 that these matrices are always similar to diagonal matrices.

SYMMETRIC AND HERMITIAN MATRICES

Definition 5.2.3. An $n \times n$ matrix A is **symmetric** if and only if $A = A^t$ and A is **hermitian**[3] if and only if $A = \overline{A}^t$.

Since the conjugate of a real number is the number itself, we see that a real symmetric matrix is also hermitian.

Example 5.2.2. Consider the following matrices:

$$A = \begin{bmatrix} 1 & 3 & 3 \\ 3 & -2 & -1 \\ 3 & -1 & 4 \end{bmatrix}, B = \begin{bmatrix} 1 & i & 1-i \\ -i & 2 & 2+i \\ 1+i & 2-i & -1 \end{bmatrix}, \text{ and } C = \begin{bmatrix} i & 2 & -i \\ 2 & 1 & 3 \\ i & 3 & -3 \end{bmatrix}.$$

The matrix A is symmetric (the entries that are symmetric about the main diagonal are equal), and since A is a real matrix, A is also hermitian. The matrix B is not symmetric since the entry in row 1 column 2 is i and the entry in row 2 column 1 is $-i$, but B is hermitian since entries that are symmetric about the main diagonal are conjugates of each other. Notice that for a matrix to be hermitian, the entries on the main diagonal must equal their conjugates, that is they must be real. Because of this, the matrix C is not a hermitian matrix. C is also not symmetric.

$\qquad \square$

Theorem 5.2.5. *The eigenvalues of a hermitian (and so also a real symmetric) matrix A are all real.* *(Hermitian and real symmetric matrices have real eigenvalues.)*

Proof. We make heavy use of properties of the conjugate and transpose operations that were established in exercises:

$$\overline{AB} = \overline{A}\ \overline{B} \text{ (Exercise 5.1.20)}$$

---
[3]Hermitian matrices are named for the French mathematician Charles Hermite (1822-1901).

$$(AB)^t = B^t A^t \text{ (Exercise 4.1.16)}$$

Notice also that

$$\overline{\overline{A}} = A \text{ and } (A^t)^t = A$$

for any matrix A. If $AX = \lambda X$, then $\overline{X}^t AX = \lambda \overline{X}^t X$. Now $\overline{X}^t X = X \cdot X$ is real and $\overline{X}^t AX$ is real since

$$\overline{\left(\overline{X}^t AX\right)^t} = \overline{X}^t \overline{A}^t \overline{\overline{X}}^{tt}.$$

Hence λ must be real. □

Example 5.2.3. Let $A = \begin{bmatrix} -1 & 2 \\ 2 & 1 \end{bmatrix}$. Then A is symmetric and so we know that the eigenvalues are real and we can find real eigenvectors. Calculating the characteristic polynomial we get $p_A(\lambda) = |A - \lambda I| = \lambda^2 - 5$. The eigenvalues are $\lambda_1 = \sqrt{5}$ and $\lambda_2 = -\sqrt{5}$. To find an eigenvector corresponding to λ_1 we must find a nonzero solution of

$$(A - \lambda_1 I)X_1 = \begin{bmatrix} -1 - \sqrt{5} & 2 \\ 2 & 1 - \sqrt{5} \end{bmatrix} \begin{bmatrix} x \\ y \end{bmatrix} = 0.$$

Radicals are always unpleasant to deal with, so let us think carefully about the solution. Since λ_1 is an eigenvalue, $A - \lambda_1 I$ is a singular matrix (the determinant must be 0). If this matrix is singular, it must have rank 1. This means that the second row is a multiple of the first and so if the first row times $\begin{bmatrix} x \\ y \end{bmatrix}$ gives zero, so must the second. To find a solution, then, we forget about the second row and concentrate on the first. To make the solution vector, we use "the old switch the entries and change one sign trick" on the first row:

$$X_1 = \begin{bmatrix} x \\ y \end{bmatrix} = \begin{bmatrix} 2 \\ 1 + \sqrt{5} \end{bmatrix}.$$

A similar calculation will find the second eigenvector.

□

Section 5.2 Exercises

For each of the matrices in Exercises 1 - 8, find eigenvalues and corresponding eigenvectors.

1. $\begin{bmatrix} 2 & -1 \\ 0 & -2 \end{bmatrix}$

2. $\begin{bmatrix} -1 & 2 \\ 0 & -1 \end{bmatrix}$

3. $\begin{bmatrix} 4 & 1 \\ -10 & -3 \end{bmatrix}$

4. $\begin{bmatrix} -5 & 6 \\ -6 & 7 \end{bmatrix}$

5. $\begin{bmatrix} 1 & 2 & -1 \\ 0 & 1 & 3 \\ 0 & 0 & 2 \end{bmatrix}$

6. $\begin{bmatrix} -2 & 3 & 1 \\ 0 & -1 & 2 \\ 0 & 0 & 3 \end{bmatrix}$

7. $\begin{bmatrix} 3 & 2 & 4 \\ 1 & 2 & 2 \\ -1 & -1 & -1 \end{bmatrix}$

8. $\begin{bmatrix} 3 & -2 & -2 \\ 0 & -1 & 0 \\ 1 & -2 & 0 \end{bmatrix}$

9. Show that $\begin{bmatrix} 1 & 2 \\ 2 & 5 \end{bmatrix}$ and $\begin{bmatrix} 1 & 2 \\ -1 & 0 \end{bmatrix}$ are not similar.

10. Prove that A is singular if and only if 0 is an eigenvalue of A.

11. Let $A = \begin{bmatrix} 1 & 2 \\ 0 & 2 \end{bmatrix}$. Compute $p_A(A)$. (Note: If $p_A(\lambda) = a_n\lambda^n + \ldots + a_1\lambda + a_0$, we define $p_A(A) = a_nA^n + \ldots + a_1A + a_0I$.)

12. Let $X_1, \ldots, X_k$ be eigenvectors of A corresponding to the eigenvalue λ, and let $X = a_1X_1 + \ldots + a_kX_k$. Show that if $X \neq 0$ then X is an eigenvector of A corresponding to λ.

    In Exercises 13 - 17, find a 2×2 matrix A over the real numbers that satisfies the given condition. Note that there may be several matrices that satisfy the condition or none.

13. A has eigenvalues -1 and 3.

14. The characteristic polynomial of A is $(1 - \lambda)^2$.

15. A has only one eigenvalue, -2, and the vector $\begin{bmatrix} 1 \\ -1 \end{bmatrix}$ is an eigenvector for this eigenvalue.

16. 0 is the only eigenvalue of A and every eigenvector is a scalar multiple of $\begin{bmatrix} 2 \\ -1 \end{bmatrix}$.

17. The complex number $1 + i$ is an eigenvalue of A.

5.3   Similarity to a Diagonal Matrix

It is not difficult to characterize those matrices that are similar to diagonal matrices. The characterization involves eigenvalues and eigenvectors of the matrices. Recall from Section 5.2 that similar matrices have the same characteristic polynomial and so the same eigenvalues. We will need two results - the proofs are left as exercises.

Lemma 5.3.1. *If $X_1, \ldots, X_k$ are linearly independent $n \times 1$ column vectors and S is a nonsingular $n \times n$ matrix, then $SX_1, \ldots, SX_k$ are linearly independent.*

Proof. Exercise 17.   □

Lemma 5.3.2. *If A is similar to B, say $B = S^{-1}AS$, and X is an eigenvector of B corresponding to λ, then SX is an eigenvector of A corresponding to λ.*

Proof. Exercise 18.   □

Using these two lemmas we can characterize matrices that are similar to a diagonal matrix.

NECESSARY AND SUFFICIENT CONDITIONS

Theorem 5.3.3. *Let A be an $n \times n$ matrix. Then A is similar to a diagonal matrix D if and only if A has n linearly independent eigenvectors. In fact, if $X_1, \ldots, X_n$ are linearly independent eigenvectors of A corresponding to the eigenvalues $\lambda_1, \ldots, \lambda_n$ and $S = [X_1, \ldots, X_n]$, then $S^{-1}AS = D = diag(\lambda_1, \ldots, \lambda_n)$.*

(A matrix is similar to a diagonal matrix iff it has n independent eigenvectors.)

Proof. Assume A is similar to a diagonal matrix $D = \mathrm{diag}(\lambda_1, \ldots, \lambda_n)$. Then $S^{-1}AS = D$ for some nonsingular matrix S. The eigenvalues of D are $\lambda_1, \ldots, \lambda_n$ and the transposes of the standard basis vectors

$$E_1^t = \begin{bmatrix} 1 \\ 0 \\ \vdots \\ 0 \end{bmatrix}, E_2^t = \begin{bmatrix} 0 \\ 1 \\ \vdots \\ 0 \end{bmatrix}, \ldots, E_n^t = \begin{bmatrix} 0 \\ 0 \\ \vdots \\ 1 \end{bmatrix}$$

are the corresponding eigenvectors. We see that D has n linearly independent eigenvectors. By Lemma 5.3.1, $SE_1^t, \ldots, SE_n^t$ are n linearly independent vectors, and by Lemma 5.3.2, $SE_1^t, \ldots, SE_n^t$ are eigenvectors of A corresponding to $\lambda_1, \ldots, \lambda_n$.

Now let $X_1, \ldots, X_n$ be linearly independent eigenvectors of A corresponding to the eigenvalues $\lambda_1, \ldots, \lambda_n$. Let $S = [X_1, \ldots, X_n]$. Using block multiplication, we see that

$$\begin{aligned}
S^{-1}AS &= S^{-1}A[X_1, \ldots, X_n] \\
&= S^{-1}[AX_1, \ldots, AX_n] \\
&= S^{-1}[\lambda_1 X_1, \ldots, \lambda_n X_n] \\
&= [S^{-1}\lambda_1 X_1, \ldots, S^{-1}\lambda_n X_n] \\
&= [\lambda_1 S^{-1}X_1, \ldots, \lambda_n S^{-1}X_n] \\
&= [S^{-1}X_1, \ldots, S^{-1}X_n]\mathrm{diag}(\lambda_1, \ldots, \lambda_n) \\
&= S^{-1}[X_1, \ldots, X_n]\mathrm{diag}(\lambda_1, \ldots, \lambda_n) \\
&= S^{-1}S\mathrm{diag}(\lambda_1, \ldots, \lambda_n) \\
&= \mathrm{diag}(\lambda_1, \ldots, \lambda_n),
\end{aligned}$$

and so A is similar to a diagonal matrix. $\square$

The above theorem says that an $n \times n$ matrix A is similar to a diagonal matrix exactly when the matrix has n linearly independent eigenvectors, and from the proof of the theorem we see that it is the S matrix made up of the eigenvectors that does the transformation.

A previous theorem gives us a special case in which this theorem applies.

Corollary 5.3.4. *Let A be an $n \times n$ matrix and assume $p_A(\lambda) = |A - \lambda I|$ has n distinct roots $\lambda_1, \ldots, \lambda_n$. Then A is similar to a diagonal matrix.*

(A matrix with n distinct eigenvalues is similar to a diagonal matrix.)

Proof. Let $X_1, \ldots, X_n$ be eigenvectors corresponding to the distinct eigenvalues $\lambda_1, \ldots, \lambda_n$. Then by Theorem 5.2.4, $X_1, \ldots, X_n$ are linearly independent and so by the previous theorem, A is similar to a diagonal matrix. $\square$

Corollary 5.3.5. *Let A be an $n \times n$ matrix and assume $p_A(\lambda)$ factors completely. Let $\lambda_1, \ldots, \lambda_k$ be the distinct characteristic roots and assume the multiplicity of λ_i is m_i. Then if $m_i = n - rank(A - \lambda_i I)$ for $i = 1, \ldots, k$, then A is similar to a diagonal matrix.*

(A matrix in which the multiplicity of each eigenvalue is the number of independent eigenvectors is similar to a diagonal matrix.)

Proof. We have $m_1 + \ldots + m_k = n$ and each eigenvalue λ_i has m_i linearly independent eigenvectors, say,

$$X_1^{(i)}, X_2^{(i)}, \ldots, X_{m_i}^{(i)}$$

since the dimension of the solution space of $(A - \lambda_i I)X = 0$ is $n - rank(A - \lambda_i I) = m_i$ by Theorem 2.9.1. Now a linear combination of these eigenvectors is either 0 or another eigenvector associated with the same eigenvalue by Exercise 5, Section 5.2. Now assume that a linear combination of the eigenvectors is zero, say

$$\sum_{j=1}^{k} \sum_{i=1}^{m_i} a_{ij} X_i^{(j)} = 0.$$

Rewriting this linear combination, we get

$$\sum_{i=1}^{m_1} a_{i1} X_i^{(1)} + \ldots + \sum_{i=1}^{m_k} a_{ik} X_i^{(k)} = 0.$$

All of the coefficients are 1's and each nonzero term is an eigenvector for one of the distinct eigenvalues $\lambda_1, \ldots, \lambda_k$. Since these nonzero eigenvectors are linearly independent, all terms must equal zero. We see that

$$\sum_{i=1}^{m_i} a_{ij} X_i^{(j)} = 0$$

for $j = 1, \ldots, k$. But the vectors in the above sum are linearly independent and so we conclude that: $a_{1j} = \ldots = a_{m_j j} = 0$ for $j = 1, \ldots, k$. It follows that the eigenvectors are linearly independent, and so, applying Theorem 5.3.3, we see that A is similar to a diagonal matrix. $\square$

"Diagonalizing" a matrix gives a nice review of our earlier work: We calculate the determinant $|A - \lambda I|$ to find the characteristic polynomial and solve it to find the eigenvalues. We then find a basis for the solution space of the homogeneous systems of equations $(A - \lambda I)X = 0$ for each eigenvalue λ. Theorem 5.3.3 tells us whether we will succeed in diagonalizing. To illustrate this process, consider the following example.

Example 5.3.1. Let $A = \begin{bmatrix} 0 & 1 \\ 2 & -1 \end{bmatrix}$. Then $p_A(\lambda) = |A - \lambda I| = \begin{vmatrix} -\lambda & 1 \\ 2 & -1-\lambda \end{vmatrix}$, which becomes $-\lambda(-1-\lambda) - 2 = \lambda^2 + \lambda - 2 = (\lambda - 1)(\lambda + 2)$. Let $\lambda_1 = 1, \lambda_2 = -2$, and compute the corresponding eigenvectors:

$$(A - \lambda_1 I)X = \begin{bmatrix} -1 & 1 \\ 2 & -2 \end{bmatrix} \begin{bmatrix} x_1 \\ x_2 \end{bmatrix} = \begin{bmatrix} 0 \\ 0 \end{bmatrix} ; \text{ take } X_1 = \begin{bmatrix} 1 \\ 1 \end{bmatrix}$$

$$(A - \lambda_2 I)X = \begin{bmatrix} 2 & 1 \\ 2 & 1 \end{bmatrix} \begin{bmatrix} x_1 \\ x_2 \end{bmatrix} = \begin{bmatrix} 0 \\ 0 \end{bmatrix} ; \text{ take } X_2 = \begin{bmatrix} 1 \\ -2 \end{bmatrix}$$

Using Theorem 5.3.3, let $S = \begin{bmatrix} X_1 & X_2 \end{bmatrix} = \begin{bmatrix} 1 & 1 \\ 1 & -2 \end{bmatrix}$. Then $S^{-1} = \begin{bmatrix} 2/3 & 1/3 \\ 1/3 & -1/3 \end{bmatrix}$, and

$$S^{-1}AS = \begin{bmatrix} 1 & 0 \\ 0 & -2 \end{bmatrix}.$$

$\square$

Section 5.3 Exercises

In Exercises 1 - 6, determine whether the matrix is similar to a diagonal matrix.

1. $\begin{bmatrix} -1 & 2 \\ 0 & 2 \end{bmatrix}$

2. $\begin{bmatrix} 1 & 0 \\ 3 & 2 \end{bmatrix}$

3. $\begin{bmatrix} 1 & 3 & 2 \\ 0 & 1 & 1 \\ 0 & 0 & 2 \end{bmatrix}$

4. $\begin{bmatrix} -1 & 2 & 0 \\ 0 & 1 & 2 \\ 0 & 0 & 3 \end{bmatrix}$

5. $\begin{bmatrix} 1 & 0 & 1 \\ 0 & 1 & 0 \\ 0 & 0 & 2 \end{bmatrix}$

6. $\begin{bmatrix} 1 & 0 & 0 \\ 0 & 1 & 1 \\ 0 & 0 & 1 \end{bmatrix}$

In Exercises 7 - 12, find a nonsingular matrix S such that $S^{-1}AS$ is a diagonal matrix.

7. $\begin{bmatrix} -1 & 2 \\ 0 & 2 \end{bmatrix}$

8. $\begin{bmatrix} 1 & 2 \\ 2 & 1 \end{bmatrix}$

9. $\begin{bmatrix} -1 & 0 \\ 2 & 2 \end{bmatrix}$

10. $\begin{bmatrix} -1 & 2 & 1 \\ 0 & 3 & 2 \\ 0 & 0 & 1 \end{bmatrix}$

11. $\begin{bmatrix} 1 & 0 & 0 \\ 2 & 2 & 2 \\ 0 & 0 & 3 \end{bmatrix}$

12. $\begin{bmatrix} 1 & 0 & 1 \\ 0 & 1 & 0 \\ 1 & 0 & 1 \end{bmatrix}$

In Exercises 13 - 15, show that the matrix is not similar to a diagonal matrix.

13. $\begin{bmatrix} 1 & 1 \\ 0 & 1 \end{bmatrix}$

14. $\begin{bmatrix} 1 & 2 & 1 \\ 0 & 1 & 1 \\ 0 & 0 & 2 \end{bmatrix}$

15. $\begin{bmatrix} 1 & 0 & 0 \\ 1 & 1 & 1 \\ 0 & 0 & 1 \end{bmatrix}$

16. Let A be a 3×3 matrix with 1 an eigenvalue of multiplicity 2 and assume that

$$A - 1I = \begin{bmatrix} 1 & 1 & 1 \\ 1 & 1 & 1 \\ 1 & 1 & 1 \end{bmatrix}.$$

Is A similar to a diagonal matrix?

17. Prove Lemma 5.3.1.

18. Prove Lemma 5.3.2.

19. Recall the definition of similarity: An $n \times n$ matrix A is similar to the $n \times n$ matrix B if and only if there is a nonsingular $n \times n$ matrix S with $B = S^{-1}AS$. Prove the following properties of the similarity relation for $n \times n$ matrices $A, B,$ and C:

   (a) A is similar to A. (Similarity is **reflexive**.)

   (b) If A is similar to B, then B is similar to A. (Similarity is **reflexive**.)

   (c) If A is similar to B and B is similar to C, then A is similar to C. (Similarity is **transitive**.)

   (We say that similarity is an **equivalence relation** since it is reflexive, symmetric, and transitive.)

5.4   Page Rank (Optional)

In this section, we investigate one of the important uses of eigenvectors and eigenvalues: the Page Rank[4] algorithm used by Google to sort search results.

   Imagine an enormous database of all web pages that includes key words from those pages. When you run a search, the software extracts those web pages that contain the key words you searched for. But how does it decide the order in which to list them? They are arranged according to their **Page Rank**, which measures their relative importance by keeping track of the other pages that link to them. The idea is that the more pages there are linking to a given page, the more important it is and therefore the more likely it is to contain what you are seeking.

   To develop this system, we will need to establish some assumptions.

1. The PageRank of a web page depends on the number of pages linking to it.

2. The PageRank of a web page depends on the PageRanks of the pages linking to it.

3. The PageRank of a web page is evenly divided among the pages it links to.

4. The PageRank of a web page is calculated iteratively, starting with equal page ranks for all pages.

---

[4]Although one might reasonably think that the Page in "Page Rank" refers to webpages, it is actually named for Larry Page, one of the founders of Google and the inventor of the algorithm.

Consider the very small internet shown below; it has only four pages.

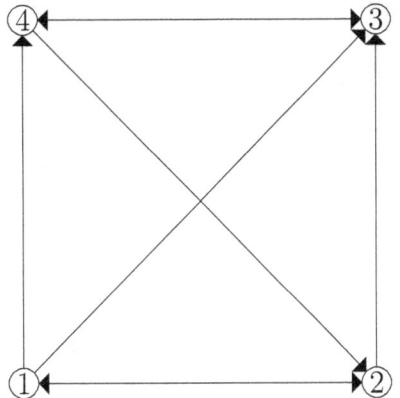

The arrows indicate links; for example, the arrow from 1 to 3 indicates that page 1 links to page 3 (but page 3 does not link back to page 1). As there are four pages, the initial ranking is $PR_0 = (0.25, 0.25, 0.25, 0.25)^T$, where the jth component of the vector corresponds to web page j and we use PR_k to record the page rank after k clicks. The values in PR_0 correspond to the probability that a random surfer would start at a particular page. To find PR_1, we apportion each page's rank equally among those pages it links to:

$$PR_1(1) = \frac{1}{2}PR_0(2)$$

$$PR_1(2) = \frac{1}{3}PR_0(1) \qquad\qquad \frac{1}{2}PR_0(4)$$

$$PR_1(3) = \frac{1}{3}PR_0(1) \quad \frac{1}{2}PR_0(2) \qquad \frac{1}{2}PR_0(4)$$

$$PR_1(4) = \frac{1}{3}PR_0(1) \qquad\qquad PR_0(3)$$

For example, since page 1 links to three other pages, each of those pages receives 1/3 of the Page Rank of page 1. At this first iteration, that is 1/3 of 0.25, but that 0.25 is recorded in the equations above as PR_0 to make it easier to generalize later.

Notice now that this system can more conveniently be represented as a matrix equation: if we let

$$A = \begin{bmatrix} 0 & 1/2 & 0 & 0 \\ 1/3 & 0 & 0 & 1/2 \\ 1/3 & 1/2 & 0 & 1/2 \\ 1/3 & 0 & 1 & 0 \end{bmatrix},$$

we find $PR_1 = APR_0$. Thus, the result of clicking a second time is $PR_2 = APR_1 = A^2PR_0$, and, in general, $PR_n = A^nPR_0$. In addition, we can see that A is stochastic! The column of A corresponding to page j indicates the fraction of j's PageRank that is apportioned to each other page. Since all of its PageRank is apportioned, the sum of each column is 1 (and all entries are non-negative). We have a number of useful tools.

Theorem 5.4.1. *If A is stochastic, so is A^n for any n.*

Proof. If $X = \begin{bmatrix} p_1 \\ \vdots \\ p_n \end{bmatrix}$ is a column vector whose entries sum to one (a stochastic vector) and

$A = \begin{bmatrix} A_1 & \cdots & A_n \end{bmatrix}$, where the entries in each A_j sum to 1, then $AX = p_1 A_1 + \ldots + p_n A_n$ (a column vector). Since the entries in each A_j sum to 1, the entries in AX sum to $p_1 + \ldots + p_n = 1$, too, so AX is a stochastic vector whenever A is a stochastic matrix and X is a stochastic vector. Therefore, $A^2 = AA = A \begin{bmatrix} A_1 & \cdots & A_n \end{bmatrix} = \begin{bmatrix} AA_1 & \cdots & AA_n \end{bmatrix}$ is stochastic since each column is a stochastic vector. Similarly, A^n is stochastic for each positive integer n. $\square$

Theorem 5.4.2. *If A is stochastic, then A has an eigenvalue of 1.*

Proof. A stochastic matrix A has columns that sum to 1, so A^T has rows that sum to 1.

Thus $A^T \begin{bmatrix} 1 \\ \vdots \\ 1 \end{bmatrix} = \begin{bmatrix} 1 \\ \vdots \\ 1 \end{bmatrix}$, and thus 1 is an eigenvalue of A^T. Since A and A^T have the same eigenvalues, A has an eigenvalue of 1 as well. $\square$

The following is also true, although we will not prove it here.

Theorem 5.4.3. *If A is stochastic, then $\lim_{n \to \infty} A^n$ exists.*

This is important because it means that our PageRanks come to a steady state after a large number of clicks: If we let $B = \lim_{n \to \infty} A^n$, we have $PR = \lim_{n \to \infty} A^n PR_0 = BPR_0$. Since $AB = B$, we have $A(BPR_0) = BPR_0$, so BPR_0 is the eigenvector of A corresponding to the eigenvalue of 1 that we know A has. Since it is also the steady-state distribution vector, it is actually the PageRank we're looking for! That is: in order to find the PageRank of our mini-internet, we need only find the eigenvector of A corresponding to the eigenvalue 1.

We proceed as usual using the matrix A started above:

$$(A - I)X = \begin{bmatrix} -1 & 1/2 & 0 & 0 \\ 1/3 & -1 & 0 & 1/2 \\ 1/3 & 1/2 & -1 & 1/2 \\ 1/3 & 0 & 1 & -1 \end{bmatrix} \begin{bmatrix} x_1 \\ x_2 \\ x_3 \\ x_4 \end{bmatrix}$$

$$= \begin{bmatrix} 0 \\ 0 \\ 0 \\ 0 \end{bmatrix}.$$

Solving by inspection, we see that $X = (1, 2, 3, 10/3)^t$ serves as our eigenvector. Scaling so that the entries sum to 1, we find $X = (0.1071, 0.2143, 0.3214, 0.3571)^t$. Thus, page 1 is considered the least important page, and page 4 is considered the most important.

Section 5.4 Exercises

1. For the stochastic matrix $A = \begin{bmatrix} 0.4 & 0.8 \\ 0.6 & 0.2 \end{bmatrix}$, find the stochastic eigenvector associated with the eigenvalue 1.

2. For the stochastic matrix $A = \begin{bmatrix} 0 & 0.7 & 0.2 \\ 0.1 & 0 & 0.8 \\ 0.9 & 0.3 & 0 \end{bmatrix}$, find the stochastic eigenvector associated with the eigenvalue 1.

3. Using the mini-internet shown, find the page ranks of all of the pages.

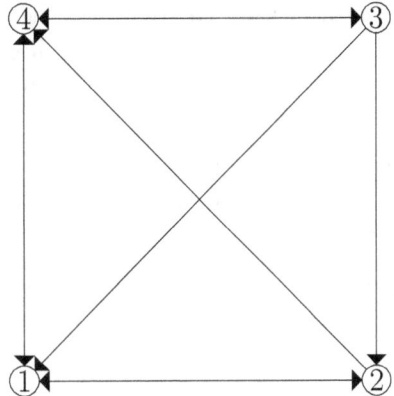

4. Using the mini-internet shown, find the page ranks of all of the pages.

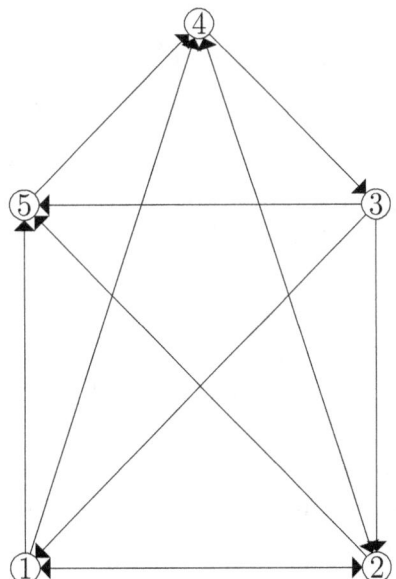

5.5   Symmetric and Hermitian Matrices

We are now in a position to begin the proof that every hermitian or real symmetric matrix is similar to a diagonal matrix. This is a deep and important result with applications in physics as shown in Section 5.8, as well as in other areas of mathematics, for example in the simplification of quadratic forms as explained in Section 5.7.

The proof that a hermitian matrix is similar to a diagonal matrix is somewhat difficult and indirect. We begin by proving that any matrix is similar, in a special way, to an upper triangular matrix. Upper triangular matrices were mentioned earlier in Section 1.7. Recall that an $n \times n$ matrix $A = [a_{ij}]$ is called **upper triangular** if and only if $a_{ij} = 0$ for $i > j$. Note that in a square matrix of order n, $a_{11}, \ldots, a_{nn}$ are the entries on the diagonal, $\{a_{ij} | i < j\}$ is the set of entries above the diagonal and $\{a_{ij} | i > j\}$ is the set of entries below the diagonal. Thus A is upper triangular if and only if all entries below the diagonal are zero. Likewise a matrix is **lower triangular** if and only if the entries above the diagonal are zero.

The special form of similarity mentioned above is related to orthonormal sets of vectors. Recall that we began the discussion in this chapter with the thought that we would like to better understand the action of a linear transformation by finding a basis relative to which the transformation was represented by a simple matrix. The basis that we will construct will, in fact, be orthonormal.

ORTHOGONAL AND UNITARY MATRICES

The special type of similarity mentioned above involves a nonsingular matrix whose inverse has special properties: A nonsingular matrix P is **orthogonal** if and only if $P^{-1} = P^t$ and P is **unitary** if and only if $P^{-1} = \overline{P}^t$. So a nonsingular matrix is orthogonal provided its inverse is its transpose and a nonsingular matrix is unitary in the event that the inverse of the matrix is the conjugate of its transpose. In general, orthogonal matrices occur in situations involving the real numbers and unitary matrices occur in the complex case.

The above terminology comes from the following: Assume $\{X_1, \ldots, X_n\}$ is an orthonormal set of real n-dimensional column vectors. Then for $i \neq j$, $X_i \cdot X_j = X_i^t X_j = 0$ and $X_i \cdot X_i = X_i^t X_i = 1$. Let P be the matrix formed using these column vectors as the columns of P; that is, $P = [X_1, \ldots, X_n]$. Then

$$P^t P = \begin{bmatrix} X_1^t \\ \vdots \\ X_n^t \end{bmatrix} [X_1 \ldots X_n]$$

$$= \begin{bmatrix} X_1^t X_1 & X_1^t X_2 & \ldots & X_1^t X_n \\ \vdots & \vdots & \ddots & \vdots \\ X_n^t X_1 & X_n^t X_2 & \ldots & X_n^t X_n \end{bmatrix}$$

$$= \begin{bmatrix} X_1 \cdot X_1 & X_1 \cdot X_2 & \ldots & X_1 \cdot X_n \\ \vdots & \vdots & \ddots & \vdots \\ X_n \cdot X_1 & X_n \cdot X_2 & \ldots & X_n \cdot X_n \end{bmatrix}$$

$$= \begin{bmatrix} 1 & 0 & \ldots & 0 \\ \vdots & \vdots & \ddots & \vdots \\ 0 & 0 & \ldots & 1 \end{bmatrix}$$

$$= I_n.$$

Thus $P^t = P^{-1}$. Likewise, if $\{X_1, \ldots, X_n\}$ is an orthogonal set of complex vectors and $U = [X_1 \ldots X_n]$ then U is a unitary matrix. It follows that since the vectors

$$\begin{bmatrix} 1/\sqrt{2} \\ 1\sqrt{2} \end{bmatrix} \quad \text{and} \quad \begin{bmatrix} -1/\sqrt{2} \\ 1/\sqrt{2} \end{bmatrix}$$

are orthogonal unit vectors, the matrix

$$P = \begin{bmatrix} 1/\sqrt{2} & -1/\sqrt{2} \\ 1/\sqrt{2} & 1/\sqrt{2} \end{bmatrix}$$

is an orthogonal matrix and

$$P^{-1} = P^t = \begin{bmatrix} 1/\sqrt{2} & 1/\sqrt{2} \\ -1/\sqrt{2} & 1/\sqrt{2} \end{bmatrix}.$$

Theorem 5.5.1. *(a) A product of orthogonal matrices is an orthogonal matrix.*

(b) A product of unitary matrices is a unitary matrix.

Proof. Exercise 20. □

Let A and B be $n \times n$ matrices. We say that A is **orthogonally similar** to B if and only if there is an orthogonal matrix P with $B = P^t A P$. Likewise A is **unitarily similar** to B provided there is a unitary matrix U with $B = \overline{U}^t A U$. We will see that any real matrix whose characteristic polynomial factors completely is orthogonally similar to an upper triangular matrix.

THE UPPER TRIANGULARIZATION PROCESS

As with ordinary similarity, if A is orthogonally similar to B, then B is orthogonally similar to A and likewise for unitary similarity. In addition, if A is orthogonally similar (or unitarily similar) to B, then A is also similar to B.

If a matrix B is upper triangular, then the determinant of B is the product of the entries on the diagonal:

$$\begin{vmatrix} b_{11} & b_{12} & \ldots & b_{1n} \\ 0 & b_{22} & \ldots & b_{2n} \\ \vdots & \vdots & \ddots & \vdots \\ 0 & 0 & \ldots & b_{nn} \end{vmatrix} = b_{11}b_{22}\ldots b_{nn}.$$

It follows that $p_B(\lambda) = |B - \lambda I| = (b_{11} - \lambda)(b_{22} - \lambda)\ldots(b_{nn} - \lambda)$ and so the eigenvalues of B lie on the diagonal. Also, we see that the characteristic polynomial of B factors completely. Since similar matrices have the same characteristic polynomial, we see that a necessary condition for a matrix to be similar to an upper triangular matrix is that the characteristic polynomial of the matrix factors completely. Of course, over the field of complex numbers every polynomial of positive degree factors completely.

The following theorem tells us about the upper triangularization process. The proof of the theorem is "constructive" in that it tells us how to construct both the upper triangular matrix and the matrix that performs the similarity transformation.

Theorem 5.5.2 (Schur's Theorem). *Let A be an $n \times n$ matrix.*

(a) *If A is a real matrix and the characteristic polynomial of A factors completely over $\mathbb{R}$, then A is orthogonally similar to a real upper triangular matrix.*

   (A real matrix is orthogonally similar to an upper triangular matrix if the characteristic polynomial factors.)

(b) *If A is a complex matrix, then A is unitarily similar to an upper triangular matrix.*   *(A complex matrix is unitarily similar to an upper triangular matrix.)*

Proof. The same proof works in each case. In (b), note that the characteristic polynomial factors completely.

The proof is by induction on the order n of the matrix A. The theorem is seen to be true for $n = 1$. Assume that the theorem is true for matrices of order $(n - 1) \times (n - 1)$.

Let A be $n \times n$ and let $\lambda_1, \ldots, \lambda_n$ be the eigenvalues of A. Let X_1 be an eigenvector corresponding to λ_1 and choose $X_2, \ldots, X_n$ such that $X_1, \ldots, X_n$ is a linearly independent set of n-dimensional column vectors. Apply the Gram-Schmidt Process to $X_1, \ldots, X_n$ and obtain an orthonormal set $Y_1, \ldots, Y_n$ of vectors. Since $Y_1 = (1/\|X_1\|)X_1$, Y_1 is also an eigenvector of A corresponding to λ_1. Let $P_1 = [Y_1, \ldots, Y_n]$. Then in case (a) P_1 is orthogonal and in case (b), P_1 is unitary by previous remarks. Also P_1 is nonsingular. Now

$$P_1^{-1}AP_1 = P_1^{-1}[AY_1, \ldots, AY_n]$$
$$= P_1^{-1}[\lambda_1 Y_1, AY_2, \ldots, AY_n]$$
$$= [\lambda_1 P_1^{-1}Y_1, P_1^{-1}AY_2, \ldots, P_1^{-1}AY_n]$$
$$= \left[\begin{array}{c|c} \lambda_1 & B_1 \\ \hline 0 & A_1 \end{array}\right],$$

where B_1 and A_1 are some matrices. The entries in the first column are $\lambda_1, 0, \ldots, 0$ since $P_1^{-1}Y_1$ is the first column of $P_1^{-1}P_1 = I_n$.

Since similar matrices have the same characteristic polynomials, we get

$$p_A(\lambda) = (\lambda_1 - \lambda)p_{A_1}(\lambda)$$

by expanding down the first column. Since $p_A(\lambda)$ factors completely, it follows that the characteristic polynomial of A_1 factors completely. Hence A_1 satisfies the hypothesis of the theorem and so by the induction assumption, there exists an orthogonal or unitary matrix P_2 such that $P_2^{-1}A_1 P_2$ is upper triangular. We "imbed" P_2 in the lower right-hand corner of the identity and call the matrix P_3:

$$P_3 = \left[\begin{array}{c|c} 1 & 0 \\ \hline 0 & P_2 \end{array}\right]$$

This will be the matrix that finally transforms $P_1^{-1}AP_1$ into upper triangular form. Notice that P_3 is orthogonal. Let $P = P_1 P_3$. Then we have

$$P^{-1}AP = P_3^{-1}P_1^{-1}AP_1 P_3$$
$$= P_3^{-1}\left[\begin{array}{c|c} \lambda_1 & B_1 \\ \hline 0 & A_1 \end{array}\right]P_3$$
$$= \left[\begin{array}{c|c} 1 & 0 \\ \hline 0 & P_2^{-1} \end{array}\right]\left[\begin{array}{c|c} \lambda_1 & B_1 \\ \hline 0 & A_1 \end{array}\right]\left[\begin{array}{c|c} 1 & 0 \\ \hline 0 & P_2 \end{array}\right]$$
$$= \left[\begin{array}{c|c} \lambda_1 & B_2 \\ \hline 0 & P_2^{-1}A_1 P_2 \end{array}\right].$$

Since $P_2^{-1}A_1 P_2$ is upper triangular, it follows that $P^{-1}AP$ is upper triangular. Since P_2 is orthogonal (or unitary in case ii)), it follows that P_3 is orthogonal (or unitary, respectively) and so by induction the theorem is proved. $\square$

Applying the above theorem, it is now easy to prove that a real symmetric matrix is similar to a diagonal matrix - in fact, orthogonally similar!

DIAGONALIZATION OF SYMMETRIC AND HERMITIAN MATRICES

A diagonal matrix D is clearly a symmetric matrix. If a matrix A is orthogonally similar to a diagonal matrix, say $P^t AP = D$, then it must also be symmetric, for $P^t AP = D =$

$D^t = (P^t A P)^t = P^t A^t P^{tt} = P^t A^t P$, and since P is nonsingular, it can be canceled to leave $A = A^t$. As a corollary to the above theorem, we can see that the converse is also true. This important result is sometimes called the Principal Axes Theorem.

Corollary 5.5.3 (Principle Axes Theorem). *Let A be a real $n \times n$ matrix. Then A is orthogonally similar to a diagonal matrix if and only if A is symmetric.*
(A real symmetric matrix is orthogonally similar to a diagonal matrix.)

Proof. Assume that A is symmetric. By Theorem 5.2.5, the eigenvalues of a real symmetric matrix are all real. It follows that the characteristic polynomial of A factors completely and so Schur's Theorem (Theorem 5.5.2) may be applied. Now A is symmetric, so $A = A^t$. By the theorem, there is an orthogonal matrix P (Recall that $P^{-1} = P^t$) with $P^t A P$ upper triangular. Now $(P^t A P)^t = P^t A^t P^{tt} = P^t A P$. Thus $P^t A P$ is symmetric and since a symmetric upper triangular matrix is diagonal, the corollary follows.

To see that the converse is true, assume that A is orthogonally similar to a diagonal matrix, say $D = P^t A P$ is diagonal. Then $D = P^t A P = D^t = (P^t A P)^t = P^t A^t P^{tt} = P^t A^t P$. Solving we see that $A = A^t$. $\qquad\square$

A similar result can be proved for hermitian matrices.

Corollary 5.5.4. *If A is an $n \times n$ complex matrix, then A is hermitian if and only if A is unitarily similar to a real diagonal matrix.*   *(A hermitian matrix is unitarily similar to a real diagonal matrix.)*

Since an $n \times n$ symmetric matrix A is similar to a diagonal matrix, the matrix must have n linearly independent eigenvectors. We saw in Section 5.3 that the column vectors of the matrix S that diagonalizes a given matrix are eigenvectors of that matrix. Since the symmetric matrix A is orthogonally similar to a diagonal matrix, we see that A must have n eigenvectors that form an orthonormal set. Similar statements are true for hermitian matrices. The following theorem states that eigenvectors corresponding to distinct eigenvalues are automatically orthogonal.

> **Theorem 5.5.5.** *Let A be a hermitian (or real symmetric) matrix and let λ_1 and λ_2 be distinct eigenvalues of A with X_1 and X_2 the corresponding eigenvectors. Then X_1 and X_2 are orthogonal.*
>
> *(For a hermitian matrix, eigenvectors corresponding to distinct eigenvalues are orthogonal.)*

Proof. Assume A is real symmetric. We have $A X_1 = \lambda_1 X_1$ and $A X_2 = \lambda_2 X_2$. Then, taking the transpose of each side we see that $(A X_1)^t = X_1^t A^t = \lambda_1 X_1^t$ and so, multiplying on the right by X_2, $X_1^t A^t X_2 = \lambda_1 X_1^t X_2$. Since $A^t = A$, $X_1^t A X_2 = \lambda_1 X_1^t X_2$. Likewise, since $A X_2 = \lambda_2 X_2$, we obtain $X_1^t A X_2 = \lambda_2 X_1^t X_2$ by multiplying on the left by X_1^t. Subtracting, we get $0 = \lambda_1 X_1^t X_2 - \lambda_2 X_1^t X_2 = (\lambda_1 - \lambda_2) X_1^t X_2 = (\lambda_1 - \lambda_2) X_1 \cdot X_2$. Since $\lambda_1 \neq \lambda_2$, we have $X_1 \cdot X_2 = 0$. $\qquad\square$

From Theorem 5.5.5, it follows that to "orthogonalize" a set of linearly independent eigenvectors of a hermitian matrix, we need only to "orthogonalize" each of the sets of eigenvectors corresponding to each eigenvalue.

Example 5.5.1. (a) Let $A = \begin{bmatrix} 7 & -2 & 1 \\ -2 & 10 & -2 \\ 1 & -2 & 7 \end{bmatrix}$. Notice that A is a real symmetric matrix and therefore it is orthogonally similar to a real diagonal matrix. The characteristic polynomial is $p_A(\lambda) = -\lambda^3 - 24\lambda^2 + 180\lambda - 432$ and the eigenvalues are $\lambda_1 = 6, \lambda_2 = 6, \lambda_3 = 12$. Corresponding eigenvectors are

$$X_1 = \begin{bmatrix} 1 \\ 0 \\ -1 \end{bmatrix}, X_2 = \begin{bmatrix} 1 \\ 1 \\ 1 \end{bmatrix}, X_3 = \begin{bmatrix} 1 \\ -2 \\ 1 \end{bmatrix}.$$

By Theorem 5.5.5, the vectors X_1 and X_2 will be orthogonal to X_3. We chose X_1 and X_2 so that they were also orthogonal, and so it follows that all three vectors X_1, X_2, and X_3 are orthogonal. It suffices, then, to normalize the vectors. We get

$$Y_1 = \begin{bmatrix} 1/\sqrt{2} \\ 0 \\ -1/\sqrt{2} \end{bmatrix}, Y_2 = \begin{bmatrix} 1/\sqrt{3} \\ 1/\sqrt{3} \\ 1/\sqrt{3} \end{bmatrix}, Y_3 = \begin{bmatrix} 1/\sqrt{6} \\ -2/\sqrt{6} \\ 1/\sqrt{6} \end{bmatrix}.$$

Then $\{Y_1, Y_2, Y_3\}$ is an orthonormal set of vectors, and if $P = [Y_1 Y_2 Y_3]$, we see that $P^{-1}AP = \text{diag}(6, 6, 12)$.

(b) Let us illustrate the method of "upper triangularization" given in the proof of Theorem 5.5.2. The method proceeds step-by-step one column at a time. Let us consider an example in which the first column has been reduced. Consider the matrix

$$A = \begin{bmatrix} 1 & 2 & 1 \\ 0 & 1 & 0 \\ 0 & 3 & 1 \end{bmatrix}.$$

By inspection we can see that $p_A(\lambda) = |A - \lambda I| = (1 - \lambda)^3$. As in the proof of the theorem, we let

$$A_1 = \begin{bmatrix} 1 & 0 \\ 3 & 1 \end{bmatrix}$$

and we try to find an orthogonal matrix P_2 such that $P_2^{-1}A_1P_2$. is upper triangular. It is easy to see that $\begin{bmatrix} 0 \\ 1 \end{bmatrix}$ is an eigenvector. We add on the orthogonal vector $\begin{bmatrix} 1 \\ 0 \end{bmatrix}$ and let $P_2 = \begin{bmatrix} 0 & 1 \\ 1 & 0 \end{bmatrix}$. As in the proof, we imbed P_2 in the identity matrtix and obtain

$$P_3 = \begin{bmatrix} 1 & 0 & 0 \\ 0 & 0 & 1 \\ 0 & 1 & 0 \end{bmatrix}.$$

This is the orthogonal matrix that upper triangularizes A; in fact,

$$P_3^{-1}AP_3 = \begin{bmatrix} 1 & 1 & 2 \\ 0 & 1 & 3 \\ 0 & 0 & 1 \end{bmatrix}.$$

Note that in the above product, $P_3^{-1} = P_3 = R_{23} = C_{23}$ and so the product can be computed using a row operation and a column operation.

(c) Consider the matrix

$$A = \begin{bmatrix} 1 & 1 & 0 \\ 2 & -1 & 2 \\ 1 & -2 & 2 \end{bmatrix}.$$

We want to orthogonally upper triangularize A. Computing the characteristic polynomial, we can see that $p_A(\lambda) = |A - \lambda I| = -\lambda(1 - \lambda)^2$. Now

$$A - 1I = \begin{bmatrix} 0 & 1 & 0 \\ 2 & -2 & 2 \\ 1 & -2 & 1 \end{bmatrix}$$

We choose an eigenvector X_1 and two vectors X_2, X_3 orthogonal to X_1 :

$$X_1 = \begin{bmatrix} 1 \\ 0 \\ -1 \end{bmatrix}, X_2 = \begin{bmatrix} 0 \\ 1 \\ 0 \end{bmatrix}, X_3 = \begin{bmatrix} 1 \\ 0 \\ 1 \end{bmatrix}.$$

Normalizing and forming the matrix P_1 we get

$$P_1 = \frac{1}{\sqrt{2}} \begin{bmatrix} 1 & 0 & 1 \\ 0 & \sqrt{2} & 0 \\ -1 & 0 & 1 \end{bmatrix}.$$

Computing $P_1^{-1}AP_1$ we obtain

$$P_1^{-1}AP_1 = \begin{bmatrix} 1 & \dfrac{3\sqrt{2}}{2} & -1 \\ 0 & -1 & 2\sqrt{2} \\ 0 & -\dfrac{\sqrt{2}}{2} & 2 \end{bmatrix}.$$

The first column has been upper triangularized, and we consider the 2×2 matrix in the lower right hand corner:

$$A_1 = \begin{bmatrix} -1 & 2\sqrt{2} \\ -\dfrac{\sqrt{2}}{2} & 2 \end{bmatrix}.$$

We know that $\lambda = 0$ is an eigenvalue and that the matrix has rank one. We take an eigenvector and add another vector that is orthogonal:

$$\begin{bmatrix} 2\sqrt{2} \\ 1 \end{bmatrix}, \begin{bmatrix} -1 \\ 2\sqrt{2} \end{bmatrix}.$$

Notice that in forming the eigenvector the entries in row one of the matrix were used with the order changed and one sign changed. The same trick gives the second vector. Normalizing, forming the 2×2 matrix, and embedding in the identity matrix we get P_3:

$$P_3 = \begin{bmatrix} 1 & 0 & 0 \\ 0 & \dfrac{2\sqrt{2}}{3} & -\dfrac{1}{3} \\ 0 & \dfrac{1}{3} & \dfrac{2\sqrt{2}}{3} \end{bmatrix}.$$

Finally, we compute $P_3^{-1}P_1^{-1}AP_1P_3$:

$$P_3^{-1}P_1^{-1}AP_1P_3 = \begin{bmatrix} 1 & \dfrac{-13}{3} & -\dfrac{\sqrt{2}}{6} \\ 0 & 0 & \dfrac{35\sqrt{2}}{18} \\ 0 & 0 & 1 \end{bmatrix}.$$

Notice that the eigenvalues appear on the diagonal in the order chosen. In calculating P_1 we chose the eigenvalue 1 and in P_3 we used 0. The remaining eigenvalue is 1, and so on the diagonal we have $1, 0, 1$. We could have made different choices and gotten a different listing of eigenvalues on the diagonal. In general, the upper triangular matrix obtained is not unique.

$\square$

Section 5.5 Exercises

For each of the matrices A in Exercises 1 - 8, find an orthogonal matrix P such that P^tAP is upper triangular. Note that neither P nor P^tAP is unique so that there may be more than one correct answer.

1. $\begin{bmatrix} 1 & 0 \\ 2 & 1 \end{bmatrix}$

2. $\begin{bmatrix} 1 & 1 \\ 1 & 1 \end{bmatrix}$

3. $\begin{bmatrix} 0 & 0 \\ 1 & 1 \end{bmatrix}$

4. $\begin{bmatrix} 1 & 1 \\ 2 & 0 \end{bmatrix}$

5. $\begin{bmatrix} 1 & 0 & 0 \\ 0 & 2 & 0 \\ 0 & 1 & -1 \end{bmatrix}$

6. $\begin{bmatrix} 1 & 0 & 0 \\ 0 & 1 & 0 \\ 1 & 0 & 1 \end{bmatrix}$

7. $\begin{bmatrix} 1 & 2 & 0 \\ 0 & 1 & 0 \\ 1 & 1 & 1 \end{bmatrix}$

8. $\begin{bmatrix} 1 & 3 & 0 \\ 0 & 1 & 0 \\ 1 & 1 & 2 \end{bmatrix}$

For each of the symmetric matrices A in Exercises 9 - 12, find an orthogonal matrix P such that $P^t A P$ is diagonal.

9. $\begin{bmatrix} 3 & -1 \\ -1 & 3 \end{bmatrix}$ 10. $\begin{bmatrix} 0 & -1 \\ -1 & 0 \end{bmatrix}$ 11. $\begin{bmatrix} 1 & 0 & 0 \\ 0 & 1 & 1 \\ 0 & 1 & 1 \end{bmatrix}$ 12. $\begin{bmatrix} 0 & 0 & 1 \\ 0 & 1 & 0 \\ 1 & 0 & 0 \end{bmatrix}$

For each of the matrices in Exercises 13 - 16, determine by inspection whether the matrix is similar to a diagonal matrix, orthogonally similar to a diagonal matrix, orthogonally similar to an upper triangular matrix, or unitarily similar to an upper triangular matrix.

13. $\begin{bmatrix} 1 & 0 \\ 2 & 2 \end{bmatrix}$ 14. $\begin{bmatrix} 1 & 0 & 0 \\ 3 & 2 & 0 \\ 1 & -2 & 3 \end{bmatrix}$ 15. $\begin{bmatrix} 1 & 0 & 1 \\ 0 & 1 & 0 \\ 1 & 0 & 1 \end{bmatrix}$ 16. $\begin{bmatrix} 1 & 2 & -1 \\ 2 & 3 & 2 \\ -1 & 2 & 1 \end{bmatrix}$

17. Find a unitary matrix U such that $U^{-1}AU$ is an upper triangular complex matrix, where

$$A = \begin{bmatrix} 1 & -1 & -1 \\ 1 & -1 & 0 \\ 1 & 0 & -1 \end{bmatrix}.$$

18. Give an example of a hermitian matrix that is not symmetric.

19. Give an example of a symmetric matrix that is not hermitian.

20. Prove Theorem 5.5.1.

21. Find a unitary matrix U such that $U^{-1} \begin{bmatrix} 1 & i \\ -i & 1 \end{bmatrix} U$ is diagonal.

5.6   The Cayley-Hamilton Theorem

Let A be an $n \times n$ matrix over a scalar field F and $p(x) = a_0 + a_1 x + \ldots + a_m x^m$ be a polynomial over F. For a scalar $r, p(r)$ is defined by

$$p(r) = a_0 + a_1 r + \ldots + a_n r^n.$$

Notice that the combination on the right is defined. If one attempts to replace the scalar r by a matrix A, notice that

$$a_0 + a_1 A + \ldots + a_n A^n$$

is not defined since a_0 is a scalar, but $a_1 A$ is a matrix. This problem is easily avoided: We define $p(A)$ by $p(A) = a_0 I + a_1 A + \ldots + a_m A^m$, where I is the $n \times n$ identity matrix and $A^k = A \cdot A \ldots A$ (k factors).

MATRICES AND POLYNOMIALS

We raise the following question: given a matrix A, is there a nontrivial polynomial p with $p(A) = 0$. If this is the case, we say that A "satisfies the polynomial equation $p(x) = 0$," or less formally, that A "satisfies the polynomial p." We will see that the investigation of this question will prove to be fruitful. The existence of such a polynomial is easily established: The vector space of all $n \times n$ matrices has dimension n^2 and so any $n^2 + 1$ matrices are linearly dependent. Thus $I, A, A^2, \ldots, A^{n^2}$ are linearly dependent and so $a_0 I + a_1 A + \ldots + a_{n^2} A^{n^2} = 0$ (the $n \times n$ zero matrix) for some scalars $a_0, \ldots, a^{n^2}$. It follows that A satisfies the polynomial $p(x) = a_0 + a_1 x + \ldots + a_n x^{n^2}$; that is, $p(A) = 0$. So we see that A satisfies a polynomial of degree n^2 or less. The following theorem shows that "less" is always true.

> **Theorem 5.6.1** (Cayley-Hamilton). *If A is a square matrix and $p_A(\lambda) = |A - \lambda I|$ its characteristic polynomial, then $p_A(A) = 0$.*   *(A matrix satisfies its characteristic polynomial.)*

Proof. Let $p_A(\lambda) = |A - \lambda I| = a_0 + a_1 \lambda + \ldots + a_n \lambda^n$ be the characteristic polynomial of A. Consider $\mathrm{Adj}(A - \lambda I)$. Since the adjoint is computed by cofactors, each entry of $\mathrm{Adj}(A - \lambda I)$ is a polynomial in λ of degree at most $n - 1$. Because of this, we may write

$$\mathrm{Adj}(A - \lambda I) = A_0 + A_1 \lambda + \ldots + A_{n-1} \lambda^{n-1},$$

where $A_0, \ldots, A_{n-1}$ are $n \times n$ matrices. Now we know that

$$(A - \lambda I)(\mathrm{Adj}(A - \lambda I)) = |A - \lambda I| I = p_A(\lambda) I.$$

Substituting the expressions for $\mathrm{Adj}(A - \lambda I)$ and $p_A(\lambda)$, we get

$$
\begin{aligned}
(A - \lambda I)\mathrm{Adj}(A - \lambda I) &= AA_0 + AA_1 \lambda + \ldots + AA_{n-1}\lambda^{n-1} - A_0\lambda - A_1\lambda^2 - \ldots - A_{n-1}\lambda^n \\
&= p_A(\lambda) I \\
&= a_0 I + a_1 I \lambda + \ldots + a_n \lambda^n.
\end{aligned}
$$

Equating coefficients of like powers of λ, we get

$$
\begin{aligned}
a_n I &= -A_{n-1} \\
a_{n-1} I &= AA_{n-1} - A_{n-2} \\
&\vdots \\
a_1 I &= AA_1 - A_0 \\
a_0 I &= AA_0.
\end{aligned}
$$

Multiply the first equation by A^n, the second by A^{n-1}, etc., and add. We get

$$a_n I A^n + \ldots + a_1 I A + a_0 I = a_n A^n + \ldots + a_1 A + a_0 I$$
$$= p_A(A)$$
$$= A^n(-A_{n-1}) + A^{n-1}(AA_{n-1} - A_{n-2})$$
$$+ \ldots + A(AA_1 - A_0) + IAA_0$$
$$= 0$$

since all terms cancel out.   □

The Cayley-Hamilton Theorem is named for two of the founders of matrix theory and linear algebra: the English mathematician Arthur Cayley (1821 - 1895) and the Irish mathematician Sir William Rowan Hamilton (1805 - 1865). Using the Cayley-Hamilton Theorem, we can write powers of an $n \times n$ matrix A in terms of the first $n - 1$ powers of A.

Example 5.6.1. Let

$$A = \begin{bmatrix} 1 & -1 \\ 2 & 3 \end{bmatrix}$$

Then $p_A(x) = (1-x)(3-x)+2 = x^2 - 4x + 5$. (Since we are thinking more about polynomials and less about eigenvalues at this point, we will let x represent the indeterminate in the polynomial under discussion rather than λ. Don't be troubled by the change.) By the Cayley-Hamilton Theorem, $p_A(A) = A^2 - 4A + 5I = 0$, so $A^2 = 4A - 5I$. Now we may write

$$A^3 = AA^2$$
$$= 4A^2 - 5A$$
$$= 4(4A - 5I) - 5A$$
$$= 16A - 5A - 20I$$
$$= 11A - 20I$$

$$A^4 = AA^3$$
$$= A(11A - 20I)$$
$$= 11A^2 - 20A$$
$$= 11(4A - 5I) - 20A$$
$$= 24A - 55I,$$

etc. In addition, the Cayley-Hamilton Theorem can be used to find the inverse of a nonsingular matrix. For example with A as above, $p_A(x) = x^2 - 4x + 5$, so that $A^2 - 4A + 5I = 0$. From this we see that $A^2 - 4A = -5I$, or $(A - 4I)A = -5I$, or $(-1/5)(A - 4I)A = I$.

It follows that $A^{-1} = -\dfrac{1}{5}(A - 4I) = -\dfrac{1}{5}\begin{bmatrix} -3 & -1 \\ 2 & -1 \end{bmatrix}$.

□

Certainly the characteristic polynomial tells us something about a matrix - the roots are the eigenvalues of the matrix. The Cayley-Hamilton tells us that the matrix satisfies this polynomial and it is not hard to see that if a matrix satisfies a given polynomial, then it satisfies any polynomial which contains the given polynomial as a factor. Perhaps there is a "least" polynomial that the matrix satisfies.

THE MINIMUM POLYNOMIAL

If $p(x) = a_0 + a_1 x + \ldots + a_n x^n$ is a polynomial of degree n, then a_n is called the **leading coefficient** of $p(x)$. If $a_n = 1$, $p(x)$ is called a **monic polynomial**. For example, the leading coefficient in $2x^2 - 3x + 7$ is 2 and $7 + 2x + x^2$ is monic, but $7 + 2x - x^2$ is not monic. Also note that if a is the leading coefficient of $p(x)$ (which implies that $a \neq 0$), then $(1/a)p(x)$ is monic. If A is an $n \times n$ matrix, then the leading coefficient of $p_A(x)$ is $(-1)^n$, so $p_A(x)$ is monic for n even and $-p_A(x)$ is monic for n odd. Now $\pm p_A(A) = 0$ and so A satisfies a monic polynomial. We make the following definition.

Definition 5.6.1. Let A be an $n \times n$ matrix. The **minimum polynomial** of A is the monic polynomial $m_A(x)$ of least degree such that $m_A(A) = 0$.

By the preceding remarks such a polynomial must exist, for A satisfies one monic polynomial and so A satisfies a monic polynomial of least degree.

Theorem 5.6.2. *If A is a square matrix and $p(x)$ a polynomial with $p(A) = 0$, then $m_A(x)$ is a factor of $p(x)$; that is, $p(x) = m_A(x)q(x)$ for some polynomial $q(x)$. In particular, $m_A(x)$ is a factor of $p_A(x)$.*

(The minimum polynomial is a factor of the characteristic polynomial.)

Proof. Use the division algorithm to divide $m_A(x)$ into $p(x)$ and obtain $p(x) = m_A(x)q(x) + r(x)$, where the degree of $r(x)$ is less than the degree of $m_A(x)$ or $r(x) = 0$. Then $p(A) = m_A(A)q(A) + r(A) = 0 + r(A)$, and so $r(A) = 0$. Now if $r(x) \neq 0$, then let $a \neq 0$ be the leading coefficient of r. Dividing by a we get $(1/a)r(A) = 0$ and $(1/a)r(x)$ is a monic polynomial of degree less than that of $m_A(x)$. This is a contradiction, so $r(x)$ must be the zero polynomial. It follows that $m_A(x)$ is a factor of $p(x)$. □

This theorem gives a method for finding the minimum polynomial of a matrix: Calculate the characteristic polynomial $p_A(x)$ and factor it into monic factors

$$p_A(x) = (-1)^n p_1(x) \ldots p_k(x).$$

Now just try all possibilities.

Example 5.6.2. Let

$$A = \begin{bmatrix} 1 & 0 & 2 \\ 0 & 1 & 0 \\ 0 & 0 & 2 \end{bmatrix}.$$

Then $p_A(x) = (1-x)^2(2-x)$. The monic factors of $p_A(x)$ are $x-2, x-1, (x-1)^2, (x-2)(x-1)$, and $(x-2)(x-1)^2$. Now try them: We can see that $x-2$ and $x-1$ fail; that is, $A-2I \neq 0$ and $A-1I \neq 0$. Next, try $(x-1)^2$:

$$(A-I)^2 = \begin{bmatrix} 0 & 0 & 2 \\ 0 & 0 & 0 \\ 0 & 0 & 1 \end{bmatrix}^2 = \begin{bmatrix} 0 & 0 & 2 \\ 0 & 0 & 0 \\ 0 & 0 & 1 \end{bmatrix}.$$

No good! (But how interesting - A matrix that equals its square! Does this happen often?)
Now try $(x-2)(x-1)$:

$$(A-2I)(A-I) = \begin{bmatrix} -1 & 0 & 2 \\ 0 & -1 & 0 \\ 0 & 0 & 0 \end{bmatrix} \begin{bmatrix} 0 & 0 & 2 \\ 0 & 0 & 0 \\ 0 & 0 & 1 \end{bmatrix} = \begin{bmatrix} 0 & 0 & 0 \\ 0 & 0 & 0 \\ 0 & 0 & 0 \end{bmatrix}.$$

Thus A satisfies the monic polynomial $(x-2)(x-1)$ and it follows that $m_A(x) = (x-2)(x-1)$.

$\square$

Corollary 5.6.3. *The minimum polynomial of an $n \times n$ matrix A is unique.*

Proof. Let both $m_1(x)$ and $m_2(x)$ have the properties of a minimum polynomial: both m_1 and m_2 are monic and of least degree. By Theorem 5.6.2, m_1 is a factor of m_2 and conversely. Assume $m_1(x) = m_2(x)q(x)$, where q is some polynomial. Now m_1 and m_2 must have the same degree (both have the least degree). It follows that $q(x) = a$ is a constant polynomial. Now consider the leading coefficients on each side of the equality: $m_1(x) = a(m_2)(x)$. Since both m_1 and m_2 are monic, $a = 1$ and $m_1 = m_2$. $\square$

Some of the trial and error in finding $m_A(x)$ can be eliminated by using the following theorem.

> **Theorem 5.6.4.** *Let A be an $n \times n$ matrix with minimum and characteristic polynomials $m_A(x)$ and $p_A(x)$, respectively. If $x-c$ is a factor of $p_A(x)$, then $x-c$ is a factor of $m_A(x)$.* (The minimum polynomial contains all of the linear factors corresponding to the distinct eigenvalues.)

Proof. As before, use the division algorithm to write:

$$m_A(x) = q(x)(x-c) + m_A(c) \tag{5.6.1}$$

Note that by the remainder theorem, the remainder is $m_A(c)$.
Substitute A into Equation 5.6.1 to obtain

$$m_A(A) = q(A)(A-cI) + m_A(c)I \tag{5.6.2}$$

Since $x-c$ is a factor of $p_A(x)$, c is an eigenvalue of A. By definition, $AX = cX$ for some nonzero $n \times 1$ column vector X. Applying Equation 5.6.2 to X, we get

$$m_A(A)X = q(A)(A-cI)X + m_A(c)X \tag{5.6.3}$$

But $AX = cX$ implies $(A - cI)X = AX - cIX = 0$ and $m_A(A) = 0$. From Equation 5.6.3 we
see that $0 = 0 + m_A(c)X$, and since $X \neq 0$, $m_A(c) = 0$. It follows that $m_A(x) = (x-c)q(x)$. $\square$

Applying Theorem 5.6.4 in finding the minimum polynomial of the matrix A in the
previous example, we see that since $p_A(x) = -(x - 1)^2(x - 2)$, the only possibilities for
$m_A(x)$ are $(x - 1)(x - 2)$ and $(x - 1)^2(x - 2)$. We try the first and it works, so
$$m_A(x) = (x - 1)(x - 2).$$

Section 5.6 Exercises

1. Let A be a 3×3 matrix with $p_A(\lambda) = -\lambda^3 + 2\lambda^2 + 3\lambda + 2$. Express A^{-1} and A^4 as
   linear combinations of powers of A.

2. Let A be a 3×3 matrix with eigenvalues $1, -2$, and -1. Express A^{-1} and A^4 as linear
   combinations of powers of A.

3. Evaluate A^6 and A^{-1} using the Cayley-Hamilton Theorem, where $A = \begin{bmatrix} 1 & -1 \\ 2 & 3 \end{bmatrix}$.

4. Using the Cayley-Hamilton Theorem, find the inverse of $\begin{bmatrix} 1 & 0 & 2 \\ 0 & -1 & 3 \\ 0 & 0 & 2 \end{bmatrix}$.

5. Let A be a 3×3 matrix with $p_A(\lambda) = -\lambda^3 + 3\lambda^2 - 2\lambda$. Find the minimum polynomial
   of A.

6. Find the minimum polynomial of $\begin{bmatrix} 1 & 0 & 0 \\ 2 & 1 & 0 \\ 3 & -1 & 1 \end{bmatrix}$.

7. Show that similar matrices have the same minimum polynomial.

8. Find the characteristic polynomial and the minimum polynomials of the 5×5 diagonal
   matrix $\text{diag}(1, 1, 1, 2, 2)$.

9. Find the characteristic polynomial and the minimum polynomials of the 2×2 matrix
$$\begin{bmatrix} 1 & 1 \\ 0 & 1 \end{bmatrix}.$$

10. Find the characteristic polynomial and the minimum polynomials of the 3×3 matrix
$$\begin{bmatrix} 1 & 1 & 0 \\ 0 & 1 & 1 \\ 0 & 0 & 1 \end{bmatrix}.$$

11. Describe the minimum polynomial of an arbitrary $n \times n$ diagonal matrix $D = \mathrm{diag}(a_1, a_2, \ldots, a_n)$. (Hint: Assume the characteristic polynomial is given by $p_D(\lambda) = (-1)^n(\lambda - b_1)^{m_1} \ldots (\lambda - b_k)^{m_k}$, where $b_1, \ldots, b_k$ are the distinct eigenvalues.)

12. Let A be an $n \times n$ matrix and assume that A is similar to a diagonal matrix. Find the minimum polynomial of A. (Hint: See Exercises 7 and 11.)

5.7   Quadratic Forms (optional)

In calculus one studies general quadratic equations of the form

$$ax^2 + bxy + cy^2 + dx + ey = f \tag{5.7.1}$$

and attempts to classify the curve represented by one of these equations. Recall that if the equation can be put into one the forms

$$\begin{aligned} x^2 + b'y &= c' \\ a'x^2 + b'y^2 &= c' \\ a'x + b'y^2 &= c' \end{aligned} \tag{5.7.2}$$

then it is relatively easy to identify the curve. Now two modifications are generally used to change an equation of the form of Equation 5.7.1 into one of the forms of Equations 5.7.2. First the term involving xy, if it exists, is eliminated by rotating the axes through an angle α, where

$$\cot 2\alpha = \frac{a - c}{b},$$

to obtain new coordinates x_1, y_1, satisfying

$$\begin{aligned} x &= x_1 \cos \alpha - y_1 \sin \alpha \\ y &= x_1 \sin \alpha + y_1 \cos \alpha \end{aligned} \tag{5.7.3}$$

Having performed this operation, the resulting equation will be free of terms involving the cross product xy. A translation of the form

$$x_2 = x_1 + h$$
$$y_2 = y_1 + k$$

will then produce an equation of the form of Equation 5.7.2. Recall that the translation can be easily found by completing the square on the terms involving x and the terms involving y.

Notice that Equation 5.7.3 can be expressed in the form of a matrix equation

$$\begin{bmatrix} x \\ y \end{bmatrix} = \begin{bmatrix} \cos \alpha & -\sin \alpha \\ \sin \alpha & \cos \alpha \end{bmatrix} \begin{bmatrix} x_1 \\ x_2 \end{bmatrix}, \tag{5.7.4}$$

and observe that the coefficient matrix in Equation 5.7.4 is an orthogonal matrix. Remembering that an orthogonal similarity transformation diagonalizes a symmetric matrix, we look for some connection in this situation. A symmetric matrix arises as follows: consider the "quadratic" part of Equation 5.7.1 and let

$$A = \begin{bmatrix} a & \dfrac{b}{2} \\ \dfrac{b}{2} & c \end{bmatrix} \quad \text{and } X = \begin{bmatrix} x \\ y \end{bmatrix}.$$

Then we see that

$$
\begin{aligned}
X^t A X &= [x y] \begin{bmatrix} a & \dfrac{b}{2} \\ \dfrac{b}{2} & c \end{bmatrix} \begin{bmatrix} x \\ y \end{bmatrix} \\
&= [x y] \begin{bmatrix} ax + \dfrac{b}{2} y \\ \dfrac{b}{2} x + cy \end{bmatrix} \\
&= ax^2 + \dfrac{b}{2} xy + \dfrac{b}{2} xy + cy^2 \\
&= ax^2 + bxy + cy^2.
\end{aligned}
$$

Of course, A is a symmetric matrix.

One wonders about higher dimensional analogs of the above procedure. In particular, what sort of rotation simplifies a quadratic equation in three variables $x, y,$ and z. We present the general theory.

QUADRATIC FORMS IN N VARIABLES

A **quadratic form** is a function q of n real variables $x_1, \ldots, x_n$ of the form

$$
\begin{aligned}
q &= q(x_1, \ldots, x_n) \\
&= a_{11} x_1^2 + a_{12} x_1 x_2 + \ldots + a_{1n} x_1 x_n \\
&+ a_{21} x_2 x_1 + a_{22} x_2^2 + \ldots + a_{2n} x_2 x_n \\
&+ \ldots + a_{n1} x_n x_1 + a_{n2} x_n x_2 + \ldots + a_{nn} x_n^2
\end{aligned}
\tag{5.7.5}
$$

where $A = [a_{ij}]$ is an $n \times n$ real symmetric matrix. Since A is symmetric, $a_{ij} = a_{ji}$ for all i, j, with $i \neq j$ and so the terms $a_{ij} x_i x_j$ and $a_{ji} x_j x_i$ are equal. It follows that q can be expressed in the form

$$
\begin{aligned}
q &= a_{11} x_1^2 + 2a_{12} x_1 x_2 + \ldots + 2a_{1n} x_1 x_n \\
&+ a_{22} x_2^2 + 2a_{23} x_2 x_3 + \ldots + 2a_{2n} x_2 x_n \\
&\quad \vdots \\
&+ a_{nn} x_n^2.
\end{aligned}
$$

We say that q is the quadratic form associated with the matrix A. There is a "complex number" version of the quadratic form that involves a hermitian matrix A and a definition similar to Equations 5.7.5. We will not consider these "hermitian forms"; however, the theory is similar to that of the "real quadratic forms".

Let us now use matrix notation for the quadratic form given in Equations 5.7.5. Let $A = [a_{ij}]$ be the $n \times n$ real symmetric matrix in the definition and let X be the column vector

$$X = \begin{bmatrix} x_1 \\ \vdots \\ x_n \end{bmatrix}$$

Then we have

$$X^t A X = [x_1 \ldots x_n] \begin{bmatrix} a_{11} & \cdots & a_{1n} \\ \vdots & \ddots & \vdots \\ & & \end{bmatrix} \begin{bmatrix} x_1 \\ \vdots \\ x_n \end{bmatrix}$$

$$= [x_1 \ldots x_n] \begin{bmatrix} a_{11}x_1 + \ldots + a_{1n}x_n \\ \vdots \\ a_{n1} + \ldots + a_{nn}x_n \end{bmatrix}$$

$$= a_{11}x_1^2 + a_{12}x_1x_2 + \ldots + a_{1n}x_1x_n$$
$$+ a_{21}x_2x_1 + a_{22}x_2^2 + \ldots + a_{2n}x_2x_n$$
$$\vdots$$
$$+ a_{n1}x_nx_1 + a_{n2}x_nx_2 + \ldots + a_{nn}x_n^2$$
$$= q(x_1, \ldots, x_n)$$
$$= q.$$

Finally then,

$$q = X^t A X. \tag{5.7.6}$$

DIAGONALIZING QUADRATIC FORMS

We say that a quadratic form $q = q(x_1, \ldots, x_n)$ is in **diagonal form** when

$$q = a_{11}x_1^2 + a_{22}x_2^2 + \ldots + a_{nn}x_n^2$$

That is, none of the "cross terms" $x_ix_j, i \neq j$, appear.

It is an easy matter now to see that any quadratic form can be put into diagonal form by an appropriate change of variables. Let q, A and X be as in Equation 5.7.6. Then A is a real symmetric matrix and so by the Principal Axes Theorem (Theorem 5.5.3) there is an orthogonal matrix S with

$$S^t A S = \text{diag}(\lambda_1, \ldots, \lambda_n) = D, \tag{5.7.7}$$

where $\lambda_1, \ldots, \lambda_n$ are the eigenvalues of A and are real numbers. Now solving Equation 5.7.7 for A we obtain

$$A = SDS^t \tag{5.7.8}$$

using the fact that $S^{-1} = S^t$ so that $SS^t = S^tS = I_n$. Substituting Equation 5.7.8 into Equation 5.7.6 we see that

$$q = q(x_1, \ldots, x_n) = X^tAX = X^t(SDS^t)X = (X^tS)D(S^tX) = (S^tX)^tD(S^tX)$$

Now, if we let

$$Y = \begin{bmatrix} y_1 \\ \vdots \\ y_n \end{bmatrix} = S^tX$$

or $X = SY$, we get

$$q = q(y_1, \ldots, y_n) = Y^tDY = \lambda_1 y_1^2 + \ldots + \lambda_n y_n^2,$$

which is a quadratic form in diagonal form.

This establishes the following:

Theorem 5.7.1. *Let $q = X^tAX$ be a quadratic form. Then there is an orthogonal matrix S such that if*

$$Y = \begin{bmatrix} y_1 \\ \vdots \\ y_n \end{bmatrix} = S^tX,$$

then $q = \lambda_1 y_1^2 + \ldots + \lambda_n y_n^2$, where $\lambda_1, \ldots, \lambda_n$ are the eigenvalues of A.

(A quadratic form can be put into diagonal form by an orthogonal matrix.)

Example 5.7.1. Consider the quadratic equation

$$2x^2 + 2y^2 + 2z^2 - 2xz = 4 \tag{5.7.9}$$

The left-hand side of this equation is a quadratic form and it can be represented by

$$q = \begin{bmatrix} x & y & z \end{bmatrix} \begin{bmatrix} 2 & 0 & -1 \\ 0 & 2 & 0 \\ -1 & 0 & 2 \end{bmatrix} \begin{bmatrix} x \\ y \\ z \end{bmatrix}.$$

Let us reduce q to diagonal form. Let

$$A = \begin{bmatrix} 2 & 0 & -1 \\ 0 & 2 & 0 \\ -1 & 0 & 2 \end{bmatrix}.$$

Calculating the characteristic polynomial of A and solving we find that the eigenvalues of A are $\lambda_1 = 1$, $\lambda_2 = 2$, and $\lambda_3 = 3$. Corresponding eigenvectors for A are

$$X_1 = \begin{bmatrix} 1 \\ 0 \\ 1 \end{bmatrix}, X_2 = \begin{bmatrix} 0 \\ 1 \\ 0 \end{bmatrix}, \text{ and } X_3 = \begin{bmatrix} 1 \\ 0 \\ -1 \end{bmatrix}.$$

Notice that X_1, X_2, and X_3 are orthogonal (necessarily since A is symmetric and the eigenvalues are distinct). We normalize the eigenvectors and construct S :

$$S = \begin{bmatrix} \frac{1}{\sqrt{2}} & 0 & \frac{1}{\sqrt{2}} \\ 0 & 1 & 0 \\ \frac{1}{\sqrt{2}} & 0 & -\frac{1}{\sqrt{2}} \end{bmatrix}.$$

Now using the variables x_1, y_1, z_1 given by

$$\begin{bmatrix} x_1 \\ y_1 \\ z_1 \end{bmatrix} = S^t \begin{bmatrix} x \\ y \\ z \end{bmatrix} = \begin{bmatrix} \frac{1}{\sqrt{2}} & 0 & \frac{1}{\sqrt{2}} \\ 0 & 1 & 0 \\ \frac{1}{\sqrt{2}} & 0 & -\frac{1}{\sqrt{2}} \end{bmatrix} \begin{bmatrix} x \\ y \\ z \end{bmatrix},$$

we get by Theorem 5.7.1 that

$$q = \lambda_1 x_1^2 + \lambda_2 y_1^2 + \lambda_3 z_1^2 = x_1^2 + 2y_1^2 + 3z_1^2.$$

It follows that Equation 5.7.9 becomes

$$x_1^2 + 2y_1^2 + 3z_1^2 = 4$$

and so the surface defined is an ellipsoid.

$\square$

EIGENVALUES AND POSITIVE-DEFINITENESS

Let $q = q(x_1, \ldots, x_n)$ be a quadratic form. Then q is called **positive** provided

$$q(x_1, \ldots, x_n) \geq 0$$

for all $x_1, \ldots, x_n$ and q is **positive definite** provided $q(x_1, \ldots, x_n) > 0$ for all $x_1, \ldots, x_n$, $(x_1, \ldots, x_n) \neq (0, \ldots, 0)$. Similarly, q is **negative (negative definite)** provided

$$q(x_1, \ldots, x_n) \leq 0$$

for all $x_1, \ldots, x_n$ (respectively, $q(x_1, \ldots, x_n) < 0$ for all $x_1, \ldots, x_n, (x_1, \ldots, x_n) \neq (0, \ldots, 0)$). If q assumes both positive and negative values, then q is called **indefinite**. A quadratic form q is **positive semidefinite (negative semidefinite)** if q is positive (negative) but not positive definite (negative definite).

The following are easy examples of the various types of quadratic forms:

$$x_1^2 + x_2^2 \text{ is positive definite;}$$
$$x_1^2 - 2x_1x_2 + 2x_2^2 = (x_1 - x_2)^2 \text{ is positive semidefinite;}$$
$$-x_1^2 - x_2^2 \text{ is negative definite;}$$
$$-(x_1 - x_2)^2 \text{ is negative semidefinite;}$$
$$x_1^2 - x_2^2 \text{ is indefinite.}$$

Using Theorem 5.7.1, it is easy to characterize the above types of quadratic forms.

Theorem 5.7.2. *Let q be a quadratic form with associated real symmetric matrix A. Then*

(a) q is positive semidefinite if and only if the eigenvalues of A are all non-negative.

(b) q is positive definite if and only if the eigenvalues of A are all positive.

(c) q is negative semidefinite if and only if the eigenvalues of A are nonpositive.

(d) q is negative if and only if the eigenvalues of A are all negative.

(e) q is indefinite if and only if A has both positive and negative eigenvalues.

(The conditions of positive and negative definiteness are related to the signs of the eigenvalues.)

Proof. Using Theorem 5.7.1, q can be expressed in the form $q = \lambda_1 y_1^2 + \ldots + \lambda_n y_n^2$. Clearly, if $\lambda_1, \ldots, \lambda_n = 0$, then $q = 0$. It follows that (a) holds. The other parts follow in a similar manner. $\square$

Quadratic forms and their properties of being positive or negative definite arise naturally in determining conditions on a function of several variables that guarantee that the function has a maximum or a minimum at a given point. Let $y = f(x_1, \ldots, x_n)$ be a real-valued function of n real variables. Let us simplify the situation with the following observation: f has a maximum (minimum) at the point $(a_1, \ldots, a_n)$ if and only if the function $g(x_1, \ldots, x_n) = f(x + a_1, \ldots, x + a_n)$ has a maximum (resp. minimum) at $(0, 0, \ldots, 0)$. We will attempt to give conditions under which $f(x_1, \ldots, x_n)$ will obtain a maximum or minimum at $(0, \ldots, 0)$.

FINDING MAXIMA AND MINIMA

Let us assume that $f(x_1, \ldots, x_n)$ has a Taylor expansion about the point $(0, \ldots, 0)$. This implies the existence and continuity of partial derivatives of orders 1, 2 and 3. We use f_i for the partial derivative of f with respect to x_i and f_{ij} or $(f_i)_j$ for the second partial of f with respect to x_i then x_j. Recall that because of the continuity of the partial derivatives, $f_{ij} = f_{ji}$.

A necessary condition for the existence of a maximum or minimum at $(0, \ldots, 0)$ is that $f_i(0, 0, \ldots, 0) = 0$ for $i = 1, \ldots, n$. Let us use the notation $a_{ij} = f_{ij}(0, 0, \ldots, 0)$.

Now according to Taylor's Theorem,

$$f(x_1, x_2, \ldots, x_n) = f(0, 0, \ldots, 0) + R_1 + R_2 + R_3 \tag{5.7.10}$$

where R_1 involves the first partials and linear terms x_i, R_2 involves second partials and products $x_i x_j$ and R_3 is the remainder involving third partials and products of the form $x_i x_j x_k$. We assume that the first partials are zero so that $R_1 = 0$. Rewriting Equation 5.7.10 we get

$$f(x_1, x_2, \ldots, x_n) - f(0, 0, \ldots 0) = R_2 + R_3.$$

We must assume that for values of $x_1, \ldots, x_n$ close to zero, the products $x_i x_j x_k$ will be small so that R_3 is small and that the sign of $f(x_1, \ldots, x_n) - f(0, \ldots, 0)$ will be the same as that of R_2. Let us write out R_2 using $a_{ij} = f_{ij}(0, \ldots, 0)$:

$$R_2 = \frac{1}{2} a_{11} x_1^2 + \ldots + a_{1n} x_1 x_n + a_{21} x_2 x_1 + a_{22} x_2^2 + \ldots + a_{n1} x_n x_1 + \ldots + a_{nn} x_n^2.$$

Now $a_{ij} = a_{ji}$ by the continuity of the second partials, and so R_2 is a quadratic form with associated symmetric matrix

$$A = \frac{1}{2}[a_{ij}] = \frac{1}{2}[f_{ij}(0, 0, \ldots, 0)]. \tag{5.7.11}$$

With these assumptions, we see that

1. If R_2 is positive definite, then $f(x_1, \ldots, x_n) - f(0, \ldots, 0)$ will be positive for values of $x_1, \ldots, x_n$ close to zero so that f will have a minimum at $(0, \ldots, 0)$;

2. If R_2 is negative definite, then $f(x_1, \ldots, x_n) - f(0, \ldots, 0)$ will be negative for values of $x_1, \ldots, x_n$ close to zero so that f will have a maximum at $(0, \ldots, 0)$.

3. If R_2 is indefinite, then f will have neither a maximum or minimum at $(0, \ldots, 0)$.

To determine the nature of R_2, one can apply Theorem 5.7.2 and consider the eigenvalues of the above matrix A.

Example 5.7.2. Let us investigate the function

$$f(x_1, x_2, x_3) = x_1^2 + x_1 x_2 + x_3^2$$

for maxima and minima. Setting first partials equal to zero we get

$$f_1 = 2x_1 + x_2 = 0$$
$$f_2 = x_1 = 0$$
$$f_3 = 2x_3 = 0$$

and so the only critical point is $(0,0,0)$. Calculating second partials at $(0,0,0)$, we get $f_{11} = 2, f_{12} = 1, f_{13} = 0, f_{22} = 0, f_{23} = 0$, and $f_{33} = 2$. Using Equation 5.7.11, the matrix A associated with the quadratic form R_2 is

$$A = \frac{1}{2} \begin{bmatrix} 2 & 1 & 0 \\ 1 & 0 & 0 \\ 0 & 0 & 2 \end{bmatrix}.$$

To determine whether R_2 is positive or negative definite or indefinite, we calculate $p_A(x)$ and solve for the eigenvalues:

$$p_A(x) = -(x - 2)(x^2 - 2x - 1).$$

Solving, we get eigenvalues

$$\lambda_1 = 2, \lambda_2 = \frac{2 + \sqrt{5}}{2}, \text{ and } \lambda_3 = \frac{2 - \sqrt{5}}{2}.$$

We can see that $\lambda_1, \lambda_2 > 0$, while $\lambda_3 < 0$. By Theorem 5.7.2, we see that A is indefinite so that f has no maxima or minima.

$\square$

Section 5.7 Exercises

1. Consider the equation $5x^2 + 4xy + 2y^2 = 9$. Find the angle of rotation that will eliminate the term $4xy$. Find new coordinates x_1, y_1 and substitute to obtain the new equation. (You will perhaps need to consult a calculus book for the half-angle formulas in order to find $\cos \alpha$ and $\sin \alpha$.)

2. Given the equation $2x^2 + 8x + y^2 - 2y = 12$, complete the square in order to find a translation that will eliminate the x and y terms.

3. Which of the following are quadratic forms

   (a) $x^2 + xy$

   (b) $x^2 + 2xy + y^2$

   (c) $x^2 + 4xyz + y^2 + z^2$

   (d) $-x^2 - z^2 + y$

   (e) $x^2 + 2x + y^2$

   (f) $x^3 - z^2$

4. Given the quadratic form, $q = 2x^2 + xy - y^2$, find matrices A and X, with A symmetric, so that $q = X^t A X$.

5. Given the quadratic form, $q = 2x^2 + xy - y^2 + 4xz + 3z^2$, find matrices A and X, with A symmetric, so that $q = X^t A X$.

6. Let $q = x_1^2 + x_2^2 + x_3^2 + 2x_2 x_3$. Find a symmetric matrix A such that $q = X^t A X$, where

$$X = \begin{bmatrix} x_1 \\ x_2 \\ x_3 \end{bmatrix}.$$

7. Let $q = 3x_1^2 + x_2^2 + x_3^2 + 4x_2x_3$. Find a matrix A such that $q = X^tAX$, where $X = \begin{bmatrix} x_1 \\ x_2 \\ x_3 \end{bmatrix}$.

8. Find an orthogonal matrix S such that the change of variables $Y = S^tX$ reduces q to a diagonal form, where q is the quadratic form of Exercise 6.

9. Find an orthogonal matrix S such that the change of variables $Y = S^tX$ reduces q to a diagonal form, where q is the quadratic form of Exercise 7.

10. Is q in Exercise 6 positive definite, negative definite, etc.?

11. Classify the curve given by $x_1^2 - 4x_1x_2 - 2x_2^2 = 4$.

12. Investigate the following function for maxima and minima:

$$f(x_1, x_2) = 2x_1^2 + x_1x_2 + x_2^2.$$

13. Investigate the following function for maxima and minima:

$$f(x_1, x_2, x_3) = x_1^2 - x_2x_3 - 2x_2^2.$$

5.8   Applications - The Inertia Tensor and Principal Axes (optional)

In most calculus courses, vector-valued functions of a real variable and the calculus of such functions are introduced. This topic was discussed briefly in Section 4.4, and the reader may find it helpful to review that section. If $R = R(t)$ is vector-valued function of the real variable t, then one can express R in terms of its coordinate functions:

$$R(t) = (x(t), y(t), z(t)),$$

where x, y and z are real-valued functions of t. The **derivative** of $R(t)$ is denoted by $R'(t)$ or $\dfrac{d}{dt}(R(t))$ and is defined by

$$R'(t) = \frac{d}{dt}(R(t)) = (x'(t), y'(t), z'(t)).$$

As is the case with scalar valued functions, the derivative measures rate of change and is used to define velocity and acceleration.

VELOCITY AND ACCELERATION

An important application of vector-valued functions is in describing the motion of a particle in three-dimensional space. We let $R(t)$ denote the **position vector** of given particle in space at time t; that is, $R(t)$ is the vector from the origin 0 of a given coordinate system to the point $(x(t), y(t), z(t))$ at which the particle is located at the time t. The **velocity vector** $V(t)$ of the particle is defined by $V(t) = R'(t)$ and the **acceleration vector** is given by $A(t) = V'(t) = R''(t)$.

Recall that in Section 4.4, the angular velocity vector $\Omega = \Omega(t)$ was introduced and it was shown that

$$V = \Omega \times R. \tag{5.8.1}$$

In this section we continue the investigation of angular motion.

For a moving particle of mass m and velocity vector V, the **momentum** P of the particle is defined by

$$P = mV. \tag{5.8.2}$$

According to Newton's second law, the rate of change of momentum of a particle is proportional to and in the direction of the force impressed on the body, and so with the correct choice of units we have $F = mA$, which is known as Newton's law of motion.

ANGULAR MOTION

For angular motion; that is, the rotation of a particle about an axis, the **angular momentum** L is defined by

$$L = R \times P, \tag{5.8.3}$$

where R is the position vector and P is the momentum vector defined in 5.8.2. The time rate of change of L is called the **torque** and is denoted by N. Torque is the angular version of the notion of force, and as with the linear situation we have

$$\frac{dL}{dt} = N. \tag{5.8.4}$$

Using Equations 5.8.1 and 5.8.2, we write L as a function of Ω as follows:

$$L = R \times P = R \times (mV) = m(R \times V) = m(R \times (\Omega \times R)). \tag{5.8.5}$$

Let $L = L(\Omega)$ and notice that $L(c\Omega) = c(m(R \times (\Omega \times R))) = cL(\Omega)$ for any scalar c, and

$$
\begin{aligned}
L(\Omega_1 + \Omega_2) &= m(R \times (\Omega_1 + \Omega_2) \times R) \\
&= m(R \times ((\Omega_1 \times R) + (\Omega_2 \times R))) \\
&= m(R \times (\Omega_1 \times R) + R \times (\Omega_2 \times R)) \\
&= m(R \times (\Omega_1 \times R)) + m(R \times (\Omega_2 \times R)) \\
&= L(\Omega_1) + L(\Omega_2)
\end{aligned}
$$

for any two vectors Ω_1 and Ω_2. From this we see that L is a linear transformation defined on 3-dimensional vectors Ω and having 3-dimensional images $L(\Omega)$; that is, L is a linear transformation from $\mathbb{R}^3$ into $\mathbb{R}^3$.

Now let us assume that we have a rigid body consisting of n particles with masses $m_1, \ldots, m_n$ and position vectors $R_1, \ldots, R_n$. The masses are constant, but the position vectors are functions of time. Let us assume that we have a fixed coordinate system with $\mathbf{i}, \mathbf{j}, \mathbf{k}$, being an orthonormal basis as in section 4.4. We will use the notation

$$R_k = R_k(t) = (x_k(t), y_k(t), z_k(t)) = x_k(t)\mathbf{i} + y_k(t)\mathbf{j} + z_k(t)\mathbf{k}. \tag{5.8.6}$$

To calculate the angular momentum of the system of particles, we add the momenta of each of the particles, using Equation 5.8.5 and obtain

$$L(\Omega) = \sum_{k=1}^{n} m_k(R_k \times (\Omega \times R_k)). \tag{5.8.7}$$

Since a sum of linear transformations is again a linear transformation, we can see that the angular momentum function of the rigid system of n particles is also a linear transformation.

THE INERTIA TENSOR

Since L is a linear transformation, it is represented by a matrix, say I_T, relative to the $\mathbf{i}, \mathbf{j}, \mathbf{k}$ basis. It follows that $L(\Omega) = I_T\Omega$ for any angular velocity vector Ω. The matrix I_T is called the **inertia tensor** of the system of particles. In general, Ω is a function of time t, so that I_T is also a function of time, but at any time t, we can calculate I_T in terms of the masses $m_1, \ldots, m_k$ and the position vectors $R_1, \ldots, R_k$.

We let

$$I_T = \begin{bmatrix} I_{xx} & I_{xy} & I_{xz} \\ I_{yx} & I_{yy} & I_{yz} \\ I_{zx} & I_{zy} & I_{zz} \end{bmatrix}$$

and notice that using the representation

$$\mathbf{i} = \begin{bmatrix} 1 \\ 0 \\ 0 \end{bmatrix}, \mathbf{j} = \begin{bmatrix} 0 \\ 1 \\ 0 \end{bmatrix}, \text{ and } \mathbf{k} = \begin{bmatrix} 0 \\ 0 \\ 1 \end{bmatrix},$$

$$L(\mathbf{i}) = I_T\mathbf{i} = \begin{bmatrix} I_{xx} & I_{xy} & I_{xz} \\ I_{yx} & I_{yy} & I_{yz} \\ I_{zx} & I_{zy} & I_{zz} \end{bmatrix} \begin{bmatrix} 1 \\ 0 \\ 0 \end{bmatrix}$$

so that

$$L(\mathbf{i}) \cdot \mathbf{i} = I_{xx}, L(\mathbf{j}) \cdot \mathbf{j} = I_{yy}, L(\mathbf{k}) \cdot \mathbf{k} = I_{zz} \tag{5.8.8}$$

Likewise $L(\mathbf{j}) \cdot \mathbf{i} = I_{xy}$, etc.

From 5.8.7, we get

$$L(\Omega) = \sum_{k=1}^{n} m_k(R_k \times (\Omega \times R_k)),$$

and applying the vector identity

$$A \times (B \times C) = B(A \cdot C) - C(A \cdot B)$$

(see Exercise 12, Section 4.4), we get

$$L(\Omega) = \sum_{k=1}^{n} m_k(\Omega(R_k \cdot R_k) - R_k(R_k \cdot \Omega)) \tag{5.8.9}$$

Using Equations 5.8.8 and 5.8.9, we calculate the entries of I_T at any fixed time t:

$$
\begin{aligned}
I_{xx} &= L(\mathbf{i}) \cdot \mathbf{i} \\
&= \sum_{k=1}^{n} m_k(\mathbf{i}(R_k \cdot R_k) - R_k(R_k \cdot \mathbf{i})) \cdot \mathbf{i} \\
&= \sum_{k=1}^{n} m_k(\mathbf{i} \cdot \mathbf{i})(R_k \cdot R_k) - (R_k \cdot \mathbf{i})(R_k \cdot \mathbf{i}) \\
&= \sum_{k=1}^{n} m_k((R_k \cdot R_k) - (R_k \cdot \mathbf{i})^2) \\
&= \sum_{k=1}^{n} m_k(x_k^2 + y_k^2 + z_k^2 - x_k^2) \\
&= \sum_{k=1}^{n} m_k(y_k^2 + z_k^2).
\end{aligned}
$$

Likewise,

$$
\begin{aligned}
I_{xy} &= L(\mathbf{i}) \cdot \mathbf{j} \\
&= \sum_{k=1}^{n} m_k(\mathbf{i}(R_k \cdot R_k) - R_k(R_k \cdot \mathbf{i})) \cdot \mathbf{j} \\
&= \sum_{k=1}^{n} m_k((\mathbf{i} \cdot \mathbf{j})(R_k \cdot R_k) - (R_k \cdot \mathbf{j})(R_k \cdot \mathbf{i})) \\
&= \sum_{k=1}^{n} m_k(-x_k y_k) \\
&= -\sum_{k=1}^{n} m_k(x_k y_k)
\end{aligned}
$$

and

$$I_{yx} = L(\mathbf{j}) \cdot \mathbf{i}$$

$$= \sum_{k=1}^{n} m_k(\mathbf{j}(R_k \cdot R_k) - R_k(R_k \cdot \mathbf{j})) \cdot \mathbf{i}$$

$$= \sum_{k=1}^{n} m_k(\mathbf{j} \cdot \mathbf{i}(R_k \cdot R_k) - (R_k \cdot \mathbf{i})(R_k \cdot \mathbf{j}))$$

$$= \sum_{k=1}^{n} m_k(-x_k y_k)$$

$$= -\sum_{k=1}^{n} m_k(x_k y_k),$$

so we see that $I_{xy} = I_{yx}$.

Similar calculations produce the other entries of I_T. We can show that

$$
\begin{aligned}
I_{xx} &= \sum_{k=1}^{n} m_k(y_k^2 + z_k^2) \\
I_{yy} &= \sum_{k=1}^{n} m_k(x_k^2 + z_k^2) \\
I_{zz} &= \sum_{k=1}^{n} m_k(x_k^2 + y_k^2)
\end{aligned}
\tag{5.8.10}
$$

and

$$
\begin{aligned}
I_{xy} &= -\sum_{k=1}^{n} m_k(x_k y_k) = I_{yx} \\
I_{xz} &= -\sum_{k=1}^{n} m_k(x_k z_k) = I_{zx} \\
I_{yz} &= -\sum_{k=1}^{n} m_k(y_k z_k) = I_{zy}.
\end{aligned}
\tag{5.8.11}
$$

The diagonal entries in Equations 5.8.10 are called the **moments of inertia** and the remaining entries in Equations 5.8.11 are called the **products of inertia**, and it is now clear why we have introduced this complicated topic.

Theorem 5.8.1. *The inertia tensor I_T of a rigid body of particles is a symmetric matrix.*
(The inertia tensor is symmetric.)

We know a lot about symmetric matrices from the theory in Section 5.5. According to Corollary 5.5.3, any real symmetric $n \times n$ matrix A is orthogonally similar to a diagonal matrix and according to Theorem 5.2.5, the entries on the diagonal are real. Applying these results to I_T we get an orthogonal matrix S and real numbers e_1, e_2, e_3 such that

$$S^t I_T S = \begin{bmatrix} e_1 & 0 & 0 \\ 0 & e_2 & 0 \\ 0 & 0 & e_3 \end{bmatrix}.$$

PRINCIPAL AXES

If E_1, E_2, E_3, are the column vectors of S so that $S = [E_1, E_2, E_3]$, then E_1, E_2, E_3 are mutually orthogonal and each is a unit vector. Also $S^t I_T S = D = \text{diag}(e_1, e_2, e_3)$ so that $I_T S = SD$ or

$$
\begin{aligned}
I_T S &= I_T [E_1, E_2, E_3] \\
&= [I_T E_1, I_T E_2, I_T E_3] \\
&= [E_1, E_2, E_3] \begin{bmatrix} e_1 & 0 & 0 \\ 0 & e_2 & 0 \\ 0 & 0 & e_3 \end{bmatrix} \\
&= [e_1 E_1, e_2 E_2, e_3 E_3].
\end{aligned}
$$

We see that $I_T E_i = e_i E_i$ for $i = 1, 2, 3$, so that each E_i is an eigenvector. By the results of Section 3.4, $D = \text{diag}(e_1, e_2, e_3)$ is the matrix for the linear transformation L relative to the basis E_1, E_2, E_3. Relative to E_1, E_2, E_3, the inertia tensor I_T is the diagonal matrix D. The vectors E_1, E_2 are E_3 are called **principal axes** of the body.

Theorem 5.8.2 (Principal Axes Theorem). *Given a rigid body of n particles and a fixed time t, there are three mutually orthogonal unit vectors E_1, E_2, E_3 such that relative to $E_1, E_2,$ and E_3 the inertia tensor I_T is a diagonal matrix; that is, all products of inertia are zero.* (Every rigid body has three principal axes.)

The above theorem is also called the Principal Axes Theorem. The principal axes are, of course, convenient in that the inertia tensor is a diagonal matrix, but there is a physical significance to principal axes which we will establish.

Recall from calculus that the center of mass of system of particles with masses m_i and position vectors (x_i, y_i, z_i) is a point $(\overline{x}, \overline{y}, \overline{z})$ where

$$
\overline{x} = \frac{\sum_{k=1}^{n} m_k x_k}{m}, \overline{y} = \frac{\sum_{k=1}^{n} m_k y_k}{m}, \text{ and } \overline{z} = \frac{\sum_{k=1}^{n} m_k z_k}{m};
$$

and where

$$
m = \sum_{k=1}^{n} m_k
$$

and where we are using the same notation for the n particles as used earlier. Let us now make assumptions regarding the specific coordinate system chosen for the presumed rigid body of n particles. First we assume that the origin 0 of the coordinate system is at the center of mass. If this is the case, $\overline{x} = \overline{y} = \overline{z} = 0$ and so we have

$$
\frac{\sum_{k=1}^{n} m_k x_k}{m} = \frac{\sum_{k=1}^{n} m_k y_k}{m} = \frac{\sum_{k=1}^{n} m_k z_k}{m} \tag{5.8.12}
$$

Next, assume that a certain time, say $t = 0$, the coordinate system is taken to be the principal axes E_1, E_2, E_3; then the inertia tensor is a diagonal matrix, say

$$I_T = \begin{bmatrix} e_1 & 0 & 0 \\ 0 & e_2 & 0 \\ 0 & 0 & e_3 \end{bmatrix}. \tag{5.8.13}$$

Let us fix the coordinate system and adopt the notation $\mathbf{i} = E_1, \mathbf{j} = E_2, \mathbf{k} = E_3$. Now assume that the body rotates about the principal axis $E_3 = \mathbf{k}$. Then the angular velocity can be expressed in the form

$$\Omega = \Omega(t) = f(t)\mathbf{k} = \begin{bmatrix} 0 \\ 0 \\ f(t) \end{bmatrix}.$$

Assume that at time t, the body has rotated through an angle $\alpha = \alpha(t)$ radians. Note that

$$\alpha(t) = \int_0^t f(x)dx.$$

Now at time $t \neq 0$, we have

$$\begin{array}{rcl} z(t) & = & z(0) \\ y(t) & = & x(0)\sin\alpha + y(0)\cos\alpha \\ x(t) & = & x(0)\cos\alpha - y(0)\sin\alpha \end{array} \tag{5.8.14}$$

At $t = 0$, the relation

$$L(\Omega) = I_T\Omega \tag{5.8.15}$$

is valid with I_T given by Equation 5.8.13, but at time $t \neq 0$, the inertia tensor has changed. We want to evaluate the product

$$L(\Omega) = I_T\Omega = I_T \begin{bmatrix} 0 \\ 0 \\ f(t) \end{bmatrix} \tag{5.8.16}$$

at the time $t \neq 0$ and so we only need evaluate the two products of inertia I_{xz} and I_{yz} and the moment of inertia I_{zz}. We calculate these at time t :

$$I_{xz} = I_{xz}(t)$$

$$= \sum_{k=1}^{n} m_k x_k z_k(t)$$

$$= -\sum_{k=1}^{n} m_k(x_k(0)\cos\alpha - y_k(0)\sin\alpha)z_k(0)$$

$$= \cos\alpha \sum_{k=1}^{n}(-m_k x_k(0)z_k(0)) + \sin\alpha \sum_{k=1}^{n} m_k y_k(0)z_k(0)$$

$$= \cos\alpha I_{xz}(0) + \sin\alpha I_{yz}(0)$$

$$= 0.$$

Likewise $I_{yz} = 0$. Calculating the moment of inertia, we get

$$L(\Omega(t)) = I_T(t)\Omega(t)$$

$$= \begin{bmatrix} - & - & 0 \\ - & - & 0 \\ - & - & e_3 \end{bmatrix} \begin{bmatrix} 0 \\ 0 \\ f(t) \end{bmatrix}$$

$$= \begin{bmatrix} 0 \\ 0 \\ e_3 f(t) \end{bmatrix}$$

$$= e_3 \begin{bmatrix} 0 \\ 0 \\ f(t) \end{bmatrix}.$$

It follows that $L(\Omega(t)) = e_3\Omega(t)$, where $\Omega(t)$ is an angular velocity in the direction of the principal axis E_3. Calculating the torque N, we get

$$N = \frac{d}{dt}L(\Omega(t)) = \frac{d}{dt}e_3 \begin{bmatrix} 0 \\ 0 \\ f(t) \end{bmatrix} = e_3 \begin{bmatrix} 0 \\ 0 \\ f'(t) \end{bmatrix},$$

and so N and $\Omega(t)$ lie in the same direction.

The physical implication of the above observation is that a rigid system of particles rotates "freely" about any one of its three principal axes. Put more precisely, if Ω is an angular velocity vector in the direction of a principal axis, then the torque N resulting from Ω lies in the same direction as Ω. The only change in Ω is a change in magnitude, not in the direction of Ω.

Example 5.8.1. Consider four points in space of mass 2 each. Let the points be located at $(0, 0, 0, 0)$, $(1, 0, 0, 0)$, $(0, 1, 0, 0)$,and $(1, 1, 0, 0)$. The points are the corners of a square and we feel sure, since the points have the same mass, that the center of mass is at the intersection of the diagonals of the square and that the principal axes should lie along the diagonals. Let us perform the calculations. The mass is

$$m = \sum_{k=1}^{4} m_k = 2 + 2 + 2 + 2 = 8,$$

$\overline{x}$ is obtained by

$$\overline{x} = \frac{\displaystyle\sum_{k=1}^{4} m_k x_k}{m} = \frac{1 \cdot 2 + 1 \cdot 2}{8} = \frac{1}{2},$$

and similarly, $\overline{y} = 1/2$ and $\overline{z} = 0$. The center of mass is located at the point $(1/2, 1/2, 0)$. To calculate the inertia tensor

$$I_T = \begin{bmatrix} I_{xx} & I_{xy} & I_{xz} \\ I_{yx} & I_{yy} & I_{yz} \\ I_{zx} & I_{zy} & I_{zz} \end{bmatrix},$$

we use Formulas 5.8.10 and 5.8.11. The moments of inertia are

$$I_{xx} = \sum_{k=1}^{4} m_k(y_k^2 + z_k^2) = 2(0^2 + 0^2 + 1^2 + 1^2) = 2 \cdot 2 = 4,$$

and similarly, $I_{yy} = 4$ and $I_{zz} = 8$. Calculating the products of inertia, we obtain

$$I_{xy} = -\sum_{k=1}^{4} m_k(x_k y_k) = -2(0 + 0 + 0 + 1) = -2,$$

$$I_{xz} = -\sum_{k=1}^{4} m_k(x_k z_k) = -2(0 + 0 + 0 + 0) = 0,$$

and likewise $I_{yz} = 0$. We see that

$$I_T = \begin{bmatrix} 4 & -2 & 0 \\ -2 & 4 & 0 \\ 0 & 0 & 8 \end{bmatrix}.$$

To find the principal axes, we need to find the eigenvalues and eigenvectors. The characteristic polynomial of I_T is $p_{I_T} = (8 - \lambda)(\lambda^2 - 8\lambda + 12) = (8 - \lambda)(6 - \lambda)(2 - \lambda)$. Thus, the eigenvalues are 2, 6, and 8. Calculating corresponding eigenvectors we get

$$\begin{bmatrix} 1 \\ -1 \\ 0 \end{bmatrix}, \begin{bmatrix} 1 \\ 1 \\ 0 \end{bmatrix}, \text{ and } \begin{bmatrix} 0 \\ 0 \\ 1 \end{bmatrix}.$$

Normalizing, we obtain the principal axes:

$$\begin{bmatrix} 1/\sqrt{2} \\ -1/\sqrt{2} \\ 0 \end{bmatrix}, \begin{bmatrix} 1/\sqrt{2} \\ 1/\sqrt{2} \\ 0 \end{bmatrix}, \text{ and } \begin{bmatrix} 0 \\ 0 \\ 1 \end{bmatrix}.$$

□

Section 5.8 Exercises

1. Let the position vector of a moving particle at any time t be given by

$$R(t) = (\cos \pi t, \sin \pi t, t).$$

   Find the location of the particle when $t = 0$ and $t = 1$. Find the velocity and acceleration vectors at any time t. Describe the path followed by the particle.

2. Let the position vector of a particle that is rotating about the z-axis be given by $R(t) = (\cos \pi t, \sin \pi t, 0)$ at any time t. Find the location of the particle when $t = 0$ and $t = 1$. Find the velocity and acceleration vectors at any time t. Describe the path followed by the particle. Calculate the angular velocity vector Ω and the cross product $\Omega \times R$.

3. Suppose that the particle in Exercise 2 has mass 3. Find the momentum vector P and calculate the angular momentum L.

4. Let a rigid system of three particles of masses 3, 2, and 4 be positioned at the points $(-1, 1, 0), (0, 1, 1)$ and $(-2, 0, 2)$, respectively. Calculate the inertia tensor for the system.

5. Find the center of mass for the system of particles in Exercise 4.

6. Consider a rigid system of two particles in space. The first particle is located at the point $(2, 0, 0)$ and is of mass 3, the second is located at $(-3, 0, 0)$ and has mass 2. Show that the center of mass is at the origin. Calculate the inertia tensor. Find principal axes.

7. Consider a rigid system of three particles in space. Each particle has mass 3. One particle is located at the point $(1, 1, 0)$, the second is located at $(1, 0, 0)$, and the third is at $(0, 1, 0)$. Find the center of mass. Calculate the inertia tensor and find principal axes.

Chapter 6

THE JORDAN CANONICAL FORM AND APPLICATIONS

6.1 Introduction

In this chapter we elaborate upon the investigations into similarity which were begun in Chapter 5 and bring our understanding of the matter to a satisfactory and elegant conclusion in the presentation of the "Jordan[1] canonical form." This term refers to a special form that a matrix may be transformed into under similarity.

We saw in Chapter 5 that the similarity transformation of a matrix into a special form is of interest from the point of view of applications and that problems of transforming a matrix under similarity are quite interesting in themselves. The diagonalization of symmetric matrices was applied to quadratic forms in Section 5.7 and to the inertia tensor in Section 5.8. We will see in Section 6.3 that the Jordan canonical form is of use in solving systems of differential equations.

It would be convenient if every real matrix were orthogonally similar to a diagonal matrix, but unfortunately, it is only the symmetric matrices that have this property. In problems involving similarity, say similarity to an upper triangular matrix, factorization of the characteristic polynomial is always a stumbling block and so any result must carry along the necessary assumptions regarding it. It has been proved that there is no "quadratic formula" type method for solving polynomial equations of degree five and larger, and so we can feel sure that this factorization must be assumed separately. Is there a best result that can be stated, with reasonable assumptions, regarding similarity? An answer will soon appear.

In this section, we will review the theory and methods developed in Chapter 5 regarding similarity, diagonalization, eigenvalues and eigenvectors, and the characteristic and minimum polynomials. In addition, we will consider several examples and present the definition of the Jordan block, a fundamental unit in the discussion that follows.

---

[1]It is named for the French mathematician Camille Jordan (1838-1922).

REVIEW

Let A be an $n \times n$ matrix. We will use the notation of Chapter 5 for the characteristic and minimum polynomials of A, and we will rely on the definitions of eigenvalue and eigenvector from that same chapter.

6.1.1   Summary of Previous Results:

(a) The eigenvalues of A are the roots of the characteristic polynomial, $p_A(\lambda)$, of A. To find an eigenvector of A corresponding to the eigenvalue λ_0, one finds a solution of the homogeneous system $(A - \lambda_0 I)X = 0$. (Theorem 5.2.1)

(b) A is similar to a diagonal matrix if and only if A has n linearly independent eigenvectors. If $X_1, \ldots, X_n$ are independent eigenvectors corresponding to the eigenvalues $\lambda_1, \ldots, \lambda_n$ and $S = [X_1 \ldots X_n]$, then S is nonsingular and $S^{-1}AS = \mathrm{diag}(\lambda_1, \ldots, \lambda_n)$. (Theorem 5.3.3)

(c) If A is a real matrix and $p_A(\lambda)$ factors completely or A is a complex matrix, then A is orthogonally similar (resp., unitarily similar) to an upper triangular matrix. The eigenvalues of an upper triangular matrix are the entries on the diagonal. (Schur's Theorem, Theorem 5.5.2)

(d) A is a real symmetric matrix if and only if A is orthogonally similar to a diagonal matrix. In this case, the eigenvalues of A are real. (Corollary 5.5.3 (the Principal Axes Theorem and Theorem 5.2.5)

(e) A is a hermitian matrix over the complex numbers if and only if A is unitarily similar to a diagonal matrix and the eigenvalues of A are real. (Corollary 5.5.4 and Theorem 5.2.5)

(f) Eigenvectors corresponding to distinct eigenvalues of A are linearly independent. If A is hermitian or real symmetric, then eigenvectors corresponding to distinct eigenvalues are orthogonal. (Theorem 5.2.4 and Theorem 5.5.5)

(g) A matrix satisfies its characteristic polynomial; that is, $p_A(A) = 0$. (Theorem 5.6.1 (the Cayley-Hamilton Theorem))

(h) The monic polynomial $m_A(\lambda)$ of least degree satisfying $m_A(A) = 0$ is the minimum polynomial of A. If $p_A(\lambda) = (a_1 - \lambda)^{m_1} \cdots (a_s - \lambda)^{m_s}$, then $m_A(\lambda) = (a_1 - \lambda)^{n_1} \cdots (a_s - \lambda)^{n_s}$, where $1 \le n_i \le m_i$ for $i = 1, \ldots, s$. That is, $m_A(\lambda)$ is a factor of $p_A(\lambda)$ and $m_A(\lambda)$ contains each of the linear factors of $p_A(\lambda)$. (Definition, Theorem 5.6.2, and Theorem 5.6.4)

(i) Similar matrices have the same minimum and characteristic polynomials. In particular, similar matrices have the same eigenvalues. (Theorem 5.2.3 and Exercise 7 of section 5.6)

Having reviewed these facts from Chapter 5, let us consider some easy examples to gain some experience with and appreciation for the theory we have just reviewed.

6.1.2 Examples

Example 6.1.1. (a) Let $A = \begin{bmatrix} 1 & 1 \\ 0 & 2 \end{bmatrix}$. Then A is upper triangular and so the eigenvalues of A are 1 and 2 (See part (c) of the Summary). Since the eigenvalues of A are distinct, the corresponding eigenvectors are linearly independent (See part (f) of the Summary), and so A is similar to the diagonal matrix $\text{diag}(1,2)$ by part (b) of the Summary.

(b) Let $A = \begin{bmatrix} 1 & 1 \\ 0 & 1 \end{bmatrix}$. Then the only eigenvalue of A is 1 and it has multiplicity 2. Now $A - 1I$ has rank 1. It follows that there are at most $2 - 1 = 1$ independent eigenvectors, and so, A is not similar to a diagonal matrix (See part (b) of the Summary).

(c) Let $A = \begin{bmatrix} 0 & 1 \\ -1 & 0 \end{bmatrix}$. Then $p_A(\lambda) = \lambda^2 + 1$ and so the characteristic polynomial does not factor completely over the real numbers. It follows that A is not similar to a diagonal matrix over the real numbers. However, considering A as a matrix over the complex numbers, $p_A(\lambda)$ factors as $p_A(\lambda) = (\lambda - i)(\lambda + i)$. Thus A has two distinct complex eigenvalues and so there is a complex matrix S with:

$$S^{-1}AS = \begin{bmatrix} i & 0 \\ 0 & -i \end{bmatrix}$$

(d) Let $A = \begin{bmatrix} 3 & 1 \\ 1 & 7 \end{bmatrix}$. Then A is symmetric and so A is similar to a real diagonal matrix (See Summary part (d)).

(e) Let $A = \begin{bmatrix} 1 & 1-i \\ 1+i & 7 \end{bmatrix}$. Then A is a hermitian matrix and so A is similar to a real diagonal matrix (See Summary part (e)).

$\square$

We will consider now the fundamental elements that make up the Jordan canonical form of a matrix.

JORDAN BLOCKS

The reader might recall that in both the "diagonalization" process and the "upper triangularization" process, the order in which the eigenvalues occurred on the diagonal of the resulting matrix was arbitrary in that any order desired could be obtained. The order could be controlled by choosing the eigenvectors in the proper order. So, for example, if a is an eigenvalue of A of multiplicity m, one could arrange to have a appear in the first m entries of the resulting similar upper triangular matrix.

We consider now a special type of matrix that has a single eigenvalue. We will see in Section 6.2 that any matrix (with $p_A(\lambda)$ factoring completely) is similar to a matrix with these special matrices on the diagonal.

A **Jordan block** is an $m \times m$ matrix J of the form

$$J = \begin{bmatrix} a & 1 & 0 & \ldots & 0 & 0 & 0 \\ 0 & a & 1 & \ldots & 0 & 0 & 0 \\ 0 & 0 & a & \ldots & 0 & 0 & 0 \\ \vdots & \vdots & \vdots & \ddots & \vdots & \vdots & \vdots \\ 0 & 0 & 0 & \ldots & a & 1 & 0 \\ 0 & 0 & 0 & \ldots & 0 & a & 1 \\ 0 & 0 & 0 & \ldots & 0 & 0 & a \end{bmatrix}.$$

We say that a is the eigenvalue associated with J, and we see that in the matrix J, each entry on the diagonal is an a and each entry on the "superdiagonal" (the entries above the diagonal) is a 1. All other entries are 0. For example,

$$[3], \begin{bmatrix} 2 & 1 \\ 0 & 2 \end{bmatrix}, \text{ and } \begin{bmatrix} -1 & 1 & 0 \\ 0 & -1 & 1 \\ 0 & 0 & -1 \end{bmatrix}$$

are Jordan blocks, but

$$\begin{bmatrix} 2 & 1 \\ 0 & 1 \end{bmatrix} \text{ and } \begin{bmatrix} -1 & 1 & 0 \\ 0 & -1 & 0 \\ 0 & 0 & -1 \end{bmatrix}$$

are not Jordan blocks. It is not difficult to calculate the minimum and characteristic polynomials for a Jordan block.

Theorem 6.1.1. *Let J be an $m \times m$ Jordan block with eigenvalue a. Then $p_J(\lambda) = (-1)^m(\lambda - a)^m$ and $m_J(\lambda) = (\lambda - a)^m$.*

(For a Jordan block the characteristic and minimum polynomials are equal, except possibly for sign.)

Proof. Since J is upper triangular, we see that

$$p_J(\lambda) = |J - \lambda I| = (a - \lambda)^m = (-1)^m(\lambda - a)^m.$$

By previous results, we know that $m_J(\lambda)$ is a factor of $p_J(\lambda)$ and so $m_J(\lambda) = (\lambda - a)^k$, where $1 \leq k \leq m$ and k is the least integer satisfying

$$m_J(J) = (J - aI)^k = 0.$$

Now

$$J - aI = \begin{bmatrix} 0 & 1 & 0 & \ldots & 0 & 0 \\ 0 & 0 & 1 & \ldots & 0 & 0 \\ 0 & 0 & 0 & \ldots & 0 & 0 \\ \vdots & \vdots & \vdots & \ddots & \vdots & \vdots \\ 0 & 0 & 0 & \ldots & 0 & 1 \\ 0 & 0 & 0 & \ldots & 0 & 0 \end{bmatrix}$$

and we see that

$$(J - aI)^2 = \begin{bmatrix} 0 & 0 & 1 & 0 & \cdots & 0 & 0 \\ 0 & 0 & 0 & 1 & \cdots & 0 & 0 \\ 0 & 0 & 0 & 0 & \cdots & 0 & 0 \\ \vdots & \vdots & \vdots & \vdots & \ddots & \vdots & \vdots \\ 0 & 0 & 0 & 0 & \cdots & 0 & 1 \\ 0 & 0 & 0 & 0 & \cdots & 0 & 0 \\ 0 & 0 & 0 & 0 & \cdots & 0 & 0 \end{bmatrix}$$

$$\vdots$$

$$(J - aI)^{m-1} = \begin{bmatrix} 0 & \cdots & 0 & 1 \\ 0 & \cdots & 0 & 0 \\ \vdots & \vdots & \ddots & \vdots \\ 0 & \cdots & 0 & 0 \end{bmatrix}$$

$$(J - aI)^m = (J - aI)(J - aI)^{m-1}$$
$$= 0.$$

From this we see that $m_J(\lambda) = (\lambda - a)^m$. $\qquad\square$

Now let us consider the eigenvectors associated with an $m \times m$ Jordan block J with eigenvalue a. Since

$$J - aI = \begin{bmatrix} 0 & 1 & 0 & \cdots & 0 & 0 \\ 0 & 0 & 1 & \cdots & 0 & 0 \\ \vdots & \vdots & \vdots & \ddots & \vdots & \vdots \\ 0 & 0 & 0 & \cdots & 0 & 1 \\ 0 & 0 & 0 & \cdots & 0 & 0 \end{bmatrix},$$

it is not hard to see that the first $m - 1$ row vectors are linearly independent and that $J - aI$ has rank $m - 1$. From this we see that J has only $m - (m - 1) = 1$ linearly independent eigenvectors. Let us find conditions under which an $m \times m$ matrix A is similar to a Jordan block.

Let A be an $m \times m$ matrix and assume A is similar to the Jordan block J with a on the diagonal. Then since similar matrices have the same characteristic polynomials, $p_A(\lambda) = (a - \lambda)^m$. Let S be the nonsingular matrix with $S^{-1}AS = J$ and assume $S = [X_1 \ldots X_m]$, where X_j is the j-th column vector of S. Then we get $AS = SJ$ and so

$$A[X_1 \ldots X_m] = [X_1 \ldots X_m] \begin{bmatrix} a & 1 & 0 & \cdots & 0 \\ 0 & a & 1 & \cdots & 0 \\ 0 & 0 & a & \cdots & 0 \\ \vdots & \vdots & \vdots & \ddots & \vdots \\ 0 & 0 & 0 & \cdots & a \end{bmatrix}.$$

It follows that

$$\begin{bmatrix} AX_1 & AX_2 & \dots & AX_m \end{bmatrix} = \begin{bmatrix} aX_1 & X_1 + aX_2 & X_2 + aX_3 & \dots & X_{m-1} + aX_m \end{bmatrix}$$

and so

$$AX_1 = aX_1$$
$$AX_2 = X_1 + aX_2$$
$$\vdots$$
$$AX_m = X_{m-1} + aX_m.$$

Rewriting this we obtain

$$(A - aI)X_1 = 0$$
$$(A - aI)X_2 = X_1$$
$$(A - aI)X_3 = X_2$$
$$\vdots$$
$$(A - aI)X_m = X_{m-1}.$$

Notice that X_1 is an eigenvector. The other vectors $X_2, \dots, X_m$ are called **generalized eigenvectors**, and $X_1, \dots, X_m$ is called a **Jordan basis**. This proves one part of the following theorem.

Theorem 6.1.2. *An $m \times m$ matrix A is similar to an $m \times m$ Jordan block J with eigenvalue a if and only if there are independent $m \times 1$ column vectors $X_1, \dots, X_m$ satisfying*

$$(A - aI)X_1 = 0$$
$$(A - aI)X_2 = X_1$$
$$(A - aI)X_3 = X_2$$
$$\vdots$$
$$(A - aI)X_m = X_{m-1}.$$

(A Jordan block corresponds to a string of generalized eigenvectors.)

Proof. See Exercise 11. □

Example 6.1.2. Let $A = \begin{bmatrix} 3 & 1 \\ -1 & 1 \end{bmatrix}$. Then $p_A(\lambda) = (\lambda - 2)^2$, so $\lambda = 2$ is an eigenvalue of multiplicity 2. The rank of $A - 2I = \begin{bmatrix} 1 & 1 \\ -1 & -1 \end{bmatrix}$ is 1 and so there is only one independent eigenvector. It follows that A is not similar to a diagonal matrix. Let $X_1 = \begin{bmatrix} 1 \\ -1 \end{bmatrix}$. Then X_1 is an eigenvector of A associated with the eigenvalue $\lambda = 2$, and there is no other eigenvector independent from X_1. Let us attempt to find a vector X_2 so that X_1, X_2 forms a Jordan basis. We need to solve the equation $(A - 2I)X_2 = X_1$ or

$$\begin{bmatrix} 1 & 1 \\ -1 & -1 \end{bmatrix} \begin{bmatrix} x \\ y \end{bmatrix} = \begin{bmatrix} 1 \\ -1 \end{bmatrix}.$$

We see that $X_2 = \begin{bmatrix} 1 \\ 0 \end{bmatrix}$ is a solution. Now let $S = [X_1 X_2] = \begin{bmatrix} 1 & 1 \\ -1 & 0 \end{bmatrix}$. Then

$$S^{-1}AS = \begin{bmatrix} 2 & 1 \\ 0 & 2 \end{bmatrix},$$

which is a Jordan block.

$\square$

Section 6.1 Exercises

For each of the matrices in Exercises 1 - 5 determine which are similar to diagonal matrices. Give reasons for your conclusion.

1. $\begin{bmatrix} 0 & 2 \\ 1 & 1 \end{bmatrix}$

2. $\begin{bmatrix} 3 & 7 \\ 7 & -2 \end{bmatrix}$

3. $\begin{bmatrix} 1 & 0 \\ 2 & 1 \end{bmatrix}$

4. $\begin{bmatrix} -1 & 2 & 1 \\ 0 & 1 & 1 \\ 1 & 0 & 2 \end{bmatrix}$

5. $\begin{bmatrix} 1 & i & 1-i \\ -i & 3 & 1 \\ 1+i & 1 & 2 \end{bmatrix}$

6. Which of the following matrices are Jordan blocks? Give reasons.

(a) $\begin{bmatrix} 1 & 1 \\ 0 & 2 \end{bmatrix}$

(b) $[2]$

(c) $\begin{bmatrix} -1 & 1 \\ 0 & -1 \end{bmatrix}$

(d) $\begin{bmatrix} 2 & 0 \\ 0 & 2 \end{bmatrix}$

(e) $\begin{bmatrix} 1 & 0 & 0 \\ 0 & 1 & 1 \\ 0 & 0 & 1 \end{bmatrix}$

7. Find a Jordan block J that is similar to the matrix $A = \begin{bmatrix} 1 & -1 \\ 1 & 3 \end{bmatrix}$.

8. Find a Jordan block J that is similar to the matrix $A = \begin{bmatrix} 1 & 0 & 0 \\ 1 & 1 & 1 \\ 1 & 0 & 1 \end{bmatrix}$.

9. For the matrices A and J in Exercise 7, find a matrix S such that $S^{-1}AS = J$.

10. For the matrices A and J in Exercise 8, find a matrix S such that $S^{-1}AS = J$.

11. Prove the remaining part of Theorem 6.1.2.

12. Let A and $S = [X_1 X_2 X_3]$ be 3×3 matrices and assume that

$$S^{-1}AS = \begin{bmatrix} 2 & 1 & 0 \\ 0 & 2 & 1 \\ 0 & 0 & 2 \end{bmatrix}.$$

Write the relationships satisfied by the matrix A and the column vectors of S.

13. Follow the instructions in Exercises 12 assuming that

$$S^{-1}AS = \begin{bmatrix} 2 & 1 & 0 \\ 0 & 2 & 0 \\ 0 & 0 & 2 \end{bmatrix}.$$

6.2   The Jordan Canonical Form

As we have observed before, not every matrix is similar to a diagonal matrix. By Theorem 5.5.2, we know that if the characteristic polynomial of a matrix A factors completely, then A is similar to an upper triangular matrix. One wonders if this is the best result that can be obtained. The answer is "no" and in this chapter we investigate this "closest-to-diagonal" matrix that can be obtained by similarity transformations.

Let A be an $n \times n$ matrix and let S be the set of all matrices that are similar to A. If the characteristic polynomial of $A, p_A(\lambda)$, factors completely, then we know that A is similar to an upper triangular matrix U. But this upper triangular matrix U is not unique. For example, suppose $A = \begin{bmatrix} 1 & 1 \\ 0 & 2 \end{bmatrix}$. Then A itself is upper triangular, and, of course, A is similar to itself so that $A \in S$. But A has two distinct eigenvalues (1 and 2) so that A is similar to the diagonal matrix $D = \begin{bmatrix} 1 & 0 \\ 0 & 2 \end{bmatrix}$. It follows that $D \in S$ and so D is a second upper triangular matrix in S. Since A is also similar to the diagonal matrix $D' = \begin{bmatrix} 2 & 0 \\ 0 & 1 \end{bmatrix}$, one can see that "absolute" uniqueness is probably impossible to achieve.

The following theorem identifies a "closest-to-diagonal" matrix J in the class S of matrices that are similar to a given matrix A, and states that this matrix J is unique (more or less). Because of this, J is called a **canonical form**, and being named after its founder, it is called the **Jordan canonical form**. The matrix J is called the Jordan canonical form "of A," and J is said to be "in" Jordan canonical form. In this context the word "canonical" has nothing to do with church law, but rather carries the implication of "simplest" and "unique." The proof is omitted here, but outlined in Appendix F.

Theorem 6.2.1. *Let A be an $n \times n$ matrix and assume that $p_A(\lambda)$ factors completely. Then A is similar to a matrix J of the form*

$$
J = \begin{bmatrix} J_1 & 0 & \dots & 0 \\ 0 & J_2 & \dots & 0 \\ \vdots & \vdots & \ddots & \vdots \\ 0 & 0 & \dots & J_k \end{bmatrix},
$$

where $J_1, \dots, J_k$ are Jordan blocks. The matrix J is unique except for the order of the blocks $J_1, \dots, J_k$, which can occur in any order.

(If the characteristic polynomial factors completely, the matrix is similar to a matrix in Jordan form.)

The above theorem is an "existence" theorem in that it states the existence of a quantity, J in this case, but offers no help in finding it. There are not many parameters in the make-up of the matrix J. We need to know how many blocks, the size of the blocks, and the eigenvalue associated with the blocks. These parameters may often be determined by investigating properties of the original matrix A.

6.2.1 PROPERTIES OF THE JORDAN FORM

It is, in general, difficult to find the Jordan canonical form of a matrix, but knowledge of certain elementary facts simplifies the task. In the following discussion we will assume that A is an $n \times n$ matrix and the characteristic polynomial of A factors completely, say $p_A(\lambda) = (a_1 - \lambda)^{m_1} \dots (a_s - \lambda)^{m_s}$, where $a_1, \dots, a_s$ are distinct. Further, let the minimum polynomial of A be $m_A(\lambda) = (\lambda - a_1)^{n_1} \dots (\lambda - a_s)^{n_s}$. Let J be the Jordan canonical form of A, and assume $J_1, \dots, J_k$ are the Jordan blocks of J.

Since J and A are similar they have the same characteristic polynomial, and since J is upper triangular, the eigenvalues of J lie on the diagonal. From this it is easy to see that the following theorem is true.

Theorem 6.2.2. *The sum of the orders of the blocks in which a_i occurs on the diagonal is m_i; that is, a_i occurs on the diagonal of J m_i times.*

(An eigenvalue of multiplicity m occurs m times on the diagonal of the Jordan form.)

Now let S be a nonsingular matrix such that $S^{-1}AS = J$, or $AS = SJ$. If $S = [X_1 \dots X_n]$, where X_j is the j-th column of S, then $X_1, \dots, X_n$ are linearly independent and we have

$$
\begin{aligned}
AS &= \begin{bmatrix} AX_1 & \dots & AX_n \end{bmatrix} \\
&= SJ \\
&= \begin{bmatrix} X_1 & \dots & X_n \end{bmatrix} \begin{bmatrix} J_1 & \dots & 0 \\ \vdots & \ddots & \vdots \\ 0 & \dots & J_k \end{bmatrix} \\
&= \begin{bmatrix} b_1 X_1 & X_1 + b_1 X_2 & \dots & X_{r-1} + b_1 X_r & b_2 X_{r+1} & X_{r+1} + b_2 X_{r+2} & \dots \end{bmatrix},
\end{aligned}
$$

where b_i is the eigenvalue associated with J_i and J_1 is $r \times r$. Note that $b_1 = b_2 = a_1$ is possible. This relabeling is necessary because a_1 could show up in both J_1 and J_2, for example.

If we let $A_i = A - b_i I$ and if we equate the columns of AS and SJ, we have

$$
\begin{aligned}
AX_1 &= b_1 X_1 & \implies & & A_1 X_1 &= 0 & & & A_2 X_{r+1} &= 0 \\
AX_2 &= X_1 + b_1 X_2 & \implies & & A_1 X_2 &= X_1 & & & A_2 X_{r+2} &= X_{r+1} \\
&\;\;\vdots & & & &\;\;\vdots & &\;\;\vdots & &\;\;\vdots \\
AX_r &= X_{r-1} + b_1 X_r & \implies & & A_1 X_r &= X_{r-1} & & & \text{etc.}
\end{aligned}
$$

A basis of the above form is called a **Jordan basis**. From the above computation one sees that $X_1, X_{r+1}, \ldots$ are linearly independent eigenvectors and there is one of them for each Jordan block of J. We have shown the following theorem:

Theorem 6.2.3. *The number of blocks associated with the eigenvalue a_i is equal to the number of linearly independent eigenvectors associated with a_i.*

(There is a block in J for each independent eigenvector.)

Although it is harder to see, the following also holds:

Theorem 6.2.4. *The order of the largest block associated with a_i is n_i, the exponent of $(\lambda - a_i)$ in $m_A(\lambda)$.*

(The largest block with a given eigenvalue is multiplicity of the eigenvalue in the minimum polynomial.)

Proof. By assumption, $m_A(\lambda) = (\lambda - a_1)^{n_1} \ldots (\lambda - a_s)^{n_s}$ is the minimum polynomial of A and since similar matrices have the same minimum polynomial (See Exercise 7, Section 5.6), $m_J(\lambda) = m_A(\lambda)$. Notice what happens when one multiplies matrices (assuming the products are defined) that are in "block-diagonal form":

$$
\begin{bmatrix} A_1 & \cdots & 0 \\ \vdots & \ddots & \vdots \\ 0 & \cdots & A_k \end{bmatrix}
\begin{bmatrix} B_1 & \cdots & 0 \\ \vdots & \ddots & \vdots \\ 0 & \cdots & B_k \end{bmatrix}
=
\begin{bmatrix} A_1 B_1 & \cdots & 0 \\ \vdots & \ddots & \vdots \\ 0 & \cdots & A_k B_k \end{bmatrix}.
$$

Assume now that J_1 is the largest block associated with the eigenvalue a_1 and that J_1 is $r \times r$. Recall that by Theorem 6.1.1, the minimum polynomial of J_1 is $(\lambda - a_1)^r$. That is, r is the least power of $(J_1 - a_1 I)$ that is the zero matrix. Now substitute J into $m_J(\lambda)$ and apply these observations:

$$m_J(J) = (J - a_1 I)^{n_1} \dots (J - a_s I)^{n_s}$$

$$= \begin{bmatrix} J_1 - a_1 I & \cdots & 0 \\ \vdots & \ddots & \vdots \\ 0 & \cdots & J_k - a_1 I \end{bmatrix}^{n_1} \cdots \begin{bmatrix} J_1 - a_s I & \cdots & 0 \\ \vdots & \ddots & \vdots \\ 0 & \cdots & J_k - a_s I \end{bmatrix}^{n_s}$$

$$= \begin{bmatrix} (J_1 - a_1 I)^{n_1} & \cdots & 0 \\ \vdots & \ddots & \vdots \\ 0 & \cdots & (J_k - a_1 I)^{n_1} \end{bmatrix} \cdots \begin{bmatrix} (J_1 - a_s I)^{n_s} & \cdots & 0 \\ \vdots & \ddots & \vdots \\ 0 & \cdots & (J_k - a_s I)^{n_s} \end{bmatrix}$$

$$= \begin{bmatrix} (J_1 - a_1 I)^{n_1} \dots (J_1 - a_s I)^{n_s} & \cdots & 0 \\ \vdots & \ddots & \vdots \\ 0 & \cdots & (J_k - a_1 I)^{n_1} \dots (J_k - a_s I)^{n_s} \end{bmatrix}.$$

Now since $m_J(J) = 0$, all blocks on the diagonal of this last matrix must be zero. It follows that

$$(J_1 - a_1 I)^{n_1} \dots (J_1 - a_s I)^{n_s} = 0.$$

But $a_1 \neq a_2, \dots, a_s$, and so each of the terms $(J_1 - a_2 I), \dots, (J_1 - a_s I)$ have nonzero entries on their diagonals. It follows that $(J_1 - a_1 I)^{n_1} = 0$, and so $n_1 \geq r$, the order of J_1. Now J_1 was assumed to be the largest block associated with a_1 and so if J_2 also has the eigenvalue a_1, $(J_2 - a_1 I)^r = 0$, as well (by Theorem 6.1.1 since r is greater than or equal to the order of J_2). But n_1 is the least integer satisfying this condition, so $n_1 \leq r$. Thus $n_1 = r$, and the theorem is proved since the order of the blocks may be arbitrarily chosen. $\square$

6.2.2 FINDING THE JORDAN FORM

While it is not in general easy to find the Jordan canonical form of a matrix, the above results provide enough information that it is possible to determine the Jordan form in certain cases. Finding the characteristic and minimum polynomials is the first step. Next, knowing the nature of the associated eigenvectors provides further clues. Often these two steps prove sufficient, but in other cases, one must try to find a Jordan basis. The following examples illustrate the various possibilities.

Example 6.2.1.   (a) Assume that A is a matrix such that:

$$p_A(\lambda) = (1 - \lambda)^3 (2 - \lambda)^2$$

$$m_A(\lambda) = (\lambda - 1)^2 (\lambda - 2).$$

Then if J is the Jordan canonical form of A, we know by Theorem 6.2.2 that 1 appears three times on the diagonal of J and 2 appears twice. By Theorem 6.2.4, the order of the largest block associated with the eigenvalue 2 is 1. From this we can see that

$$J = \begin{bmatrix} \begin{bmatrix} 1 & 1 \\ 0 & 1 \end{bmatrix} & & & 0 \\ & [1] & & \\ & & [2] & \\ 0 & & & [2] \end{bmatrix}.$$

Here we have shown the Jordan blocks that lie on the diagonal and represent the 0's that lie outside the blocks by a single 0.

(b) If A is such that

$$p_A(\lambda) = (1 - \lambda)^3(2 - \lambda)^2 = -m_A(\lambda),$$

then the largest block with a 2 on the diagonal has order 2 and the largest block with a 1 on the diagonal has order 3. From this we see that the Jordan canonical form is:

$$J = \begin{bmatrix} \begin{bmatrix} 1 & 1 & 0 \\ 0 & 1 & 1 \\ 0 & 0 & 1 \end{bmatrix} & 0 \\ 0 & \begin{bmatrix} 2 & 1 \\ 0 & 2 \end{bmatrix} \end{bmatrix}.$$

(c) Assume that A is a 4×4 matrix with

$$p_A(\lambda) = (1 - \lambda)^4 \text{ and } m_A(\lambda) = (\lambda - 1)^2.$$

By Theorem 6.2.4, the largest Jordan block associated with the eigenvalue 1 is 2×2. This leaves two possibilities. If A has three independent eigenvectors, then there are three blocks in the Jordan canonical form J and

$$J = \begin{bmatrix} \begin{bmatrix} 1 & 1 \\ 0 & 1 \end{bmatrix} & & 0 \\ & [1] & \\ 0 & & [1] \end{bmatrix}.$$

On the other hand, if A has only two independent eigenvectors, then J has two Jordan blocks on its diagonal and so

$$J = \begin{bmatrix} \begin{bmatrix} 1 & 1 \\ 0 & 1 \end{bmatrix} & 0 \\ 0 & \begin{bmatrix} 1 & 1 \\ 0 & 1 \end{bmatrix} \end{bmatrix}.$$

(d) Knowing the characteristic and minimum polynomials and the number of linearly independent eigenvectors may not be sufficient to determine the Jordan canonical form of a matrix. Suppose that A is a 7×7 matrix with

$$p_A(\lambda) = (1 - \lambda)^7 \text{ and } m_A(\lambda) = (\lambda - 1)^3.$$

If A has three linearly independent eigenvectors, then the Jordan canonical form J of A must have three Jordan blocks. Because of the minimum polynomial of A, the largest block must be 3×3. These conditions don't determine J. J could consist of a 3×3 Jordan block and two 2×2 blocks or J could have two 3×3 Jordan blocks and one 1×1 block.

(e) Let $A = \begin{bmatrix} 1 & 1 & 0 \\ 1 & 1 & 1 \\ 0 & -1 & 1 \end{bmatrix}$. Then $p_A(\lambda) = (1 - \lambda)^3$ and so $\lambda = 1$ is the only eigenvalue of A. Let us examine the eigenvectors of A in order to determine the Jordan canonical form of A. Consider

$$(A - 1I)X = \begin{bmatrix} 0 & 1 & 0 \\ 1 & 0 & 1 \\ 0 & -1 & 0 \end{bmatrix} \begin{bmatrix} x_1 \\ x_2 \\ x_3 \end{bmatrix} = 0.$$

We see that $x_2 = 0$ and $x_1 = -x_3$, and so $X_1 = \begin{bmatrix} 1 \\ 0 \\ -1 \end{bmatrix}$ is a solution and there is only one linearly independent solution. It follows from Theorem 6.2.3 that there is only one Jordan block in the Jordan canonical form J of A and so

$$J = \begin{bmatrix} 1 & 1 & 0 \\ 0 & 1 & 1 \\ 0 & 0 & 1 \end{bmatrix}.$$

To make the example interesting, let us find a matrix S such that $S^{-1}AS = J$. From the previous discussion we know that we want to take $S = [X_1 X_2 X_3]$, where X_1, X_2, X_3 is a Jordan basis; that is, $(A - I)X_1 = 0, (A - I)X_2 = X_1$ and $(A - I)X_3 = X_2$. We compute solutions to these equations:

$$(A - I)X_1 = 0; \text{ take } X_1 = \begin{bmatrix} 1 \\ 0 \\ -1 \end{bmatrix}$$

$$(A - I)X_2 = X_1; \text{ take } X_2 = \begin{bmatrix} 1 \\ 1 \\ -1 \end{bmatrix} \text{ or } \begin{bmatrix} 0 \\ 1 \\ 0 \end{bmatrix}$$

$$(A - I)X_3 = X_2; \text{ take } X_3 = \begin{bmatrix} 1 \\ 1 \\ 0 \end{bmatrix} \text{ or } \begin{bmatrix} 0 \\ 1 \\ 1 \end{bmatrix}.$$

Then X_1, X_2, X_3 are linearly independent and so they form a Jordan basis. If we let

$$S = \begin{bmatrix} 1 & 1 & 1 \\ 0 & 1 & 1 \\ -1 & -1 & 0 \end{bmatrix},$$

then

$$S^{-1} = \begin{bmatrix} 1 & -1 & 0 \\ -1 & 1 & -1 \\ 1 & 0 & 1 \end{bmatrix}$$

and $S^{-1}AS = J$.

(f) Finding a Jordan basis is not always straightforward. Suppose that

$$A = \begin{bmatrix} 3 & 1 & 0 \\ -1 & 1 & 0 \\ 1 & 1 & 2 \end{bmatrix}.$$

The characteristic polynomial is $(2 - \lambda)^3$ and we see that

$$A - 2I = \begin{bmatrix} 1 & 1 & 0 \\ -1 & -1 & 0 \\ 1 & 1 & 0 \end{bmatrix},$$

and so $A - 2I$ has rank 1. This means that there are two independent eigenvectors and therefore two Jordan blocks, a 2×2 block and a 1×1 block. To find a Jordan basis we must find X_1, X_2, X_3 satisfying

$$(A - 2I)X_1 = 0$$
$$(A - 2I)X_2 = X_1$$
$$(A - 2I)X_3 = 0$$

Solving for eigenvectors X_1 and X_3, consider

$$(A - 2I)X = \begin{bmatrix} 1 & 1 & 0 \\ -1 & -1 & 0 \\ 1 & 1 & 0 \end{bmatrix} \begin{bmatrix} x \\ y \\ z \end{bmatrix} = \begin{bmatrix} 0 \\ 0 \\ 0 \end{bmatrix}.$$

We see that $X_1 = \begin{bmatrix} 0 \\ 0 \\ 1 \end{bmatrix}$ and $X_3 = \begin{bmatrix} 1 \\ -1 \\ 0 \end{bmatrix}$ are obvious choices. Now let's find X_2. We must solve

$$(A - 2I)X_2 = X_1$$

or

$$(A - 2I)X_2 = \begin{bmatrix} 1 & 1 & 0 \\ -1 & -1 & 0 \\ 1 & 1 & 0 \end{bmatrix} \begin{bmatrix} x \\ y \\ z \end{bmatrix} = \begin{bmatrix} 0 \\ 0 \\ 1 \end{bmatrix}.$$

Considering the first and last rows, we see that there's no solution. Perhaps we should switch eigenvectors and try to solve

$$(A - 2I)X_2 = X_3$$

or

$$(A - 2I)X_2 = \begin{bmatrix} 1 & 1 & 0 \\ -1 & -1 & 0 \\ 1 & 1 & 0 \end{bmatrix} \begin{bmatrix} x \\ y \\ z \end{bmatrix} = \begin{bmatrix} 1 \\ -1 \\ 0 \end{bmatrix}.$$

Again, no solution. The theorem above states that our matrix A is similar to a matrix in Jordan form and it implies the existence of a Jordan basis. The theorem doesn't guarantee that the Jordan basis will be easy to find! We need to replace the eigenvector X_1 by another eigenvector so that $(A - 2I)X_2 = X_1$ does have a solution. Let's let

$$X_1 = \begin{bmatrix} 1 \\ -1 \\ 1 \end{bmatrix}.$$

Then X_1 is and eigenvector and X_2 can be chosen to be $\begin{bmatrix} 1 \\ 0 \\ 0 \end{bmatrix}$. The matrix $S =$

$\begin{bmatrix} 1 & 1 & 0 \\ -1 & 0 & 0 \\ 1 & 0 & 1 \end{bmatrix}$ will properly transform the matrix A to its Jordan form.

$\square$

Section 6.2 Exercises

In Exercises 1 - 6, find the Jordan canonical form of the matrix satisfying the given conditions.

1. A is a 3×3 matrix with $p_A(\lambda) = (-2 - \lambda)^3$ and $m_A(\lambda) = (\lambda + 2)^2$.

2. A is a 3×3 matrix with $p_A(\lambda) = (2 - \lambda)^3$ and $m_A(\lambda) = (\lambda - 2)^3$.

3. A is a 3×3 matrix with $p_A(\lambda) = (2 - \lambda)^3$ and $(A - 2I)^2 = 0$, but $A \neq 2I$.

4. A is a 5×5 matrix with $p_A(\lambda) = (2 - \lambda)^3(3 - \lambda)^2$ and $m_A(\lambda) = (\lambda - 2)^2(\lambda - 3)$.

5. A is a 6×6 matrix with $p_A(\lambda) = (2 - \lambda)^4(3 - \lambda)^2$ and $m_A(\lambda) = (\lambda - 2)^2(\lambda - 3)^2$. The matrix $A - 2I$ has rank 4.

6. A is a 6×6 matrix with $p_A(\lambda) = (1 - \lambda)^4(-2 - \lambda)^2$ and $m_A(\lambda) = (\lambda + 2)^2(\lambda - 1)^2$. The matrix $A - I$ has rank 3.

Find the Jordan canonical form of the matrices in Exercises 7 - 12.

7. $\begin{bmatrix} 1 & 1 \\ 1 & 1 \end{bmatrix}$

9. $\begin{bmatrix} 2 & 1 & 0 \\ -1 & 0 & 0 \\ 0 & 0 & 2 \end{bmatrix}$

11. $\begin{bmatrix} 0 & 1 & 0 \\ -8 & 6 & 1 \\ -1 & -1 & -1 \end{bmatrix}$

8. $\begin{bmatrix} 1 & 0 \\ 1 & 0 \end{bmatrix}$

10. $\begin{bmatrix} 1 & 0 & 1 \\ 1 & 1 & 1 \\ 0 & 0 & 1 \end{bmatrix}$

12. $\begin{bmatrix} 1 & 0 & 1 \\ 1 & 1 & 2 \\ 0 & 0 & 1 \end{bmatrix}$

13. Let $A = \begin{bmatrix} 1 & 1 & 0 \\ 0 & 1 & 0 \\ 1 & -1 & 2 \end{bmatrix}$. Find a matrix S such that $S^{-1}AS$ is in Jordan canonical form.

14. Find the Jordan canonical form J of $\begin{bmatrix} 1 & 0 & 1 \\ 1 & 1 & 2 \\ 0 & 0 & 1 \end{bmatrix}$ and find a nonsingular matrix S with $S^{-1}AS = J$.

15. Give an example of a 3×3 matrix A satisfying: (a) $\lambda = 2$ is the only eigenvalue of A, and (b) there are two linearly independent eigenvectors of A associated with this eigenvalue.

16. Give an example of a 3×3 matrix A satisfying: (a) $\lambda = 1$ is the only eigenvalue of A, and (b) there is only one linearly independent eigenvector of A associated with this eigenvalue.

17. Let A be a 3×3 matrix satisfying: (a) $\lambda = 1$ is the only eigenvalue of A, and (b) there are three linearly independent eigenvectors of A associated with the eigenvalue of $\lambda = 1$. Show that $A = I$.

6.3   Systems of Constant Coefficient Differential Equations (optional)

The Jordan canonical form of a matrix is of use in solving differential equations. In most elementary courses in differential equations, a general discussion of systems of first order, constant coefficient linear differential equations is not included. This omission is mainly due to the unavailability of the Jordan canonical form which is necessary in solving such systems. It seems appropriate to include a discussion of these systems here. In what follows, the reader will find some background in differential equations helpful, but not absolutely necessary.

By a **system of first order constant coefficient linear differential equations** we mean a system of the form

$$
\begin{aligned}
x_1' &= a_{11}x_1 + \ldots + a_{1n}x_n + f_1(t) \\
&\vdots \qquad\qquad\qquad\qquad \vdots \\
x_n' &= a_{n1}x_1 + \ldots + a_{nn}x_n + f_n(t)
\end{aligned}
\tag{6.3.1}
$$

Of course a **solution** of System 6.3.1 is a collection of functions $x_1(t), \ldots, x_n(t)$ that satisfy the equations on some interval.

These systems of equations arise naturally: If we consider the n-th order constant coefficient linear differential equation

$$
y^{(n)} + a_{n-1}y^{(n-1)} + \ldots + a_1y' + a_0y = g(t),
\tag{6.3.2}
$$

then this equation can be "reduced" to a system of the form 6.3.1 by making the substitutions

$$
\begin{aligned}
x_1 &= y \\
x_2 &= y' \\
&\vdots \\
x_n &= y^{(n-1)}
\end{aligned}
\tag{6.3.3}
$$

Using the substitutions 6.3.3 in 6.3.2 we obtain

$$
\begin{aligned}
x_1' &= x_2 \\
x_2' &= x_3 \\
&\vdots \\
x_{n-1}' &= x_n \\
x_n' &= -a_{n-1}x_n - \ldots - a_1x_2 - a_0x_1 + g(t),
\end{aligned}
\tag{6.3.4}
$$

which is a system of equations of the form 6.3.1.

A further natural occurrence of linear systems of differential equations arises in applying Kirchhoff's laws (see Section 1.7) to electrical networks involving inductors and capacitors, applying laws of motion to coupled spring-mass systems, and in other physical situations. To solve a first order system of linear differential equations we will need to be able to solve a single first order equation. We start with a brief review.

6.3.1   SOLVING A FIRST ORDER EQUATION

The general first order linear differential equation is an equation of the form

$$
y' + f(t)y = g(t),
\tag{6.3.5}
$$

where f and g are assumed to be continuous. This equation can be solved by using the integrating factor

$$
p(t) = e^{\int f(t)\,dt}
\tag{6.3.6}
$$

If both sides of Equation 6.3.5 are multiplied by $p(t)$, we obtain

$$y'p(t) + f(t)yp(t) = g(t)p(t). \tag{6.3.7}$$

But notice that

$$\begin{aligned} \frac{d}{dt}(yp(t)) &= y'p(t) + yp'(t) \\ &= y'p(t) + yf(t)p(t) \end{aligned} \tag{6.3.8}$$

since

$$p'(t) = e^{\int f(t)dt} = p(t)f(t).$$

Now combining 6.3.7 and 6.3.8, we get

$$\frac{d}{dt}(yp(t)) = g(t)p(t);$$

integrating, we obtain

$$yp(t) = \int g(t)p(t)dt + C$$

or

$$y = \frac{1}{p(t)} \int g(t)p(t)dt + \frac{C}{p(t)}. \tag{6.3.9}$$

The assumption of continuity of the functions $f(t)$ and $g(t)$ guarantees that the integrals in 6.3.6 and 6.3.9 exist. The solution 6.3.9 is called the **general solution** of 6.3.5 and it involves the unknown constant C. An initial condition of the form $y(a) = b$ determines the value of C and gives the unique solution that satisfies this condition.

Example 6.3.1. Consider the equation

$$y' + 2ty = t.$$

Applying the above method we multiply by the integrating factor

$$p(t) = e^{\int 2tdt} = e^{t^2}$$

and obtain

$$y'e^{t^2} + 2te^{t^2}y = te^{t^2}.$$

Rewriting this equation we get

$$\frac{d}{dt}ye^{t^2} = te^{t^2}$$

and integrating both sides with respect to t gives

$$ye^{t^2} = \int te^{t^2}dt = \frac{e^{t^2}}{2} + C.$$

It follows that

$$y = \frac{1}{2} + Ce^{-t^2}.$$

If a solution is desired that satisfies $y(0) = 2$, then we get

$$2 = \frac{1}{2} + Ce^0$$

so that $C = \frac{3}{2}$ and we obtain the specific solution

$$y = \frac{1}{2} + \frac{3}{2}e^{-t^2}.$$

$\square$

Just as matrix notation simplified the expression and solution of systems of linear equations, the same holds true for systems of differential equations.

6.3.2 MATRIX NOTATION

Let us now return to the problem of finding a solution to the system of equations 6.3.1. We introduce matrix notation to write Equations 6.3.1 in a more compact "matrix" differential equation. Let

$$X = X(t) = \begin{bmatrix} x_1(t) \\ \vdots \\ x_n(t) \end{bmatrix},$$

$$A = [a_{ij}],$$

and

$$F = F(t) = \begin{bmatrix} f_1(t) \\ \vdots \\ f_n(t) \end{bmatrix}.$$

For a "matrix" function $X(t)$ we define the derivative X' of X by

$$X'(t) = \begin{bmatrix} x_1'(t) \\ \vdots \\ x_n'(t) \end{bmatrix}.$$

Using this notation, the system 6.3.1 can then be expressed as a matrix differential equation of the form

$$X' = AX + F(t) \tag{6.3.10}$$

6.3.3   USING THE JORDAN FORM

The first step in solving the matrix equation 6.3.10 is as follows: A is similar to a matrix J in Jordan canonical form (J may have complex numbers on its diagonal), say $S^{-1}AS = J$ or $A = SJS^{-1}$. Then

$$X' = SJS^{-1}X + F(t)$$

or

$$S^{-1}X' = JS^{-1}X + S^{-1}F(t).$$

Let $Y = S^{-1}X$ and $G(t) = S^{-1}F(t)$. Then Equation 6.3.10 becomes

$$Y' = JY + G(t). \tag{6.3.11}$$

Now let $J = \begin{bmatrix} J_1 & \cdots & 0 \\ \vdots & \ddots & \vdots \\ 0 & \cdots & J_k \end{bmatrix}$, where each J_i is Jordan block. Assume that J_i is $n_i \times n_i$ and let $Y = \begin{bmatrix} Y_1 \\ \vdots \\ Y_k \end{bmatrix}$ and $G(t) = \begin{bmatrix} G_1(t) \\ \vdots \\ G_k(t) \end{bmatrix}$, where Y_i and $G_i(t)$ are $n_i \times 1$. Then we get

$$
Y' = \begin{bmatrix} Y_1' \\ \vdots \\ Y_k' \end{bmatrix}
$$
$$
= \begin{bmatrix} J_1 & \cdots & 0 \\ \vdots & \ddots & \vdots \\ 0 & \cdots & J_k \end{bmatrix} \begin{bmatrix} Y_1 \\ \vdots \\ Y_k \end{bmatrix} + \begin{bmatrix} G_1(t) \\ \vdots \\ G_k(t) \end{bmatrix}
$$
$$
= \begin{bmatrix} J_1 Y_1 + G_1(t) \\ \vdots \\ J_k Y_k + G_k(t) \end{bmatrix}.
$$

From this we see that we need only solve the systems

$$Y_i' = J_i Y_i + G_i(t) \tag{6.3.12}$$

for $i = 1, \ldots, k$, where each J_i is a Jordan block. Given the solutions Y_i, let $Y = \begin{bmatrix} Y_1 \\ \vdots \\ Y_k \end{bmatrix}$ and take $X = SY$. X is then the solution of 6.3.10 and so we need only consider a system of equations of the form

$$Z' = JZ + H(t), \tag{6.3.13}$$

where J is a Jordan block. Let $Z = \begin{bmatrix} z_1 \\ \vdots \\ z_m \end{bmatrix}$, $H(t) = \begin{bmatrix} h_1(t) \\ \vdots \\ h_m(t) \end{bmatrix}$, and

$$J = \begin{bmatrix} a & 1 & 0 & \ldots & 0 & 0 \\ 0 & a & 1 & \ldots & 0 & 0 \\ 0 & 0 & a & \ldots & 0 & 0 \\ \vdots & \vdots & \vdots & \ddots & \vdots & \vdots \\ 0 & 0 & 0 & \ldots & a & 1 \\ 0 & 0 & 0 & \ldots & 0 & a \end{bmatrix}.$$

The system 6.3.12 is then equivalent to

$$
\begin{aligned}
z_1' &= az_1 + z_2 + h_1(t) \\
&\vdots \\
z_{m-1}' &= az_{m-1} + z_m + h_{m-1}(t) \\
z_m' &= az_m + h(t).
\end{aligned}
\tag{6.3.14}
$$

The problem of solving a system of differential equations has now been reduced to solving a system associated with a single Jordan block. We tackle that problem next.

6.3.4   SOLVING WHEN JORDAN BLOCKS OCCUR

Solving systems determined by Jordan blocks as in 6.3.14 is relatively straightforward. We start from the bottom and work up: solve the last equation for z_m and substitute z_m into the next to the last equation and solve for z_{m-1}, and so forth.. At each step we need to solve a constant coefficient first order linear differential equation of the form $y' - ay = f(t)$.

To solve $y' - ay = f(t)$, we proceed as before: Find the integrating factor $e^{\int -a\,dt} = e^{-at}$. Multiply the equation by the integrating factor and get

$$y'e^{-at} - ae^{-at}y = e^{-at}f(t).$$

Notice that the left hand side of the equation is the derivative of ye^{-at}, and so we have

$$\frac{d}{dt}(ye^{-at}) = e^{-at}f(t).$$

Now we can integrate both sides to obtain

$$ye^{-at} = \int e^{-at}f(t)dt$$

and solve for y, obtaining $y = e^{at} \int e^{-at}f(t)dt$.

Example 6.3.2. Solve $Z' = \begin{bmatrix} 2 & 1 \\ 0 & 2 \end{bmatrix} Z + \begin{bmatrix} e^t \\ e^{2t} \end{bmatrix}$. We must solve the system:

(a) $z_1' = 2z_1 + z_2 + e^t$

(b) $z_2' = 2z_2 + e^{2t}$.

1. Solve (b) first: The integrating factor is e^{-2t}; multiply and get

$$z_2' e^{-2t} z_2 = \frac{d}{dt} z_2 e^{-2t} z_2 = \frac{d}{dt} z_2 e^{-2t} = e^{-2t} e^{2t} = 1.$$

Integrating, we get $z_1 e^{-2t} = \int 1 \, dt - t + C$, so that $z_2 = t e^{2t} + C e^{2t}$.

2. Substitution into (a) gives

$$z'1 = 2z_1 + t e^{2t} + C e^{2t} + e^{t};$$

now solve for z_1. The integrating factor is e^{-2t}. Multiply, manipulate and get

$$z_1' e^{-2t} - 2 e^{-2t} z_1 = \frac{d}{dt} z_1 e^{-2t} = t + C + e^{-t}.$$

Integrating, we get

$$z_1 e^{-2t} = \frac{t^2}{2} + Ct - e^{-t} + C'$$

or

$$z_1 = \frac{t^2}{2} e^{2t} + Ct e^{2t} - e^{t} + C' e^{2t}.$$

It follows that the general solution of the system of equations is:

$$z_1 = \frac{t^2}{2} e^{2t} + Ct e^{2t} - e^{t} + C' e^{2t}$$

$$z_2 = t e^{2t} + C e^{2t}.$$

If a solution satisfying the initial conditions $z_1(0) = 2$, $z_2(0) = 3$ is desired, we can solve for C and C' : $z_2(0) = 0 e^{2 \cdot 0} + C e^{2 \cdot 0} = 3$, so $C = 3$. $z_1(0) = \frac{0}{2} e^{2 \cdot 0} + C \cdot 0 \cdot e^{0} - e^{0} + C' e^{0} = 2$, so $C' = 1$ and we get the following solution of the initial value problem:

$$z_1 = \frac{t^2}{2} e^{2t} + 3t e^{2t} - e^{t} + e^{2t}$$

$$z_2 = t e^{2t} + 3 e^{2t}.$$

$\square$

The above example illustrates only one step in a rather long and complicated process. To illustrate the entire process, we will go through the details of the following example:

Example 6.3.3. Find the general solution of

$$x_1' = 3x_1 + x_2 + x_3 + 1$$
$$x_2' = 2x_1 + 2x_2 + x_3 + e^t$$
$$x_3' = -6x_1 - 3x_2 - 2x_3 + e^{2t}.$$

We first formulate the problem in matrix form:

$$X' = \begin{bmatrix} 3 & 1 & 1 \\ 2 & 2 & 1 \\ -6 & -3 & -2 \end{bmatrix} X + \begin{bmatrix} 1 \\ e^t \\ e^{2t} \end{bmatrix}$$

and calculate the Jordan canonical form J of the matrix

$$A = \begin{bmatrix} 3 & 1 & 1 \\ 2 & 2 & 1 \\ -6 & -3 & -2 \end{bmatrix}.$$

Now $p_A(\lambda) = (1 - \lambda)^3$ and $m_A(\lambda) = (\lambda - 1)^2$ since $(A - 1I)^2 = 0$. We see that

$$J = \begin{bmatrix} 1 & 1 & 0 \\ 0 & 1 & 0 \\ 0 & 0 & 1 \end{bmatrix}.$$

We will need to know the matrix S that satisfies $S^{-1}AS = J$. Recall that $S = \begin{bmatrix} X_1 & X_2 & X_3 \end{bmatrix}$, where X_1, X_2, X_3 is a Jordan basis. Because of the form of J, these basis vectors must satisfy $(A - I)X_1 = 0, (A - I)X_2 = X_1, (A - I)X_3 = 0$. After some experimentation, we get

$$X_1 = \begin{bmatrix} 1 \\ 1 \\ -3 \end{bmatrix}, X_2 = \begin{bmatrix} 0 \\ 0 \\ 1 \end{bmatrix}, \text{ and } X_3 = \begin{bmatrix} -1 \\ 1 \\ 1 \end{bmatrix}.$$

Thus $S = \begin{bmatrix} 1 & 0 & -1 \\ 1 & 0 & 1 \\ -3 & 1 & 3 \end{bmatrix}$. Also, $S^{-1} = -\dfrac{1}{2}\begin{bmatrix} -1 & -1 & 0 \\ -4 & -2 & -2 \\ 1 & -1 & 0 \end{bmatrix} = \dfrac{1}{2}\begin{bmatrix} 1 & 1 & 0 \\ 4 & 2 & 2 \\ -1 & 1 & 0 \end{bmatrix}.$

By the previous section, we must solve

$$Y' = JY + G(t),$$

where $Y = S^{-1}X$ and $G(t) = S^{-1}F(t) = \begin{bmatrix} \dfrac{1}{2}(1 + e^t) \\ 2 + e^t + e^{2t} \\ \dfrac{1}{2}(e^t - 1) \end{bmatrix}$. We let $Y = \begin{bmatrix} y_1 \\ y_2 \\ y_3 \end{bmatrix}$ and solve the

system, which you will recall involves solving the systems determined by the blocks of J. The 1×1 block gives the equation

$$y_3' = y_3 + \frac{1}{2}(e^t - 1).$$

The solution is $y_3 = \frac{1}{2}te^t + Ce^t + \frac{1}{2}$.

The 2×2 block gives the system

$$y_1' = y_1 + y_2 + \frac{1}{2}(1 + e^t)$$
$$y_2' = y_2 + 2 + e^t + e^{2t}.$$

Solving the second equation, we get

$$y_2 = -2 + te^t + e^{2t} + C'e^t.$$

Substituting into the first equation gives the equation

$$y_1' - y_1 = y_2 + \frac{1}{2}(1 + e^t) = -\frac{3}{2} + te^t + e^{2t} + \left(C' + \frac{1}{2}\right)e^t.$$

Solving this equation we get:

$$y_1 = \frac{3}{2} + \frac{t^2}{2}e^t + e^{2t} + \left(C' + \frac{1}{2}\right)te^t + C''e^t$$
$$= \frac{3}{2} + e^t\left(\frac{t^2}{2} + \left(C' + \frac{1}{2}\right)t + C''\right) + e^{2t}.$$

We have found y_1, y_2 and y_3 above and so we have determined Y. Now $Y = S^{-1}X$, so $X = SY$ and so

$$X = \begin{bmatrix} 1 & 0 & -1 \\ 1 & 0 & 1 \\ -3 & 1 & 3 \end{bmatrix} \begin{bmatrix} \frac{3}{2} + e^t\left(\frac{t^2}{2} + \left(C' + \frac{1}{2}\right)t + C''\right) + e^{2t} \\ -2 + te^t + e^{2t} + C'e^t \\ \frac{1}{2}te^t + Ce^t + \frac{1}{2} \end{bmatrix}$$

$$= \begin{bmatrix} 1 + e^t\left(\frac{t^2}{2} + C't + C'' - C\right) + e^{2t} \\ 2 + e^t\left(\frac{t^2}{2} + (C' + 1)t + C'' + C\right) + e^{2t} \\ -6 + e^t\left(-\frac{3t^2}{2} - 3C't - 3C'' + C' + C\right) - 2e^{2t} \end{bmatrix}$$

is the general solution of the original system of equations.

To find the specific solution that satisfies the initial conditions $x_1(0) = 0, x_2(0) = 1, x_3(0) = 1$, substitute $t = 0$ and solve for C, C' and C''. We obtain the following system of equations:

$$
\begin{array}{rclcll}
x_1(0) & = & 0 & = & 1 + C'' - C + 1 & \text{or} & C'' - C & = & -2 \\
x_2(0) & = & 1 & = & 2 + C'' + C + 1 & \text{or} & C'' + C & = & -2 \\
x_3(0) & = & 1 & = & -6 - 3C'' + C' + C - 2 & \text{or} & -3C'' + C' + C = 9.
\end{array}
$$

Solving the system of linear equations, we get $C = 0, C' = 3$, and $C'' = -2$. The specific solution is then

$$
X = \begin{bmatrix}
1 + e^t \left(\dfrac{t^2}{2} + 3t + -2 \right) + e^{2t} \\[2mm]
2 + e^t \left(\dfrac{t^2}{2} + 4t - 2 \right) + e^{2t} \\[2mm]
-6 + e^t \left(-\dfrac{3t^2}{2} - 9t + 9 \right) - 2e^{2t}
\end{bmatrix}.
$$

$\square$

Section 6.3 Exercises

In Exercises 1 - 4, find an integrating factor and use it to find the general solution of the given first order linear equation. Then find a specific solution which satisfies the given initial condition.

1. $y' - 2y = e^t, y(0) = 1$

2. $y' + 3y = t, y(0) = -1$

3. $y' - ty = t, y(0) = 0$

4. $y' - t^{-1}y = t^2, y(1) = -1$

In Exercises 5 and 6, express the systems in the form of a matrix equation, that is, find matrices X, A, and $F(t)$ with $X' = AX + F(t)$.

5.

$$
\begin{aligned}
x_1' &= 2x_1 + x_2 + e^t \\
x'2 &= 3x_2 + e^{2t}
\end{aligned}
$$

6.

$$
\begin{aligned}
x_1' &= 3x_1 + x_2 + \sin t \\
x_2' &= 2x_1 - 2x_2 + t
\end{aligned}
$$

7. Find the general solution of

$$
X' = \begin{bmatrix} -1 & 1 \\ 0 & -1 \end{bmatrix} X + \begin{bmatrix} t \\ e^t \end{bmatrix}
$$

and then find the specific solution that satisfies the following initial conditions: $x_1(0) = 1, x_2(0) = -2$.

8. Find the general solution of

$$X' = \begin{bmatrix} 2 & 1 \\ 0 & 2 \end{bmatrix} X + \begin{bmatrix} t^2 \\ t \end{bmatrix}$$

and then find the specific solution that satisfies the following initial conditions: $x_1(0) = -1, x_2(0) = 0$.

9. Solve the following initial value problem:

$$X' = \begin{bmatrix} 2 & 1 & 0 \\ 0 & 2 & 0 \\ 0 & 0 & 1 \end{bmatrix} X + \begin{bmatrix} 1 \\ 0 \\ 0 \end{bmatrix},$$

$x_1(0) = 0, x_2(0) = 1, x_3(0) = 0$.

10. Find the general solution of the following system of equations:

$$x_1' = 2x_1 - x_2 + x_3 + 1$$
$$x_2' = x_1 + x_3 + e^t$$
$$x_3' = x_1 - 2x_2 + 3x_3 + e^{-t}$$

11. Find the general solution of

$$X' = \begin{bmatrix} -1 & 1 \\ 4 & -1 \end{bmatrix} X.$$

Chapter 7

Linear Programming

7.1 Introduction

In business, industry and many other areas, one often encounters problems that involve the optimization of some function (that usually measures profit, efficiency, cost, etc.) by properly determining how resources should be allocated or how production should be geared. Such problems are called "mathematical programming" problems - presumably because the solution must be given in the form of a "program" for obtaining the optimum value of the function. We will consider a particular sort of mathematical programming called "linear programming" - the word linear referring to the fact that the function to be optimized will be a linear function of the variables involved.

Unlike most of the linear algebra that we have studied, linear programming is a relatively recent development. Linear programming was first studied in this country about 1947 by George B. Dantzig and several other mathematicians who were working on a project for the U. S. Air Force. Dantzig developed a method for solving linear programming problems that is called the Simplex Method. It seems that the Russian mathematician L. V. Kantorovich worked on similar programming problems a few years earlier - about 1939. Recent developments in linear programming are discussed in Section 7.3.

We will first of all consider a very simple linear programming problem in order to get some idea of what is involved: A company manufactures cars and trucks - one model of each. Its profit is \$110 on each car and \$100 on each truck. Let x and y represent the number of cars and truck manufactured per week, respectively. Since the company is always able to sell all the cars and trucks it manufactures, the profit per week will be $P = 110x + 100y$. The company, of course, wants to maximize its profits, but there are some restrictions: The same engine is used in both the cars and trucks, and because of a restriction in the factory's physical plant only 50 engines per week can be manufactured. From this we see that $x + y \leq 50$. Also because of restrictions in other parts of the factory, we have: $4x + y \leq 155, x + 2y \leq 90$. Since it is not possible to manufacture fewer than 0 cars or trucks, we have $x \geq 0$ and $y \geq 0$. The problem we must solve is:

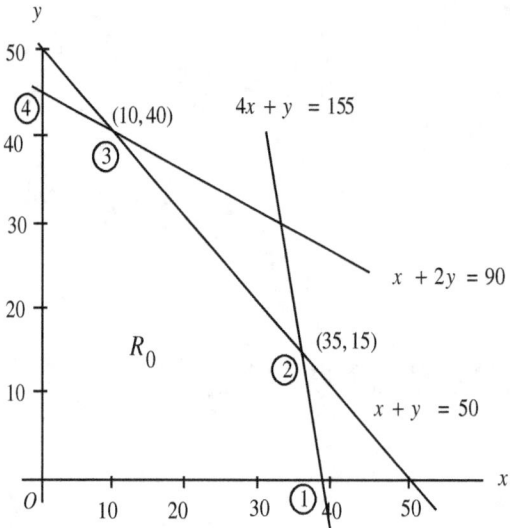

Figure 7.1: Constraints for the auto manufacturer

$$
\begin{array}{ll}
\text{Maximize} & P = 110x + 100y \qquad \text{(linear form)} \\
\text{Subject to} & 0 \le x, 0 \le y \\
 & x + y \le 50 \quad \text{(linear constraints)} \\
 & 4x + y \le 155 \\
 & x + 2y \le 90
\end{array}
\qquad (7.1.1)
$$

To solve the problem, we begin by plotting the lines that bound the set of points (x, y) that satisfy the linear constraints listed above. These lines consist of the x-axis, the y-axis and the three lines in the diagram (see Figure 7.1). We call the region determined by the linear constraints R_0. Notice that R_0 is a closed and bounded set and that P is a continuous function on R_0. By the results of calculus, P has a maximum value on R_0 and so we can be sure that an optimal value of P exists. However in a linear programming problem, the existence of an optimal value, while important, is hardly the complete solution - one must actually know the point (x_0, y_0) at which P obtains the optimal value.

We take a different approach and consider the "graph" of P as a function of x and y defined on the region R_0. P is a linear function of x and y, and so defines a plane through the origin; see Figure 7.1. We are looking for the maximum value of P on the region R_0, and so we are looking for the point on the plane determined by P that lies furthest above R_0. We argue that the maximum value of P cannot occur at an interior point of the region R_0 and that it, in fact, must occur at one of the "corners" of the region. we list the values of P at the corners:

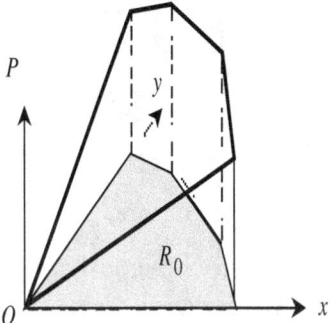

Figure 7.2: A graph of the objective function over the domain

Point	Coordinates	Value of P
1.	$(155/4, 0)$	$17050/4 = 4262.5$
2.	$(35, 15)$	5350
3.	$(10, 40)$	5100
4.	$(0, 45)$	4500

From the table above we see that the maximum profit per week is \$5350 and it can be obtained when 35 cars and 15 trucks are manufactured.

THE GENERAL PROBLEM

We propose to investigate problems of the above sort and we hope to develop methods for solving them. First, we must formulate the general problem. It seems clear that we should attempt to solve:

$$\text{Maximize} \quad z \; = \; c_1 x_1 + \ldots + c_n x_n + b_0 \tag{7.1.2}$$

$$\text{Subject to} \quad a_{11} x_1 + \ldots + a_{1n} x_n \; \leq \; b_1$$

$$\vdots$$

$$a_{m1} x_1 + \ldots + a_{mn} x_n \; \leq \; b_m \tag{7.1.3}$$

$$x_1, \ldots, x_n \; \geq \; 0.$$

The above problem is called a **linear programming** (usually abbreviated LP) problem. The function 7.1.2 is called the **objective function** and the inequalities 7.1.3 are called the **linear constraints**. A point $(x_1, \ldots, x_n)$ that satisfies the linear constraints is called a **feasible point**.

This formulation of the LP problem is more general than it might first appear:

1. To minimize z, maximize $-z$ and take the negative of the maximum value found.

2. A constraint of the form $a_1 x_1 + \ldots + a_n x_n \geq b$ that involves a "greater than" is equivalent to the condition $-a_1 x_1 - \ldots - a_n x_n \leq -b$.

3. If the condition $x_i \leq 0$ is required, the variable x_i can be replaced by $-x_i$ and the appropriate change in the a_{ij}'s can be made.

4. If one of the inequalities in 7.1.3 is an equality, we can use the following trick: $a = b$ if and only if $a \leq b$ and $-a \leq -b$. Thus, adding one equality rules out this apparent restriction.

Before going further, we should consider some extreme cases of the above problem in order to get an idea of the problems we face:

1. There may be no maximum value of z: For example, if $z = x + y$ is the objective function and the constraints are $0x + y \leq 2, 0 \leq x, y$, then the variable x is not restricted and so arbitrarily large values of z can be obtained.

2. There may be no feasible points. For example, if one tries to maximize the function $z = x + y$ subject to the constraints $x + y = -1$ and $x, y \geq 0$, it is quickly seen that there are no points that satisfy the constraints.

3. The method may be impractical. In the problem we considered at the beginning, there were two variables x and y and five linear constraints. To find all of the "corners" of the region it was necessary to solve $\begin{pmatrix} 5 \\ 2 \end{pmatrix} = 10$ pairs of linear equations in two unknowns.

If there were 10 constraints and 5 unknowns, we would have to solve $\begin{pmatrix} 10 \\ 5 \end{pmatrix} = 252$ systems of equations each with five equations in five unknowns. Furthermore, many of the solutions might not be feasible points so that each solution would have to be tested to be certain that it satisfied all of the other constraints.

SLACK VARIABLES

Inequalities are harder to deal with than equations and so we will introduce "**slack variables**" $x_{n+1}, \dots, x_{n+m}$ to change the inequalities of the general LP problem 7.1.2, 7.1.3 into equalities. The problem then becomes

$$
\begin{array}{lrcl}
\text{Maximize} & z = c_1 x_1 + \dots + c_n x_n & & \\
\text{Subject to} & a_{11} x_1 + \dots + a_{1n} x_n + x_{n+1} & = & b_1 \\
& a_{21} x_1 + \dots + a_{1n} x_n + x_{n+2} & = & b_2 \\
& \vdots & & \\
& a_{m1} x_1 + \dots + a_{mn} x_n + x_{n+m} & = & b_m \\
& x_1, \dots, x_n, \dots, x_{n+m} & = & 0
\end{array} \tag{7.1.4}
$$

It is easy to see that the two problems are equivalent. We will attempt to solve the latter problem.

Matrix notation is useful. We define

$$A = \begin{bmatrix} a_{11} & \cdots & a_{1n} & 1 & \cdots & 0 \\ \vdots & \ddots & \vdots & \vdots & \ddots & \vdots \\ a_{m1} & \cdots & a_{mn} & 0 & \cdots & 1 \end{bmatrix}, X = \begin{bmatrix} x_1 \\ \vdots \\ x_{n+m} \end{bmatrix}, B = \begin{bmatrix} b_1 \\ \vdots \\ b_m \end{bmatrix},$$

$$\text{and } C = \begin{bmatrix} c_1 & \cdots & c_n & 0 & \cdots & 0 \end{bmatrix}.$$

We identify the 1×1 matrix CX with its single scalar entry, and for two $m \times n$ matrices $A = [a_{ij}]$ and $B = [b_{ij}]$, we define $A \leq B$ if and only if $a_{ij} \leq b_{ij}$ for all i and j. With these understandings, the general linear programming problem can now be stated in the following form:

$$
\begin{aligned}
\text{Maximize} \quad & z = CX + b_0 \\
\text{Subject to} \quad & AX = B \\
& X = 0
\end{aligned}
\qquad (7.1.5)
$$

Section 7.1 Exercises

In Exercises 1-4, solve the LP problem that has the same linear constraints as those in Problem 7.1.1 but with the objective function given by:

4. $P = 100x + 200y$

5. $P = 100x + 125y$

6. $P = 75x + 40y$

7. $P = 100x + 50y$

8. Solve the following LP problem:

$$
\begin{aligned}
\text{Maximize} \quad & z = 40x + 50y \\
\text{Subject to} \quad & x, y \geq 0 \\
& x + 2y \leq 720 \\
& 5x + 4y \leq 1800 \\
& 3x + y \leq 900
\end{aligned}
$$

9. Solve the LP problem with the constraints as in Exercise 8, but with objective function $z = 50x + 40y$.

10. Consider the following LP problem:

$$
\begin{aligned}
\text{Maximize} \quad & z = 2x_1 + 4x_2 + x_3 \\
\text{Subject to} \quad & x_1 - x_2 + 3x_3 \leq 3 \\
& 2x_1 + x_2 + 3x_3 \leq 4 \\
& x_1 + 3x_2 \leq 6 \\
& x_1, x_2, x_3 \geq 0
\end{aligned}
$$

Add slack variables to put the problem in the form 7.1.4 and then express the problem in the form of a matrix LP problem of the form 7.1.5.

11. Consider the following LP problem:

$$
\begin{array}{rrcl}
\text{Maximize} & z = 2x_1 + x_2 + 2x_3 \\
\text{Subject to} & x_1 + x_2 + 2x_3 & \leq & 4 \\
& 2x_1 - x_2 + 2x_3 & \leq & 6 \\
& x_1 + 3x_2 + x_3 & \leq & 6 \\
& x_1, x_2, x_3 & \geq & 0
\end{array}
$$

Add slack variables to put the problem in the form 7.1.4 and then express the problem in the form of a matrix LP problem of the form 7.1.5.

12. What can be concluded about the following LP problem?

$$
\begin{array}{rrcl}
\text{Maximize} & P = x + y + z \\
\text{Subject to} & x + y - z & \leq & 1 \\
& x + y & \leq & 1 \\
& x, y, z & \geq & 0
\end{array}
$$

13. Consider the following LP problem:

$$
\begin{array}{rrcl}
\text{Maximize} & P = 2x + 3y + z \\
\text{Subject to} & x + y - z & \leq & 5 \\
& x + y + z & \leq & 2 \\
& x, y, z & \geq & 0
\end{array}
$$

What may be concluded?

14. Let $z = c_1 x_1 + \ldots + c_n x_n$ be a function of the n variables $x_1, \ldots, x_n$, and assume that z is defined only on the line segment from $(a_1, \ldots, a_n)$ to $(b_1, \ldots, b_n)$. (NOTE: This line consists of the set of points $\{t(a_1, \ldots, a_n) + (1 - t)(b_1, \ldots, b_n) | 0 \leq t \leq 1\}$). Prove that z is either constant on the line or takes on a maximum value at one of the endpoints of the line.

15. Suppose that we consider a different sort of optimization problem, one in which there is a circular constraint region. For example, suppose we wish to solve:

$$
\begin{array}{rrcl}
\text{Maximize} & z = 4x + 5y \\
\text{Subject to} & x, y & \geq & 0 \\
& x^2 + y^2 & \leq & 100
\end{array}
$$

Here there is a much simpler solution. Recall from calculus that the gradient of a function points in the direction of maximum increase of the value of the function. The function z above is a function of the two variables x and y and it has the gradient $\nabla z = (4, 5)$, which is a constant. To find the maximum value of z, proceed along the line through the vector $(4, 5)$ until the boundary of the circle $x^2 + y^2 = 100$ is encountered. The maximum value should be found at that point. Find the point and the corresponding value of z.

7.2 The Simplex Method in Equation Form

In order to introduce the simplex method for solving linear programming problems, we will solve the example of Section 7.1 using this method.

Recall that in Section 1.5 the notion of "pivoting" was introduced. Given a system of linear equations and a nonzero coefficient, say a, of x_j in the i-th equation, to **pivot on** a means one does the following sequence of operations: (1) multiply both sides of the i-th equation by $1/a$; (2) eliminate the variable x_j from the other equations by adding to each equation the proper multiple of the i-th equation. The coefficient a is called the **pivot**. Of course, this definition of "pivot" is completely analogous to the notion of "pivot" that was defined on entries in a matrix. Indeed pivot operations on systems of equations gave us the motivation to consider row operations and finally pivots on the entries in the augmented matrix of the system of equations.

Example 7.2.1. Given the system

$$2x_1 + 3x_2 = 6$$
$$x_1 + 2x_2 = 2,$$

to pivot on the coefficient 3 of x_2 in the first equation means to replace the system by

$$(2/3)x_1 + x_2 = 2$$
$$(-1/3)x_1 = -2.$$

$\square$

To solve the problem of the previous section using the simplex method we do the following:

1. Add slack variables and write the problem as shown:

$$
\begin{array}{rrrrrl}
x_1 + & x_2 + & x_3 & & & = 50 \\
4x_1 + & x_2 + & & x_4 & & = 155 \\
x_1 + & 2x_2 + & & & x_5 & = 90 \\
P - & 110x_1 - & 100x_2 & & & = 0
\end{array}
$$

2. Select a pivot as follows: Find the "most negative" coefficient in the last row; in this case, it's -110 and it's the coefficient of x_1. The pivot will be the coefficient of x_1 in one of the first three equations. To determine which equation, compute the ratios of the constant terms by the positive coefficients of $x_1 : 50/1, 155/4, 90/1$. Choose the

equation with the smallest ratio (155/4) and pivot on that coefficient (4) of x_1. This gives

$$
\begin{aligned}
(3/4)x_2 + \quad\quad\quad x_3 - \quad (1/4)x_4 &= 45/4 \\
x_1 + \quad (1/4)x_2 + \quad (1/4)x_4 &= 155/4 \\
(7/4)x_2 - \quad (1/4)x_4 + \quad\quad\quad x_5 &= 205/4 \\
P + (110/4 - 100)x_2 + (110/4)x_4 &= 17050/4
\end{aligned}
$$

The reader will wonder about the reason behind the strange rules! The "most negative" entry is chosen in order to increase the efficiency - in general, the final solution is obtained in fewer steps if the "most negative" entry is chosen. The reason for choosing the smallest ratio is to keep the constants on the right-hand side of the constraint equations positive (See Exercise 5). The need for this will be made clear in the next section.

3. Repeat the process: the coefficient of x_2 is the most negative in the last row and the coefficient 3/4 of x_2 in the first equation has the smallest ratio, so pivot on 3/4 and get:

$$
\begin{aligned}
x_2 + (4/3)x_3 - (1/3)x_4 &= 15 \\
x_1 - (1/4)x_3 + (1/3)x_4 &= 35 \\
-(7/4)x_3 - (37/36)x_4 + \quad x_5 &= 25 \\
P + (290/3)x_3 + (40/12)x_4 &= 5350
\end{aligned}
$$

4. Since there are no more negative coefficients in the last equation, the process stops. By inspecting the last equation, we see that the maximum value for P is 5350 (since x_3 and x_4 are both nonnegative) and this maximum occurs when $x_3 = x_4 = 0$. The first and second equations then give $x_2 = 15$ and $x_1 = 35$. It is not hard to see that the last LP problem is equivalent to the original problem, and so the problem is solved.

The above method is known as the "Simplex Method in Equation Form." As in working with systems of linear equations, we see that it is clumsy writing the variables at each step and using matrix notation simplifies the matter. Notice the operations that were performed on the equations: multiples of the first three equations were added to the other equations, the first three equations were multiplied by constants, but the last equation was not multiplied by a constant and multiples of it were not added to other equations. These are the basic operations that may be performed on linear programming problems. To put it more succinctly, pivot operations may be performed on the coefficients in the constraint equations.

Section 7.2 Exercises

In Exercises 1-4, solve the LP problems using the simplex method in equation form:

1.

$$\begin{aligned}
\text{Maximize} \quad & P = x_1 + 2x_2 \\
\text{Subject to} \quad & x_1 + x_2 \leq 7 \\
& 2x_1 + x_2 \leq 11 \\
& x_1, x_2 \geq 0
\end{aligned}$$

3.

$$\begin{aligned}
\text{Maximize} \quad & P = 50x_1 + 40x_2 \\
\text{Subject to} \quad & x_1 + 2x_2 \leq 200 \\
& 5x_1 + 4x_2 \leq 1800 \\
& 3x_1 + x_2 \leq 900 \\
& x_1, x_2 \geq 0
\end{aligned}$$

2.

$$\begin{aligned}
\text{Maximize} \quad & P = 3x_1 + 2x_2 \\
\text{Subject to} \quad & x_1 + 2x_2 \leq 9 \\
& 2x_1 + x_2 \leq 20 \\
& x_1, x_2 \geq 0
\end{aligned}$$

4.

$$\begin{aligned}
\text{Maximize} \quad & P = 30x_1 + 70x_2 \\
\text{Subject to} \quad & 2x_1 + x_2 \leq 240 \\
& 4x_1 + 7x_2 \leq 1800 \\
& 2x_1 + x_2 \leq 900 \\
& x_1, x_2 \geq 0
\end{aligned}$$

5. Consider the system of linear equations

$$ax + by = h$$
$$cx + dy = k$$

and assume that h and k are positive. Assume that $a > 0, c > 0$, and $h/a < k/c$. Show that a pivot on a results in a system of equations with positive constants.

7.3 The Simplex Method

We will investigate the general LP problem as formulated before:

$$\begin{aligned}
\text{Maximize} \quad & z = CX + b_0 \\
\text{Subject to} \quad & AX = B, X \geq 0,
\end{aligned} \qquad (7.3.1)$$

where $A = [a_{ij}]$ is $m \times n$, $C = (c_1, \ldots, c_n)$, $B = \begin{bmatrix} b_1 \\ \vdots \\ b_m \end{bmatrix}$, and $X = \begin{bmatrix} x_1 \\ \vdots \\ x_n \end{bmatrix}$.

Notice that in the matrix equation $AX = B$, we have assumed that slack variables have been added to make the inequalities become equalities, and in the assumption that A is $m \times n$, we have assumed that the slack variables are among the variables $x_1, \ldots, x_n$. This differs from the notation in Section 7.1 where we assumed that A was $m \times (n+m)$. We will call the matrix

$$\left[\begin{array}{c|c} A & B \\ \hline -C & b_0 \end{array} \right]$$

the **augmented matrix** of the LP problem.

In solving LP problems in equation form, we saw that certain operations could be performed on the equations. Just as in the solution of systems of linear equations, this leads us to row operations that may be performed on the augmented matrix of the LP problem.

ROW OPERATIONS

Theorem 7.3.1. *Assume that an arbitrary finite sequence of elementary row operations are performed on the augmented matrix of the LP problem 7.3.1, except that the last row is not added to the other rows and the last row is not switched with another row. Let the resulting matrix be*

$$\left[\begin{array}{c|c} A' & B' \\ \hline -C' & b'_0 \end{array}\right]$$

Then the LP problem:

$$\begin{aligned} Maximize \quad & z = C'X + b'0 \\ Subject\ to \quad & A'X = B', X \geq 0 \end{aligned}$$

is equivalent to the original problem 7.3.1. That is, the solution sets of $AX = B$ and $A'X = B'$ are equal and for any solution X, $z = CX + b_0 = C'X + b'0$, so that z takes on a maximum value at X_0 if and only if z' takes on a maximum value at X_0.

(Certain row operations are permissible on the augmented matrix of an LP problem.)

Proof. The matrix $[A|B]$ is row equivalent to the matrix $[A'|B']$ so that by the results of Chapter 1, $AX = B$, and $A'X = B'$ are equivalent and so have the same solution sets.

Let X_0 be a solution of $AX = B$. The bottom row of the new augmented matrix, $[-C'|b'_0]$ consists of the original bottom row $[-C|b_0]$ with multiples of the rows of $[A|B]$ added to it. We have $[-C'|b'_0] = [-C|b_0] + \begin{bmatrix} r_1 & \cdots & r_m \end{bmatrix} [A|B]$, and so

$$\begin{aligned} z' &= C'X_0 + b'_0 \\ &= (C - (r_1, \ldots, r_m)A)X_0 + (b_0 + (r_1, \ldots, r_m)B) \\ &= CX_0 + b_0 + (r_1, \ldots, r_m)(-AX_0 + B) \\ &= CX_0 + b_0 \\ &= z \end{aligned}$$

since $-AX_0 + B = 0$. It follows that z has a maximum at X_0 if and only if z' has a maximum at X_0. $\qquad\square$

In effect, this theorem states that pivots may be performed on any entry of the matrix A in the upper left-hand corner of the augmented matrix of an LP problem and the resulting matrix will be the augmented matrix of an equivalent LP problem. Just as with the problem of finding solutions of systems of linear equations, the trick is to find an "algorithm" or systematic method for arriving at an equivalent augmented matrix for which the solution of the problem is apparent.

BASIC VARIABLES AND FEASIBLE POINTS

Recall that a point $(x_1, \ldots, x_n)$ satisfying the constraint $AX = B, X = 0$ is called the **feasible point** or **feasible solution**. A feasible point that maximizes the value of z is called an **optimal feasible solution**. We wish to find a special sort of feasible solution. Assume that in A the columns of I_m appear; say column i of I_m is column j_i of $[A|B]$. If all the columns of I_m appear as columns of A, we say that A and $[A|B]$ are in **basic form**. It is then easy to find a solution to $AX = B$: Let $x_{j_i} = b_i$ for $i = 1, \ldots, m$ and let all other $x_j = 0$. The variables x_{j_i} corresponding to the columns of the identity matrix are called **basic variables** and a solution obtained in this manner is called a **basic solution**. Notice that if we began with the constraints written as "less thans" and add a slack variable to make each inequality an equality, that the resulting augmented matrix is automatically in basic form.

Example 7.3.1. Assume that:

$$[A|B] = \begin{bmatrix} 0 & 2 & 1 & 0 & 0 & 1 & 3 \\ 1 & 1 & 0 & 0 & 0 & -1 & 6 \\ 0 & 1 & 0 & 0 & 1 & 2 & 7 \\ 0 & 2 & 0 & 1 & 0 & 1 & 2 \end{bmatrix}.$$

Then $[A|B]$ is in basic form since the columns of I_4 appear as columns $1, 3, 4,$ and 5 of A. The basic variables are x_1, x_2, x_4, x_5 and the variables x_2 and x_6 are nonbasic. The resulting basic solution is $x_1 = 6, x_3 = 3, x_4 = 2, x_5 = 7, x_2 = 0,$ and $x_6 = 0$.

$\square$

In order for a basic solution to be feasible, we must have $B \geq 0$. We will make a further assumption: a system of equations $AX = B$ is called **nondegenerate** if and only if every $m \times m$ submatrix of the $m \times (n + 1)$ matrix $[A|B]$ is nonsingular. We will say that the LP problem 7.3.1 is **nondegenerate** provided that $AX = B$ is nondegenerate.

A basic solution of the nondegenerate system $AX = B$ can be obtained as follows: choose the entries in rows and columns $j_1, \ldots, j_m$ of A and let A' be the submatrix of A consisting of these rows and columns. Choose $x_{j_1}, \ldots, x_{j_m}$ so that

$$A' \begin{bmatrix} x_{j_1} \\ \vdots \\ x_{j_m} \end{bmatrix} = B$$

and choose the other x_j's to be zero. Notice that since every $m \times m$ submatrix of $[A|B]$ is nonsingular, each $x_{j_i} = |A'_i|/|A'|$, and $|A'_i|$ is the determinant of an $m \times m$ submatrix of $[A|B]$ (with columns rearranged), we have x_{j_i} nonzero. From this we see the following theorem.

> **Theorem 7.3.2.** *The basic variables in any basic solution of a nondegenerate system of equations take on nonzero values. (Basic variables have nonzero values if the system is nondegenerate.)*

Consider now the LP problem 7.3.1 and its augmented matrix

$$\left[\begin{array}{c|c} A & B \\ \hline -C & b_0 \end{array}\right].$$

By Theorem 7.3.1, certain pivot operations may be performed on this matrix. Assume that after a sequence of pivots the resulting matrix is

$$\left[\begin{array}{c|c} A' & B' \\ \hline -C' & b'_0 \end{array}\right]$$

and that the matrix $[A'|B']$ is in basic form with $x_{j_1}, \ldots, x_{j_m}$ being the basic variables and $x_{j_{m+1}}, \ldots, x_{j_n}$ the remaining variables. Assume further that the LP problem is nondegenerate and that $A' = [a'_{ij}]$, $B' = [b'_i]$, and $C' = [c'_j]$. We may assume that $c'_{j_1}, \ldots, c'_{j_m} = 0$ (for if not, a pivot on $a'_{ij_i} = 1$ would make $c'_{j_i} = 0$). In order for the basic solution to be feasible, we must have $B' > 0$. This brings us to the resolution:

RESOLUTION OF AN LP PROBLEM

Theorem 7.3.3. *Under the above assumptions we have:*

(I) *If* $-c_{j_{m+1}}, \ldots, -c_{j_n}$ *are all nonnegative, then* $x_{j_1} = b'_{j_1}, \ldots, x_{j_m} = b'_{j_m}$, *and* $x_{j_{m+1}} = \ldots = x_{j_n} = 0$ *is an optimal feasible solution with the maximum value of* z *being* b'_0 .

(If the $-C$-part contains all non-negative numbers, the the optimal value is in the lower right-hand corner.)

(II) *If one of the entries* $-c'_{j_i} (m+1 \le i \le n)$ *is negative, then we have two cases:*

(a) *If* $a'_{kj_i} = 0$ *for* $k = 1, \ldots, m$, *then there exist feasible solutions with arbitrarily large values of* z *so that* z *has no maximum value.*

(If the $-C$-part contains a negative and all entries above it are negative, then there is no largest value.)

(b) $a'_{kj_i} > 0$ *for at least one value of* $k (1 \le k \le m)$, *then a new basic solution can be found that increases the value of* z.

(If there is a positive number above a negative in the $-C$-part, then the value of z can be increased.)

Proof.   1. The objective function associated with the new augmented matrix above is $z' = c_{j_{m+1}} x_{j_{m+1}} + \ldots + c_{j_n} x_{j_n} + b'_0$. Since $c_{j_{m+1}}, \ldots, c_{j_n}$ are all negative or zero, and since for any feasible solution $x_{j_{m+1}}, \ldots, x_{j_n}$ are all nonnegative, we see that the maximum value of z' is b'_0 and that it occurs when $x_{j_{m+1}} = \ldots = x_{j_n} = 0$.

2. (a) Since $a'_{kj_1} \le 0 \, for \, k = 1, \ldots, m$, solutions of $A'X = B'$ can be found with arbitrarily large values of x_{j_i}. Considering the form z' and the fact that $c'_{j_i} > 0$, we see that arbitrarily large values of z' can be obtained.

   (b) Assume $a'_{kj_1} > 0$. If we pivot on a'_{kj_1}, a new basic feasible solution is obtained and in the resulting matrix b'_0 is replaced by

$$b_0'' = b_0' + c_{j_i}'(b_{j_i}'/a_{kj_1}').$$

Since c_{j_i}', b_{j_i}', and a_{kj_1}' are all positive, $b_0'' > b_0'$.

$\square$

We will make frequent use of this theorem in solving LP problems. The theorem tells us when a solution has been obtained, when there is no solution in that no maximum value of z exists, and how to proceed to larger values of z when they exist. Notice that part (II) (b) makes no claim that the process will ever end; that is, that a largest value of z will ever be found.

The Simplex Algorithm gives a method for solving (if a solution exists) a linear programming problem. The method involves two parts called "Phase I" and "Phase II." Theorem 7.3.3 parts I and II (a) provide a means for determining when the problem is solved. The algorithm and its phases consist of a set of instructions for solving the problem. We will not attempt to prove that the algorithm "works." Instead, we will only state it and give illustrations of its application.

The LP problem considered is the original one 7.3.1:

$$\text{Maximize} \quad z = CX + b_0$$
$$\text{Subject to} \quad AX = B, X \geq 0,$$

where $A = [a_{ij}]$ is $m \times n$, $B = [b_i]$ is $m \times 1$ and $C = [z_i]$ is $1 \times n$. The augmented matrix is

$$\left[\begin{array}{c|c} A & B \\ \hline -C & b_0 \end{array} \right].$$

Note that the assumption of nondegeneracy still holds.

Theorem 7.3.4 (The Simplex Algorithm). *If $B \geq 0$, go to Phase II.*
If $B \ngeq 0$, begin with Phase I.
Phase I:

1. *Let b_k be the lowest negative entry in B (that is, the negative in B entry with only positive entries below it.)*

2. *Choose a negative entry a_{kj} in row k (the best choice is the "most negative" entry). This determines the pivot column.*

3. *For a_{kj} and each positive element a_{ij} below a_{kj} in the pivot column, form the quotient b_i/a_{ij}. The entry a_{ij} giving the smallest quotient b_i/a_{ij} is the pivot element.*

4. *Pivot on a_{ij}.*

5. *Repeat steps 1 - 4 until either there are no negative elements in the last column or until there is a negative entry in the last column with the remainder of that row consisting of positive entries indicating that there is no feasible solution and so that the LP problem has no solution.*

If $B \geq 0$, do the following:
Phase II:

1. *Choose some entry in the bottom row that is negative, say $-c_j$ in column j (the best choice is the "most negative" entry).*

2. *For the positive entries in column j, calculate the quotients b_i/a_{ij}. The entry a_{ij} giving the smallest quotient is chosen to be the pivot.*

3. *Pivot on a_{ij}.*

4. *Repeat steps 1 - 4 until there are no negative entries in the last row, indicating that the optimal feasible solution has been found; or until there is a negative entry in the last row with all entries above it being nonpositive, indicating that z has no maximum value.*

Example 7.3.2. (Using Phase I) LP problem:

$$\begin{array}{rrcl}
\text{Maximize} & z = 40x_1 + 60x_2 \\
\text{Subject to} & 2x_1 + x_2 & \geq & 70 \\
& x_1 + x_2 & \leq & 40 \\
& x_1 + 3x_2 & \geq & 90 \\
& x_1, x_2 & \geq & 0.
\end{array}$$

The first step is to change the "$\geq$" to "$\leq$" and to add slack variables, obtaining

$$\text{Maximize} \quad z = 40x_1 + 60x_2$$
$$\text{Subject to} \quad -2x_1 - x_2 + x_3 = -70$$
$$x_1 + x_2 + x_4 = 40$$
$$-x_1 - 3x_2 + x_5 = -90$$
$$x_1, x_2, x_3, x_4, x_5 = 0.$$

Now write the augmented matrix and apply Phase I rules:

$$\begin{bmatrix} -2 & -1 & 1 & 0 & 0 & -70 \\ 1 & 1 & 0 & 1 & 0 & 40 \\ -1 & -3 & 0 & 0 & 1 & -90 \\ -40 & -60 & 0 & 0 & 0 & 0 \end{bmatrix}.$$

Phase I rules dictate that the -1 or the -3 in row 3 may be the pivot, with -3 being more efficient. Pivot on -3 and get

$$\begin{bmatrix} -5/3 & 0 & 1 & 0 & -1/3 & -40 \\ 2/3 & 0 & 0 & 1 & 1/3 & 10 \\ 1/3 & 1 & 0 & 0 & -1/3 & 30 \\ -20 & 0 & 0 & 0 & -20 & 1800 \end{bmatrix}.$$

A negative entry remains in the last column and so Phase I rules must again be followed, this time with Column 1 as the pivot column. Calculate the ratios: $-40/(-5/3)$, $10/(2/3)$, and $30/(1/3)$. The smallest is $10/(2/3)$, and so we pivot on $2/3$, obtaining

$$\begin{bmatrix} 0 & 0 & 1 & 5/2 & 5/2 & -70/3 \\ 1 & 0 & 0 & 3/2 & 3/2 & 15 \\ 0 & 1 & 0 & -1/3 & -1/3 & 20 \\ 0 & 0 & 0 & 0 & -10 & 2000 \end{bmatrix}.$$

A negative entry remains in row 1 of the last column, but it is the only negative entry in row 1. The constraint equation corresponding to row 1 is:

$$x_3 + (5/2)x_4 + (5/2)x_5 = -70/3$$

and since $x_3, x_4, x_5 \geq 0$, there can be no solution, and so the LP problem has no solution.

(a) (Using both Phases I and II) LP problem:

$$\text{Maximize} \quad z = 9x_1 + 3x_2$$
$$\text{Subject to} \quad x_1 + x_2 \leq 20$$
$$-x_1 + 2x_2 \geq 10$$
$$3x_1 + x_2 \geq 6$$
$$x_1, x_2 \geq 0.$$

Multiply the second and third inequalities by -1 (thus reversing the inequalities), add slack variables and write the augmented matrix of the LP problem. We obtain:

$$\begin{bmatrix} 1 & 1 & 1 & 0 & 0 & 20 \\ 1 & -2 & 0 & 1 & 0 & -10 \\ -3 & -1 & 0 & 0 & 1 & -6 \\ -9 & -3 & 0 & 0 & 0 & 0 \end{bmatrix}.$$

The last column has negative entries and so Phase I must be applied. Row 3 will be the pivot row and entries -3 or -1 in columns 1 or 2 may be chosen. The entry -3 in column 1 is better, but let us choose the -1 in column 2 as the pivot. Pivoting, we obtain

$$\begin{bmatrix} -2 & 0 & 1 & 0 & 1 & 14 \\ 7 & 0 & 0 & 1 & -2 & 2 \\ 3 & 1 & 0 & 0 & -1 & 6 \\ 0 & 0 & 0 & 0 & -3 & 18 \end{bmatrix}.$$

The last column has no negative entries and so Phase I need not be applied. The negative entry in the last row indicates that Phase II must be applied. The only positive ratio is $14/1$ obtained from the entry in row 1 and column 5. We pivot on this entry, obtaining

$$\begin{bmatrix} -2 & 0 & 1 & 0 & 1 & 14 \\ 3 & 0 & 2 & 1 & 0 & 30 \\ 1 & 1 & 1 & 0 & 0 & 20 \\ -6 & 0 & 3 & 0 & 0 & 60 \end{bmatrix}.$$

Notice that the constraint portion of the matrix is now in basic form with a basic solution being: $x_1 = 0, x_2 = 20, x_3 = 0, x_4 = 30$, and $x_5 = 14$, and a corresponding z value of 60. Because of the negative entry in row 4 column 1, Phase II must again be applied. The positive ratios are: $30/3$ and $20/1$ and so the 3 in row 2, column 1, is chosen as the pivot. Pivoting again, we obtain

$$\begin{bmatrix} 0 & 0 & 7/3 & 2/3 & 1 & 34 \\ 1 & 0 & 2/3 & 1/3 & 0 & 10 \\ 0 & 1 & 1/3 & -1/3 & 0 & 10 \\ 0 & 0 & 7 & 2 & 0 & 120 \end{bmatrix}.$$

The last row has no negative entries and the last column remains positive. This means that the problem has been resolved. The basic solution is: $x_1 = 10, x_2 = 10, x_3 = 0, x_4 = 0, x_5 = 34$, and the maximum value of z is 120.

$\square$

EFFICIENCY OF THE SIMPLEX METHOD

Let us consider the efficiency of the Simplex Method. In Section 4.5 we introduced the idea of computational complexity. How much memory does it take to solve a problem using a computer, and how much time does it take? While we cannot present the details here, it is appropriate to give an outline of the current situation regarding the solution of LP problems.

Theorem 7.3.3 states that if a negative number appears in the $-C$ part of the augmented matrix of an LP problem, and if there is a positive number above it, then the value of z can be increased. The Simplex Algorithm uses this fact, and states roughly: if there is a negative in the bottom row, locate the pivot element, pivot, and repeat until the negatives are gone. How many pivots must we do? We might unwisely assume that each time we perform a pivot in Phase II, the number of negatives is reduced by one. This is not the case.

Much research has been done on the question of the efficiency of methods of solving LP problems. In 1972, V. Klee and G. L. Minty (How good is the simplex method?, Inequalities III, Academic Press, New York, 1972, 159 - 179) showed that in a worst case, the simplex method is of exponential order. In 1979, L. G. Khachiyan (A polynomial algorithm in linear programming, Doklady Akademii Nauk SSSR 244:S (1979) 1093 - 1096) found the so called "ellipsoid algorithm" which has a running time proportional to $n^6 L^2$. Here n is the number of variables and L is a measure of the length of the input data.

A major breakthrough came in 1984 with the announcement of the "Karmarkar Algorithm" (N. Karmarkar, A new polynomial-time algorithm for linear programming, Combinatorica 4(4) (1984) 373 - 395.) by the mathematician Narendra Karmarkar of AT&T Bell Laboratories. The running-time of this algorithm is proportional to $n^{3.5} L^2$. Karmarkar's algorithm involves the application of a transformation to the LP problem so that a sphere is involved. We saw in Exercise 15 that it is easy, considering the gradient of the objective function, to maximize a linear function when the feasible region is spherical. Karmarkar's method takes advantage of this observation.

While the worst case behavior of the Simplex Method is not good, experience has shown that its behavior the solution of many ordinary problems is quite good. Further study of this problem will no doubt give rise to new results.

Section 7.3 Exercises

In Exercises 1 - 6, solve the linear programming problems with the given augmented matrices. Use Theorem 7.3.3 and the Simplex Method.

1.

$$\begin{bmatrix} 1 & 0 & 0 & 0 & 2 & 1 & 1 \\ 0 & 1 & 0 & 2 & 1 & -1 & 2 \\ 0 & 0 & 1 & 0 & 2 & 1 & 1 \\ 0 & 0 & 0 & 2 & 1 & 1 & 3 \end{bmatrix}$$

2.

$$\begin{bmatrix} 1 & 0 & 0 & 0 & 2 & 1 & 1 \\ 0 & 1 & 0 & 2 & -1 & 1 & 2 \\ 0 & 0 & 1 & 3 & 1 & 1 & 1 \\ 0 & 0 & 0 & 2 & -1 & 1 & 2 \end{bmatrix}$$

3.

$$\begin{bmatrix} 1 & 0 & 0 & 0 & -1 & 1 & 1 \\ 0 & 1 & 0 & 0 & -2 & 1 & 1 \\ 0 & 0 & 1 & 0 & -1 & 2 & 1 \\ 0 & 0 & 0 & 2 & -1 & -1 & 2 \end{bmatrix}$$

5.

$$\begin{bmatrix} 1 & 2 & 2 & 1 & 2 \\ 0 & 0 & 1 & -1 & -1 \\ 0 & 0 & 2 & -1 & 2 \end{bmatrix}$$

4.

$$\begin{bmatrix} 1 & 1 & 0 & 0 & 1 & 0 & 1 \\ 1 & 2 & 1 & 0 & 2 & 0 & 3 \\ 0 & 2 & 0 & 0 & 1 & 1 & 2 \\ 0 & -1 & 0 & 1 & 1 & -1 & -1 \end{bmatrix}$$

6.

$$\begin{bmatrix} 1 & 0 & 0 & 1 & -1 & -1 \\ 0 & 1 & 0 & -1 & 1 & 2 \\ 0 & 0 & 1 & 1 & 2 & 3 \\ 0 & 0 & 2 & 3 & 5 & \end{bmatrix}$$

In Exercises 7 - 10, add slack variables and write the augmented matrix of the LP problem. Then solve the given problem using the Simplex Method.

7.

$$\begin{aligned}
\text{Maximize} \quad & z = 40x_1 + 60x_2 \\
\text{Subject to} \quad & 2x_1 + x_2 \leq 70 \\
& x_1 + x_2 \leq 40 \\
& x_1 + 3x_2 \leq 90 \\
& x_1, x_2 \geq 0
\end{aligned}$$

8.

$$\begin{aligned}
\text{Maximize} \quad & z = 40x_1 + 60x_2 \\
\text{Subject to} \quad & x_1 + x_2 \leq 70 \\
& x_1 + 2x_2 \leq 40 \\
& 3x_1 + x_2 \leq 90 \\
& x_1, x_2 \geq 0
\end{aligned}$$

9.

$$\begin{aligned}
\text{Maximize} \quad & z = 5x_1 + 3x_2 + 4x_3 \\
\text{Subject to} \quad & 4x_1 + 2x_2 + 4x_3 \leq 80 \\
& 2x_1 + 2x_2 + 3x_3 \leq 50 \\
& x_1 + 3x_2 + 2x_3 \leq 40 \\
& x_1, x_2, x_3 \geq 0
\end{aligned}$$

10.

$$\begin{aligned}
\text{Maximize} \quad & z = 5x_1 + 3x_2 + 4x_3 \\
\text{Subject to} \quad & 2x_1 + x_2 + 4x_3 \leq 60 \\
& 5x_1 + 2x_2 + 2x_3 \leq 40 \\
& x_1 + 3x_2 + 2x_3 \leq 50 \\
& x_1, x_2, x_3 \geq 0
\end{aligned}$$

11. A company makes two products, X's and Y's. They make \$40 on each X and \$50 on each Y. The following chart gives the constraints on each product - each item required 3 jobs in production. How many X's and Y's should be made in order to maximize the profit?

	Job 1	Job 2	Job 3
X	1 hour	3 hours	3 hours
Y	2 hours	4 hours	1 hour
Total Hours Available	720	1800	900

12. A company makes microscopes and telescopes and can sell all they can make. The facts are listed in the following chart.

	assembly	testing	packaging	profit
microscopes	2 hr/unit	3 hr/unit	1 hr/unit	$10/unit
telescopes	4 hr/unit	1 hr/unit	2 hr/unit	$20/unit
Hours Available	50 hr/wk	60 hr/wk	25hr/wk	

Set up the LP problem to maximize profit for the company and solve it using the Simplex Algorithm.

7.4   Applications - Game Theory

The theory of games is an important and relatively new area of mathematics. Early work was done in the area by John von Neumann and the first book on the subject was Theory of Games and Economic Behavior by von Neumann and Oskar Morgenstern in 1944. We will present here only the most elementary introduction to the subject and show how linear programming can be used in solving simple problems in game theory.

Consider the game of matching pennies. As it is often played, two players, A and B, each toss a penny independently of each other, and the outcomes are compared. If the outcomes match (both head or both tails) then player B wins a penny from player A. Otherwise (that is, if one has a head and the other a tail), A wins a penny from B. This game can be summarized in the following table:

		Player B	
		H	T
	H	-1	1
Player A			
	T	1	-1

The table shows Players A's gain or loss. We will call Player A the **row player** and Player B the **column player**.

The array of numbers

$$\begin{bmatrix} -1 & 1 \\ 1 & -1 \end{bmatrix}$$

is called the **payoff matrix**. The entries represent the amount paid to the row player A for each of the four possible outcomes $(H,H), (H,T), (T,H)$ and (T,T). A negative entry represents a payoff to the column player B.

A game of this sort is called a **two-person game** since it involves only two players. The game is called **zero-sum** since the money lost by one player is gained by the other. We will consider only two-person, zero-sum games.

Matching pennies in the form presented above is not very interesting since no skill or strategy is involved. Let us change the rules slightly: Player A places a penny with either heads or tails up on the table so that B cannot see it. B then guesses whether it is heads or tails and tries to match it by placing his penny with either heads or tails up depending on the guess. Now the game involves some strategy! How should A and B make their choices in order to win the most money or lose the least? We will see shortly that the version of matching pennies that was first presented really amounts to both players playing with optimal strategy.

STRATEGIES

By a **strategy** we mean a process for deciding at each step in a game which choice is to be made. It is a basic principle in game theory one should assume that one's opponent is not only rational, but intelligent. For example, if one player decides to employ some fixed strategy, say choosing heads and tails alternately, then the opponent will detect this strategy and make use of it. For this reason fixed or determined strategy is rejected in favor of a rational, but random or mixed strategy. In this type of strategy, one decides in what proportion each choice should be made, but makes each individual choice randomly in a manner that will produce the correct proportions.

Let us return to the second version of matching pennies and take the point of view of the column player B. Player B wishes to find an optimal strategy. We let s denote the proportion of choices by B of heads and t denote the proportion of tails chosen. Then

$$s = 0, t = 0, \text{ and } s + t = 1. \tag{7.4.1}$$

Against each of Player A's choices, B must expect to pay the following amounts:

$$-s + t \text{ if } A \text{ chooses heads}$$

$$s - t \text{ if } A \text{ chooses tails.}$$

Denote the largest of these sums by h so that

$$-s + t \leq hs - t \leq h \tag{7.4.2}$$

We wish to assume that h is positive. In general this can be accomplished by making all terms in the payoff matrix nonnegative by adding a fixed constant k to each. For each strategy, including the optimal one, this would increase the expected payoff by k, but would not alter the strategy. Considering this, we add 1 to each entry in the payoff matrix and obtain

$$\begin{bmatrix} 0 & 2 \\ 2 & 0 \end{bmatrix}.$$

Now B's greatest expected loss is h, and of course, B would like to minimize h. This can be done by maximizing $M = \dfrac{1}{h}$. Let

$$u = sh, v = th, \text{ and } M = \frac{1}{h} = \frac{s+t}{h} = u + v \qquad (7.4.3)$$

Then using the new payoff matrix, 7.4.1 and 7.4.2 become

$$M = u + v$$
$$u \geq 0$$
$$v \geq 0$$
$$2v \leq 1$$
$$2u \leq 1.$$

SOLVING USING LP METHODS

To solve our original problem, we must solve the following LP problem:

$$\begin{array}{rrcl} \text{Maximize} & M = u + v \\ \text{Subject to} & u, v & \geq & 0 \\ & 2v & \leq & 1 \\ & 2u & \leq & 1. \end{array}$$

We add slack variables and write the augmented matrix

$$\begin{bmatrix} 0 & 2 & 1 & 0 & 1 \\ 2 & 0 & 0 & 1 & 1 \\ -1 & -1 & 0 & 0 & 0 \end{bmatrix}.$$

Using the simplex method we obtain the following matrix after two pivots:

$$\begin{bmatrix} 0 & 1 & \frac{1}{2} & 0 & \frac{1}{2} \\ 1 & 0 & 0 & \frac{1}{2} & \frac{1}{2} \\ 0 & 0 & \frac{1}{2} & \frac{1}{2} & 1 \end{bmatrix}.$$

The maximum value of M is thus 1, and it is obtained when $u = v = \dfrac{1}{2}$. Using 7.4.3 we get

$$h = \frac{1}{M} = 1,$$

$$s = uh = \frac{1}{2},$$

$$t = vh = \frac{1}{2}.$$

It follows that the optimal strategy is $\left(\frac{1}{2}, \frac{1}{2}\right)$ and that B's expected maximum loss is $h - 1 = 0$ (since we added 1 to each entry in the payoff matrix).

A similar procedure can be used to find the optimal strategy for A: Recall that the original payoff matrix showed Player A's (that is, the row player's) gain. If we change signs, the matrix then gives Player A's loss. Thus

$$\begin{bmatrix} 1 & -1 \\ -1 & 1 \end{bmatrix}$$

is the payoff matrix giving A's loss. As before, let (s_1, t_1) be a strategy for A and add 1 to each entry in the matrix to make all entries non-negative. We get

$$\begin{bmatrix} 2 & 0 \\ 0 & 2 \end{bmatrix}$$

and so for the two possible choices of player B, the corresponding losses for A are

$$2s_1 \text{ and } 2t_1.$$

We let h be the largest of these and obtain

$$s_1 \leq h$$
$$2t_1 \leq h.$$

As before, we minimize h by maximizing $\frac{1}{h}$ and solve the LP problem

$$\begin{aligned} \text{Maximize} \quad & M = u_1 + v_1 \\ \text{Subject to} \quad & 2u_1 \leq 1 \\ & 2v_1 \leq 1 \\ & u_1, v_1 \geq 0, \end{aligned}$$

where $u_1 = \frac{s_1}{h}$ and $v_1 = \frac{t_1}{h}$. The augmented matrix for this LP problem is

$$\begin{bmatrix} 2 & 0 & 1 & 0 & 1 \\ 0 & 2 & 0 & 1 & 1 \\ 1 & 1 & 0 & 0 & 0 \end{bmatrix}.$$

Applying the simplex method, we get

$$\begin{bmatrix} 1 & 0 & \frac{1}{2} & 0 & \frac{1}{2} \\ 0 & 1 & 0 & \frac{1}{2} & \frac{1}{2} \\ 0 & 0 & \frac{1}{2} & \frac{1}{2} & 1 \end{bmatrix}.$$

It follows that the optimal strategy for player A is

$$s_1 = \frac{1}{2}$$
$$t_1 = \frac{1}{2}$$

and A's maximum expected loss is $h - 1 = \frac{1}{M} - 1 = 0$ (since we added a 1 to each entry in the payoff matrix).

While we will not present the details here, it is not difficult to see how the above situation can be generalized. Any $m \times n$ matrix $[a_{ij}]$ can be thought of as the **payoff matrix** of a "two-person, zero-sum matrix game." The row player, player A, chooses a number i, $1 \leq i \leq m$, and the column player, player B, chooses a number $j, 1 \leq j \leq n$. The payoff matrix $[a_{ij}]$ records player A's gain. If $a_{ij} > 0$, A wins that amount from player B, and if $a_{ij} < 0$, player A pays player B the negative of that amount. Whatever A loses, B gains.

A **strategy** for the row player or player A is an m-tuple $P = (p_1, \ldots, p_m)$ and a strategy for the column player or player B is an n-tuple $Q = (q_1, \ldots, q_n)$, where $p_1 + \ldots + p_m = 1$ and $q_1 + \ldots + q_n = 1$. Optimal strategies can be found for each player by using a procedure similar to that above. One obtains an LP problem and solves it using the simplex algorithm.

Section 7.4 Exercises

1. A certain zero-sum, two-person game has payoff matrix $\begin{bmatrix} -1 & 2 \\ 3 & -2 \end{bmatrix}$. What happens if the row player chooses 2 and the column player 1? What happens if the row player chooses 1 and the column player 1?

2. A certain zero-sum, two-person game has payoff matrix $\begin{bmatrix} -1 & 2 & -2 \\ 2 & -1 & 3 \end{bmatrix}$. Describe what happens when the row player chooses 2.

3. Suppose that the payoff matrix for a certain game contains only non-negative numbers. Which player wins?

4. For the game described in Exercise 1, find an optimal strategy for the column player by setting up and solving the correct LP problem. Can this player expect to win or lose? How much?

5. For the game described in Exercise 2, find an optimal strategy for the column player by setting up and solving the correct LP problem. Can this player expect to win or lose? How much?

If a game has payoff matrix $[a_{ij}]$, then for every pair of strategies there is an expected value of the game for the row player. This expected value is the average amount that the row player expects to win per round if the assumed strategies are followed. If we assume that $(p_1, \ldots, p_m)$ is the row player's strategy and $(q_1, \ldots, q_n)$ is the column player's strategy, then the expected value is given by $\displaystyle\sum_{i=1}^{m}\sum_{j=1}^{n} a_{ij}p_iq_j$, and this equals the matrix product

$$
\begin{bmatrix} p_1 & \cdots & p_m \end{bmatrix}
\begin{bmatrix}
a_{11} & a_{12} & \cdots & a_{1n} \\
a_{21} & a_{22} & \cdots & a_{2n} \\
\vdots & \vdots & \ddots & \vdots \\
a_{m1} & a_{m2} & \cdots & a_{mn}
\end{bmatrix}
\begin{bmatrix} q_1 \\ \vdots \\ q_n \end{bmatrix}.
$$

6. What does the row player expect to win in the matching pennies game if she uses the strategy $(0.3, 0.7)$ and the column player uses the strategy $(0.4, 0.6)$?

7. What is the expected value of the game in Exercise 1 if each player makes choices with equal probability?

8. What is the expected value of the game in Exercise 2 if the row player chooses rows with equal probability, but the column player uses the strategy $(0.3, 0.4, 0.3)$?

9. What is the expected value of the game in Exercise 1 if each player uses optimal strategy?

Appendix A

Sets

A basic concept in mathematics is that of a set. We won't give a precise definition to the notion of a set, but we will use the term **set** (class, collection) to mean a well-defined collection of objects. The objects that make up the set are called **elements** and we think of the elements of a set as "being" in the set, or belonging to the set.

Sets are often denoted by capital letters $A, B, \ldots$ and the elements of sets are often denoted by lower case letters, $a, b, x, y, \ldots$. If a is an element of the set A, we write $a \in A$.

The number systems of mathematics provide us with our first examples of sets. The set of **integers** is the set consisting of the counting numbers $1, 2, 3, \ldots$ along with their negatives $-1, -2, -3, \ldots$, and 0. The set of integers is denoted by $\mathbb{Z}$. The set $\mathbb{Q}$ of **rational numbers** is the set consisting of all fractions a/b where $a, b \in Z$ and $b \neq 0$. The reader is probably already familiar with the set $\mathbb{R}$ of **real numbers**. The elements of $\mathbb{R}$ can be thought of as those numbers that can be represented by (not necessarily terminating) decimals. Finally, $\mathbb{C}$, the set of **complex numbers**, is the set of all numbers of the form $a + bi$, where $a, b \in R$ and where i is the so-called **imaginary unit**, and satisfies $i^2 = -1$.

A set A is a **subset** of set B if each element of A is an element of B. If this is the case we write $A \subseteq B$. We say A **equals** B and write $A = B$ if $A \subseteq B$ and $B \subseteq A$; that is, $A = B$ if A and B have exactly the same elements. If we identify the integer n with the fraction $n/1$, the rational number n/m with its decimal expansion and the real number r with the complex number $r + 0i$, then we have

$$\mathbb{Z} \subseteq \mathbb{Q} \subseteq \mathbb{R} \subseteq \mathbb{C}.$$

DEFINING SETS

There are several ways of specifying sets. The first method is to list the elements between braces. For example, the set A consisting of the integers $1, 2, 3$ is written $A = \{1, 2, 3\}$. The set B of even positive integers is written as $B = \{2, 4, 6, \ldots\}$ where the ellipses "$\ldots$" indicate that the list is to be continued in the obvious manner. An alternate method is demonstrated by the following examples.

$$A = \{x \in Z | 0 < x < 4\},$$
$$B = \{x | x \text{ is a positive even integer}\}$$

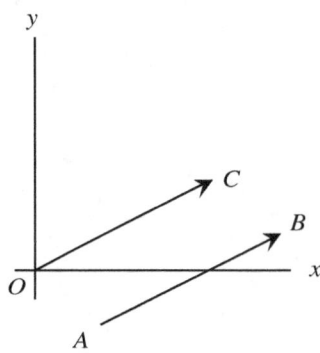

Figure A.1: Vectors $\vec{OC}$ and $\vec{AB}$

The first brace { is read "the set of all" and the vertical bar | is read "such that." Thus $\{x \in Z | 0 < x < 4\}$ is read "the set of all x in $\mathbb{Z}$ such that 0 is less than x and x is less than 4."

To negate a symbol mathematicians often put a slash / through it. For example, $x \notin A$ means x is not an element of A, $A \nsubseteq B$ means A is not a subset of B, and so on.

It is convenient to include the possibility of a set not having any elements. This set is called the **empty set** and is denoted by the symbol $\emptyset$. This set arises naturally as for example the set of all solutions of an equation that has no solutions, for example, in $|x| + 1 = 0$. Of course we are most often concerned about sets that have at least one element; such sets are called **nonempty**.

As a final example of a set, we consider the set $\mathbb{R}^2$ of 2-dimensional vectors. The reader is probably already familiar with vectors from the study of calculus or elementary physics. By a two-dimensional vector, we mean a directed line segment $\vec{AB}$, where A and B are two points in a certain fixed plane. Picture this fixed plane as having a coordinate system as in Figure A below.

Vectors having the same direction and the same magnitude are identified; that is, assumed to be equal. It follows that the vector $\vec{AB}$ can be identified with the vector $\vec{OC}$. If the point C has coordinates (x_0, y_0) then the vector is determined by (x_0, y_0), and so we will think of 2-dimensional vectors as pairs of real numbers (x_0, y_0). We define

$$R^2 = \{(x, y) | x, y \in R\}.$$

Note that equality of pairs defined by $(x, y) = (x', y')$ if and only if $x = x'$ and $y = y'$. This is a natural definition since it simply means that in a fixed coordinate system, both pairs describe the same point.

For two sets A and B, new sets can be defined from the elements of these sets. The **union** of A and B is the set $\{x | x \in A \text{ or } x \in B\}$ and is denoted by $A \cap B$. The **intersection** of A and B is the set $A \cap B = \{x | x \in A \text{ and } x \in B\}$, and, finally, the **complement** of B in A is the set $A - B = \{x | x \in A \text{ and } x \notin B\}$. In other words, the union of two sets is the set containing all of the elements from the two sets, the intersection is the set of elements common to the two sets, and the complement is the set of all elements in the one set that are not in the other set.

Suppose that $A = \{1, 2, 3, 4\}$ and $B = \{2, 4, 6\}$. Then $A \cup B = \{1, 2, 3, 4, 6\}$, $A \cap B = \{2, 4\}$, and $A - B = \{1, 3\}$. Notice that $B - A = \{6\} \neq A - B$.

Appendix A Exercises

1. List the elements of the following sets:

   (a) $\{x \in Z | x + 1 = 2\}$

   (b) $\{x \in Z | 0 < x < 6\}$

   (c) $\{x \in C | x^2 = -1\}$

   (d) $\{x \in R | |x| - 3 = 2\}$

2. Prove the following: If A, B, and C are sets and $A \subseteq B$ and $B \subseteq C$, then $A \subseteq C$.

3. Let: $A = \{2, 3, 4\}, B = \{6/2, 2, 22\}, C = \{x \in Z | x > 1\}, D = \{x | x = 2 \, or \, x = 3\}$, and $E = \{x \in Z | 1 < x < 5\}$.

   (a) Which of the above sets are equal?

   (b) Which of the above are subsets of another of the sets?

4. Let $A = \{2, 3, 4\}, B = \{1, 2, 3, 5\}$. List the elements of the sets $A \cup B$, $A - B$, and $A \cap B$.

5. Prove the following statement or disprove it by giving an example: If A, B, and C are any three sets, then $A \cup B \subseteq C$ implies $A \subseteq C$.

6. Prove the following statement or disprove it by giving an example: If A, B, and C are any three sets, then $A \cap B \subseteq C$ implies $A \subseteq C$.

Appendix B

Functions

Much of mathematics is concerned with the study of special correspondences between the elements of various sets. These special correspondences are called "functions" and defined as follows:

Definition B.1. Let A and B be sets. A **function** from A to B is a correspondence that associates with each element x in the set A a unique element y in B.

Since the element $y \in B$ is uniquely determined by x, it is convenient to give the function a name, say f, and to write $y = f(x)$ (read f of x). To indicate that f is a function from the set A to the set B, we write $f : A \to B$. We say that y or $f(x)$ is the **image** of x under the function f. The image of f is defined to be the set $\Im(f) = \{f(x) | x \in A\}$.

It is important to note that in the definition of a function f from a set A into a set B each element x of A must have a corresponding value y in the set B. That is, f must be defined on all elements in A. It is also important to note that f must be "well-defined." This part comes from the word "unique" in the definition of a function. For example, suppose that we try to define $f : \mathbb{R} \to \mathbb{R}$ by $f(x) = \sqrt{x}$. Then, since $\sqrt{-2}$ is not defined, f is not defined on all of the set $\mathbb{R}$, and so f is not a function from $\mathbb{R}$ into $\mathbb{R}$. If we attempt to define a function $f : \mathbb{Q} \to \mathbb{Q}$ by $f\left(\dfrac{m}{n}\right) = \dfrac{1}{n}$, then f is not well-defined since $\dfrac{1}{2} = \dfrac{2}{4}$, but $f\left(\dfrac{1}{2}\right) = \dfrac{1}{2}$ and $f\left(\dfrac{2}{4}\right) = \dfrac{1}{4}$.

Example B.1.   (a) Let $A = \{x \in \mathbb{R} | 0 < x\}$ be the set of positive real numbers. Let $B = \mathbb{R}$. Define $f : A \to B$ by $f(x) = 1/x$. Then f is a function from A to B.

(b) The operations on the sets of numbers $\mathbb{Z}, \mathbb{Q}, \mathbb{R}, \mathbb{C}$ provide us with different examples of functions. We can think of addition of real numbers as a function $A : \mathbb{R}^2 \to \mathbb{R}$ defined by $A((a, b)) = a + b$. Similarly multiplication may be thought of as a function $M : \mathbb{R}^2 \to \mathbb{R}$.

(c) Addition of vectors is defined as follows: Let $\vec{OA}$ and $\vec{OB}$ be the 2-dimensional vectors with endpoints (a_1, a_2) and (b_1, b_2). The sum of the two vectors is taken to be the diagonal of the completed parallelogram as in Figure 3. That is, $\vec{OA} + \vec{OB} = \vec{OC}$.

Now in this function, each pair of vectors is associated with a unique new vector - the sum of the two. Knowing that addition is a function allows certain manipulations on

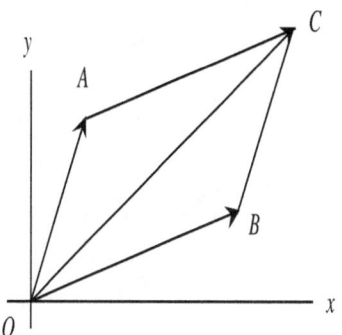

Figure B.1: The parallelogram law of vector addition

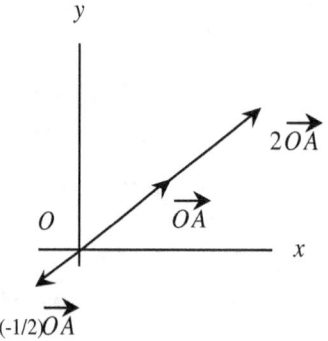

Figure B.2: Scalar multiplication

vectors. If we know that two vectors are equal, then a vector can be added to each and the results will be equal, since the addition is well-defined.

(d) Scalar multiplication furnishes an example of a function that associates with each pair $(r, \vec{OC})$, where r is a real number and $\vec{OC}$ is a 2-dimensional vector, a unique vector $r\vec{OC}$ that is defined as follows: If $r > 0$, then $r\vec{OC}$ is defined to be the vector in the direction of $\vec{OC}$ that has length equal to r times the length of $\vec{OC}$. If $r < 0$, then $r\vec{OC}$ is defined to be the vector that has the same length as $(-r)\vec{OC}$ but points in the opposite direction.

(e) Many of the functions considered in this text are correspondences between sets of vectors. For example, consider the function

$$T : R^2 \to R^2$$

defined by $T(x, y) = (x + y, x - y)$. This is an example of a special function called a linear transformation.

(f) Finally, on any set A we have the **identity function** on A. This function is denoted by 1_A and is defined by $1_A(x) = x$ for all x in the set A.

□

If we have two functions f and g both defined on the same sets, say $f : A \to B$ and $g : A \to B$, then we say that f **equals** g, and write $f = g$, if $f(x) = g(x)$ for all x in the set A. We see that equality of functions means identity.

If $f : A \to B$ is a function from the set A into the set B, we say that f is **onto** if each element of B corresponds to some element in A, and we say that f is **one-to-one** (**1-1**) if each element of B corresponds to at most one element of A. These definitions need some explanation: to say that f is onto means that given an element $y \in B$, then there is an element $x \in A$ with $y = f(x)$. In order that f be 1-1, it must be the case that no two different elements in A correspond to the same element in B. We see that for a one-to-one function, if $f(x) = f(x')$, then $x = x'$.

Let us consider some examples.

Example B.2. If $f : \mathbb{R} \to \mathbb{R}$ is defined by $f(x) = x^2$, then f is not onto since $f(x) = -1$ is never possible and f is not 1-1 since $f(2) = 4 = f(-2)$. The function $f(x) = 3x + 1$ is both 1-1 (if $3x + 1 = 3x' + 1$, then $x = x'$) and onto (for any real number r there is a real number x such that $3x + 1 = r$). The function $f(x) = x(x + 1)(x - 1)$ is not 1-1, but it is onto; the function $f(x) = e^x$ is 1-1 but not onto. (Why?)

□

If A, B, and C are sets and $f : A \to B$ and $g : B \to C$ are functions, we may define a new function $g \circ f : A \to C$ by $g \circ f(x) = g(f(x))$. This new function $g \circ f$ is indeed a function from A to C and it is called the **composite** of f and g. Notice how the composite appears to come out backwards. We have

$$A \xrightarrow{f} B \xrightarrow{g} C,$$

but the composite is $g \circ f$. This phenomenon arises as a consequence of writing the function name to the left of the variable name: $f(x)$ instead of $(x)f$. Notice that if $f : A \to B$ is any function, then $f \circ 1_A(x) = f(1_A(x)) = f(x)$, so that $f \circ 1_A = f$, and $1_B \circ f(x) = 1_B(f(x)) = f(x)$, so that $1_B \circ f = f$. We see that the identity function behaves as an identity relative to the operation of composition of functions.

We state without proof two important theorems about general functions:

Theorem B.1. *Let A, B, C, and D be nonempty sets, and let $f : A \to B, g : B \to C$, and $h : C \to D$ be functions. Then $h \circ (g \circ f) = (h \circ g) \circ f$* (Composition of functions is associative.)

Theorem B.2. *Let $f : A \to B$ be a function. Then f is 1-1 and onto if and only if there is a function $g : B \to A$ with $g \circ f = 1_A$ and $f \circ g = 1_B$.*
(A function has an inverse if and only if it is 1-1 and onto.)

The function g in the above theorem is called the **inverse** of f and is denoted by f^{-1}. The -1 is a superscript only, however; it is *not* an exponent.

Appendix B Exercises

1. Let $f : R \to R$ be defined by $f(x) = 2x + 1$. Calculate $f(5), f(-2)$. Is f 1-1? Is f onto?

2. Let $g : R \to R$ be defined by $g(x) = x^2 + 1$. Calculate $g(2), g(-2)$. Is g 1-1? Is g onto?

3. Let f and g be as in Exercises 1 and 2 above. Calculate $f \circ g(2)$ and $g \circ f(2)$.

4. Let $f : R \to R$ be defined by $f(x) = 3x - 2$. Show that f is 1-1 and onto and find the inverse function f^{-1}.

5. Let T be the function in Example B.1 (e). Show that

$$T((x_1, y_1) + (x_2, y_2)) = T(x_1, y_1) + T(x_2, y_2) \text{ and } T(r(x, y)) = rT(x, y).$$

6. Prove Theorem B.1.

7. Prove Theorem B.2.

Appendix C

Fields

The reader is probably familiar with the rational, real, and complex number systems that are studied in traditional algebra and calculus courses. Recall that $\mathbb{Q}$ is used to denote the set of rational numbers, $\mathbb{R}$ denotes the set of real numbers, and $\mathbb{C}$ is the collection of complex numbers. According to our conventions,

$$\mathbb{Q} \subseteq \mathbb{R} \subseteq \mathbb{C}.$$

These number systems have, along with a certain set of elements, two operations - addition and multiplication. Properties of these operations allow one to manipulate algebraic expressions and solve equations. We wish to establish results and develop methods that are as general as possible in that they apply to many different number systems. For this reason (not to make things more difficult) we introduce the definition of a "field."

In general, a field is an algebraic system that has the arithmetic properties common to the rational, real, and complex number systems and in order to simplify matters, the reader may assume that any field discussed is one of these three fields. This assumption will, however, carry with it a resulting loss in generality of the results and methods that are presented.

By an **operation** (also called a "**binary operation**") on a set F we mean a function that associates with each pair (a, b) of elements of F some unique element of F. If we name this function $\otimes$, it is common practice to denote the element of F that is associated with the pair (a, b) by $a \otimes b$ - using the so-called "**infix**" notation rather than the more customary "**prefix**" notation $\otimes(a, b)$. The fields mentioned above have addition and multiplication operations and we will assume that every field has such operations.

With these comments we then make the following definition:

A **field** is a nonempty set F along with an operation of addition $(+)$ and an operation of multiplication $(\cdot)$ that satisfy the following:

(A1) $(a + b) + c = a + (b + c)$ for all $a, b, c \in F$.

(A2) $a + b = b + a$ for all $a, b \in F$.

(A3) There is an element $0 \in F$ satisfying $a + 0 = a$ for all $a \in F$.

(A4) For each $a \in F$ there is an element $-a \in F$ with $a + (-a) = 0$.

(M1) $(a \cdot b) \cdot c = a \cdot (b \cdot c)$ for all $a, b, c \in F$.

(M2) $a \cdot b = b \cdot a$ for all $a, b \in F$.

(M3) There is an element $1 \in F$ with $1 \cdot a = a$ for all $a \in F$.

(M4) For each $a \in F, a \neq 0$, there is an element $a^{-1} \in F$ satisfying $a \cdot a^{-1} = 1$.

(D) For all $a, b, c \in F, a \cdot (b + c) = (a \cdot b) + (a \cdot c)$.

The above properties are well-known properties of the systems of rational, real, and complex number systems along with the usual operations of addition and multiplication. It follows that $\mathbb{Q}, \mathbb{R}$, and $\mathbb{C}$ are examples of fields.

> *Example* C.1. For a much different example of a field, let $\mathbb{Z}_2 = \{0, 1\}$ and define $+$ and $\cdot$ by
>
> $$\begin{array}{ll} 0 + 0 \ = \ 0 & 0 \cdot 1 \ = \ 1 \cdot 0 = 0 \\ 0 + 1 \ = \ 1 + 0 = 1 & 0 \cdot 0 \ = \ 0 \\ 1 + 1 \ = \ 0 & 1 \cdot 1 \ = \ 1. \end{array}$$
>
> Then Z_2 along with $+$ and $\cdot$ satisfies the hypotheses above and so Z_2 is a field, called the field of integers modulo 2.

(a) We can find a field with 3 elements. Let $Z_3 = \{0, 1, 2\}$. We define addition and multiplication operations by the following tables:

+	0	1	2
0	0	1	2
1	1	2	0
2	2	0	1

$\cdot$	0	1	2
0	0	0	0
1	0	1	2
2	0	2	1

Under these operations Z_3 is a field: the integers modulo 3.

$\square$

A field can be thought of as an environment in which a certain amount of algebra can be done. In the system of integers, any equation of the form $x + a = b$ can be solved, where a and b are integers and x is some unknown. To solve an equation of the form $ax + b = c$ for the unknown x requires the capability of dividing by the coefficient a, or multiplying by its reciprocal. This is not in general possible in the system of integers, but property (M4) guarantees that this operation may be done in any field, provided that $a \neq 0$.

Let us solve the equation $ax + b = c$ carefully assuming that a, b, and c are elements of a field F, and of course that $a \neq 0$. We list each step and the associated field property which permits it:

$$\begin{aligned}
ax + b &= c && \text{given} \\
(ax + b) + (-b) &= c + (-b) && \text{(A4)} \\
ax + (b + (-b)) &= c + (-b) && \text{(A1)} \\
ax + (0) &= c + (-b) && \text{(A4)} \\
ax &= c + (-b) && \text{(A3)} \\
a - 1(ax) &= a - 1(c + (-b)) && \text{(M4)} \\
(a - 1a)x &= a - 1(c + (-b)) && \text{(M1)} \\
1x &= a - 1(c + (-b)) && \text{(M4)} \\
x &= a - 1(c + (-b)) && \text{(M3)}
\end{aligned}$$

In the second step above, $-b$ was added to both sides of the equation. This is permitted by the assumption that addition is a function and therefore well-defined. In effect, the assumption that elements have unique images under a functional correspondence produces the familiar rule: Equals added to equals, sums equal. We see that the properties of a field are exactly what we need to be sure that we can solve linear equations. Notice that the equation $x^2 = 2$ cannot be solved over the field $\mathbb{Q}$, but it does have a solution in the field $\mathbb{R}$ – but that's another story.

Some of the familiar properties of addition and multiplication of the real and complex numbers are not listed as hypotheses for a field, but it turns out that they can be proved assuming only these hypotheses. We list them in the following theorem, the proof of which is left as an exercise.

Theorem C.1. *Let F be a field. Then for all $a, b, c \in F$, the following hold:*

(a) *If $a + b = a + c$, then $b = c$.*   *(The cancellation law of addition.)*

(b) *If $a \cdot b = a \cdot c$ and $a \neq 0$, then $b = c$.*   *(The cancellation law of multiplication.)*

(c) *$0 \cdot a = a \cdot 0 = 0$.*   *(Zero times any element is zero.)*

(d) *$-a = (-1) \cdot a$*   *($-a$ is -1 times a.)*

(e) *$-(-a) = a$*   *(The additive inverse of $-a$ is a.)*

(f) *$(a^{-1})^{-1} = a$ if $a \neq 0$*   *(The multiplicative inverse of a^{-1} is a.)*

(g) *If $a \cdot b = 0$, then $a = 0$ or $b = 0$.*   *(There are no nonzero zero divisors.)*

(h) *If $a \cdot b \neq 0$, then $(a \cdot b)^{-1} = b^{-1} \cdot a^{-1}$.* *(The inverse of a product is the product of the inverses.)*

Appendix C Exercises

1. Prove Theorem C.1.

2. Let F be a field and $a, b \in F$. Prove the following: If $(x - a)(x - b) = 0$, then $x = a$ or $x = b$.

3. We define subtraction on a field F by $a - b = a + (-b)$ for any $a, b \in F$. Prove the following for $a, b, c \in F$:

    (a) $a - (-b) = a + b$

    (b) $a - (b + c) = (a - b) - c$

    (c) $(a - b)(a + b) = a \cdot a - b \cdot b$

4. For elements a and b in a field F, we define a/b by $a/b = a \cdot b^{-1}$ provided $b \neq 0$. Show the following for $a, b, c, d \in F, b \neq 0, d \neq 0$.

    (a) $(a/b) \cdot (c/d) = (a \cdot c)/(b \cdot d)$

    (b) $a/b + c/d = (ad + bc)/(bd)$

    (c) $(a/b)/(c/d) = (ad)/(bc)$

5. Find a field with four elements. (Hint: Let $F = 0, 1, a, b$, where 0 and 1 satisfy the obvious conditions. Define addition and multiplication so that the properties of a field are satisfied).

Appendix D

Facts about polynomials

A few facts about polynomials are helpful in finding the roots of the characteristic polynomial of a matrix. Recall that a **polynomial over a field** F is an expression of the form $p(x) = a_0 + a_1 x + \ldots + a_n x_n$, where $a_0, \ldots, a_n \in F$, and that the set of $a_n \in F$, and that the set of all polynomials over F is denoted by $F[x]$. If $q(x) = b_0 + \ldots + b_m x_m$ is another polynomial over F, the product $p(x)q(x)$ is defined to be the polynomial

$$c_0 + c_1 x + \ldots + c_{n+m} x^{n+m},$$

where

$$c_k = \sum_{i+j=k} a_i b_j.$$

If $p(x) = a_0 + \ldots + a_n x_n$ and $a_n \neq 0$, then we say that p has degree n and write deg $p(x) = n$. Note that in the above product $p(x)q(x), c_{n+m} = a_n b_m$ and it follows that deg $p(x)q(x) = \deg p(x) + \deg q(x)$, provided that neither of $p(x)$ or $q(x)$ is the zero polynomial.

If $a \in F$ and $p(x)$ is as shown, we define $p(a)$ by $p(a) = a_0 + a_1 a + \ldots + a_n a^n$. We say that $a \in F$ is a **root** of $p(x)$ (or a **solution** of the equation $p(x) = 0$) if $p(a) = 0$.

In elementary high school algebra one learns "long hand division" of a polynomial by another. This is formalized as follows:

> **Theorem D.1** (The Division Algorithm). *If $p(x), s(x) \in F[x]$ and $s(x)$ is not the zero polynomial, then there exist polynomials $q(x), r(x) \in F[x]$ with $p(x) = s(x)q(x) + r(x)$ and $r(x) = 0$ or deg $r(x) < \deg s(x)$. (Long hand division works with polynomials.)*

Applying the division algorithm to the polynomials $p(x)$ and $s(x) = x - a$, we get

$$p(x) = (x - a)q(x) + r(x)$$

where $\deg r(x) < 1$ or $r(x) = 0$. Substituting a for x, we get $p(a) = r(a)$. But since $\deg r(x) = 0$, $r(x)$ is a constant polynomial and so $r(x) = p(a)$. From this observation, the following two theorems follow easily.

Theorem D.2 (The Factor Theorem). *$x - a$ is a factor of $p(x)$ (that is, $p(x) = (x-a)q(x)$ for some $q(x)$) if and only if a is a root of $p(x)$.* *($x - a$ is a factor iff a is a root.)*

Theorem D.3 (The Remainder Theorem). *If one divides $x - a$ into $p(x)$, the remainder is $p(a)$.* *(The remainder on dividing by x - a is $p(a)$.)*

A polynomial of positive degree is called **prime** if it cannot be factored as a product of polynomials of positive degree. Each polynomial of degree one (called a linear polynomial) is prime and the following important theorem tells us that these are the only prime polynomials over the complex numbers.

Theorem D.4 (The Fundamental Theorem of Algebra). *Every polynomial over the complex numbers has a root in the set of complex numbers.* *(Every polynomial has a root in $\mathbb{C}$.)*

From this theorem it follows that if $p(x) = a_0 + \ldots + a_n x_n$ is a polynomial over the complex numbers of degree n, then there are complex numbers $c_1, \ldots, c_n$ (not necessarily distinct) such that $p(x) = a_n(x - c_1) \ldots (x - c_n)$. In general, if $p(x)$ is a polynomial over a field F, we say that $p(x)$ **factors completely** (over F) if there exist $a, a_1, \ldots, a_n \in F$ such that $p(x) = a(x - a_1) \ldots (x - a_n)$. Using this terminology, then, we say that each polynomial over $\mathbb{C}$, the field of complex numbers, factors completely over $\mathbb{C}$.

In the polynomial $p(x) = a_0 + \ldots + a_n x_n$, the coefficient a_n is called the **leading coefficient**. The polynomial is called **monic** if the leading coefficient is 1; that is, if $a_n = 1$. A monic polynomial $p(x)$ factors in the form

$$p(x) = (x - c_1) \ldots (x - c_n).$$

Recall that for a complex number $z = a + bi$, the **conjugate** of z is the complex number denoted by $\overline{z}$ and defined by $\overline{z} = a - bi$. The following facts about polynomials are easily verified:

1. If $p(x)$ is a polynomial with real coefficients and z is a complex number with $p(z) = 0$, then $p(\overline{z}) = 0$.

2. $(x - z)(x - \overline{z}) = x^2 - (z + \overline{z})x + z\overline{z}$ is a polynomial with real coefficients (for any complex number z).

3. Any polynomial with real coefficients can be factored as a product of real polynomials of degree 1 and 2.

4. A polynomial of odd degree with real coefficients has a real root.

We can use the above facts to help us factor polynomials, if it is possible to do so.

D.1 Factoring Polynomials

In Chapters 5 and 6, the assumption that the characteristic polynomial factors completely is frequently made. The above remarks indicate that every polynomial (of positive degree) factors completely (that is, into linear factors) over the field of complex numbers, but over the field of real numbers, it can only be shown that every polynomial (of positive degree) factors as a product of linear and quadratic polynomials. This theory is, of course, very nice, but in order to solve problems one must actually factor a polynomial completely - not just know that such a factorization exists. Unfortunately, no foolproof plan exists for finding such a factorization. The reader may have already encountered this situation in calculus: In integrating rational functions using the method of partial fractions, it is necessary to factor the polynomial in the denominator before the partial fraction decomposition can be accomplished.

The following hints and comments should be of help:

1. In some situations the polynomial can be factored before it is "multiplied out." Factor before collecting terms, if possible.

2. If a root a is found, divide $x - a$ into a polynomial, say $p(x)$, and obtain $p(x) = (x - a)p_1(x)$. Now $p_1(x)$ is of smaller degree than p and the remaining roots of $p(x)$ are all roots of $p_1(x)$, so try to factor $p_1(x)$.

3. Hope for a "nice" root, that is, an integer or a rational number. These can sometimes be found using the following:

> **Theorem D.1** (The Rational Root Test). *Let $p(x) = a_0 + \ldots + a_n x_n$, be a polynomial with integral coefficients (that is, $a_0, \ldots, a_n$ are integers). If r/s is a rational number with r and s integers having no common divisors and $p(r/s) = 0$, then r divides a_0 and s divides a_n.*
>
> *(The only possible rational roots are a factor of the constant term over a factor of the leading coefficient.)*

To find the rational roots of a polynomial one need only check all of the quotients r/s where r divides a_0 and s divides a_n.

4. When the polynomial has been sufficiently factored that only a quadratic polynomial remains unfactored, apply the quadratic formula to obtain the remaining roots and corresponding factors.

Example D.1. (a) Suppose we wish to factor

$$p(x) = x^3 - 8x^2 + 17x - 6.$$

Applying the rational root test we see that, since the polynomial is monic, the only possible rational roots are $\pm 1, \pm 2, \pm 3, \pm 6$.

We use **synthetic division** (see a high school or college algebra text) to check:

$$\begin{array}{r|rrrr} 1 & 1 & -8 & 17 & -6 \\ & & 1 & -7 & 10 \\ \hline & 1 & -7 & 10 & 4 \end{array}$$

We see that $x = 1$ is not a root since the remainder is 4. Now we try $x = 3$:

$$\begin{array}{r|rrrr} 3 & 1 & -8 & 17 & -6 \\ & & 3 & -15 & 6 \\ \hline & 1 & -5 & 2 & 0 \end{array}$$

Since the remainder is 0, $x = 3$ is a root. We also see that the quotient has the coefficients $1, -5$, and 2, and so the polynomial factors as

$$x^3 - 8x^2 + 17x - 6 = (x - 3)(x^2 - 5x + 2).$$

Using the quadratic formula to solve the remaining quadratic we obtain roots:

$$r_1 = \frac{5 + \sqrt{17}}{2},$$

$$r_2 = \frac{5 - \sqrt{17}}{2}$$

The complete factorization is

$$x^3 - 8x^2 + 17x - 6 = (x - 3)\left(x - \frac{5 + \sqrt{17}}{2}\right)\left(x - \frac{5 + \sqrt{17}}{2}\right).$$

(b) Consider the polynomial:

$$p(x) = x^4 + 1.$$

Since $x^4 + 1 > 0$ for all real numbers x, we can see that p has no real roots and so the factorization will be into a product of two quadratics. We use a trick:

$$p(x) = x^4 + 1 = x^4 - (i^2).$$

We see that p is a difference of two squares and so can be factored in that manner:

$$p(x) = (x^2 - i)(x^2 + i).$$

We can apply the trick again:

$$p(x) = (x - \sqrt{i})(x + \sqrt{i})(x - \sqrt{-i})(x + \sqrt{-i}),$$

and using $\sqrt{i} = \dfrac{1}{\sqrt{2}} + \dfrac{1}{\sqrt{2}}i$ and $\sqrt{-i} = \dfrac{1}{\sqrt{2}} - \dfrac{1}{\sqrt{2}}i$, we get

$$p(x) =$$

$$\left(x - \left(\frac{1}{\sqrt{2}} + \frac{1}{\sqrt{2}}i\right)\right)\left(x + \frac{1}{\sqrt{2}} + \frac{1}{\sqrt{2}}i\right)\left(x - \left(\frac{1}{\sqrt{2}} - \frac{1}{\sqrt{2}}i\right)\right)\left(x + \frac{1}{\sqrt{2}} - \frac{1}{\sqrt{2}}i\right).$$

Notice that

$$(x - z)(x - \overline{z}) = x^2 - (z + \overline{z})x + z\overline{z}),$$

and so grouping factors involving roots that are conjugates produces a quadratic with real coefficients since $z + \overline{z}$ is a real number. We get

$$p(x) = (x^2 - (2/\sqrt{2})x + 1)(x^2 + (2/\sqrt{2})x + 1),$$

the factorization of $p(x)$ over the real numbers.

□

One final definition: If a is a root of the polynomial $p(x)$, the **multiplicity** of a is the integer k such that $p(x) = (x - a)^k p_1(x)$ for some $p_1(x)$, but $p(x) \neq (x - a)^{k+1}p_2(x)$ for any $p_2(x)$. We might say that a is a root of $p(x)$ "k times". For example, 2 is a root of $x^2 - 4x + 4$ of multiplicity 2; 1 is a root of $x^3 - 2x^2 + x$ of multiplicity 2 and 0 is a root of multiplicity 1.

Appendix D Exercises

In Exercises 1 - 3, factor the polynomials into prime factors over the real numbers.

1. $x^3 - 5x^2 - 8x + 12$

2. $x^3 + 9x^2 + 9x + 8$

3. $x^4 - 2x^3 + 2x^2 - 2x + 1$

   In Exercises 4-6, factor the polynomials into into a product of linear factors over the complex numbers.

4. $x^3 - 5x^2 - 8x + 12$

5. $x^3 + 9x^2 + 9x + 8$

6. $x^4 - 2x^3 + 2x^2 - 2x + 1$

7. Find a polynomial with real coefficients that has $1 + i$ as a root.

8. Does the polynomial $x^3 - 7x^2 + 2x + 1$ have a rational root? A real root?

Appendix E

Mathematical Induction

The **Principle of Mathematical Induction** gives us an important method of proof. Since we shall be using induction in proofs, it is appropriate that it be stated here. The method of induction is based on a property of the integers and it may be regarded as a theorem about the integers. The integers, $\mathbb{Z}$, form an "ordered system" in that we have the order relation $<$ defined on the integers and we have the notion of "positive integer." The set of all positive integers is denoted by $\mathbb{Z}^+$. Of course, the rational numbers $\mathbb{Q}$ and the real numbers $\mathbb{R}$ are also ordered, but the integers have an additional property not shared by these systems.

Theorem E.1 (The Well-Ordering Property). *Every nonempty subset of the set of positive integers has a least element.* *(A set of integers with positive elements has a least positive element)*

The well-ordering property is one of the axioms for the system of integers, that is, it is assumed to be true along with the other assumptions about the integers. If we think of the set of positive real numbers, we can see that this set has no least element and so the reals fail to have the well-ordering property.

Now suppose that we have a subset S of the set of positive integers and suppose that S has two properties: 1) $1 \in S$, 2) If $k \in S$ then $k + 1 \in S$. It is not hard to see that S must be exactly the set of positive integers: If S is not the set of positive integers, then there is a positive integer not in S. Since there is one such positive integer, there must be a least positive integer which is not in S. The integer that comes before this least one must be in S (the least one couldn't have been 1 by property 1)), but then by 2) the next integer is also in S. This gives a contradiction and so, $S = \mathbb{Z}^+$.

The Principle of Mathematical Induction is simply an application of the result stated in the above paragraph. Suppose that for each $n \in \mathbb{Z}^+$, S_n is a statement; that is, S_n is either true or false. Let S be defined by $S = \{n \in \mathbb{Z}^+ | S_n \text{ is true}\}$. If S has the properties 1) and 2) described above, then $S = Z^+$. But $S = \mathbb{Z}^+$ is equivalent to the statement S_n is true for all $n \in \mathbb{Z}^+$. Interpreted in this context, 1) states that S_1 is true, and 2) is equivalent to the statement, "If S_k is true for some positive integer k, then S_{k+1} is true. Thus, using the well-ordering property, we have proved the following theorem.

Theorem E.2 (The Principle of Mathematical Induction). *Let S_n be a statement for each positive integer n. Assume that the following two statements are true:*

1. S_1 is true.

2. If S_k is true for some positive integer k, then S_{k+1} is true.

Then S_n is true for all positive integers n.

Example E.1.   (a) Let S_n represent the statement

$$1^2 + 2^2 + \ldots + n^2 = \frac{n(n+1)(2n+1)}{6}.$$

Then S_1 is true since $1^2 = \dfrac{1 \cdot 2 \cdot 3}{6}$. Now assume that S_k is true for some integer k. Then

$$1^2 + 2^2 + \ldots + k^2 = \frac{k(k+1)(2k+1)}{6}.$$

Adding $(k+1)^2$ to both sides, we see that

$$1^2 + 2^2 + \ldots + k^2 + (k+1)^2 = \frac{k(k+1)(2k+1)}{6} + (k+1)^2$$
$$= (k+1)\frac{k(2k+1) + 6k + 6}{6}$$
$$= (k+1)\frac{2k^2 + 7k + 6}{6}$$
$$= (k+1)\frac{(k+2)(2k+3)}{6}.$$

We were to show that $1^2 + 2^2 + \ldots + k^2 + (k+1)^2 = \dfrac{(k+1)(k+1+1)(2(k+1)+1)}{6} = \dfrac{(k+1)(k+2)(2k+3)}{6}$, so S_{k+1} is true and by induction we know that S_n is true for all positive integers n.

(b) Induction can be used to make definitions. Let A be a $k \times k$ matrix and for a positive integer n, define A^n as follows:

  (a) Define A^1 to be A.

  (b) If A^k is defined for some positive integer k, define A^{k+1} by $A^{k+1} = A^k A$.

It is not hard to see that the statement "A^n is defined" can be proved true for all positive integers n.

(c) Using the above definition we can prove that $A^m A^n = A^{m+n}$ for all positive integers m and n. While the proof is left as an exercise, we will give a strong hint: Think of m as being a fixed integer and let S_n represent the statement, "$A_m A_n = A^{m+n}$. Notice that

$$A^m A^{n+1} = A^m A^n A = A^{m+n} A = A^{m+n+1}.$$

(d) If $a_1, \ldots, a_n$ is any list of scalars, we define $\sum_{i=1}^{k} a_i$ inductively by the following: We define $\sum_{i=1}^{1} a_i = a_1$. If $\sum_{i=1}^{k} a_i$ is defined, we define $\sum_{i=1}^{k+1} a_i = \left(\sum_{i=1}^{k} a_i \right) + a_{k+1}$. In Exercise 6, the reader is asked to prove the distributive property: $\sum_{i=1}^{k} a_i = \sum_{i=1}^{k} (aa_i)$.

$\square$

Appendix E Exercises

1. Prove that for any positive integer n,

$$1 + 2 + \ldots + n = \frac{n(n+1)}{2}.$$

2. Prove that for any positive integer n,

$$1^3 + 2^3 + \ldots + n^3 = \frac{n^2(n+1)^2}{4}.$$

3. Prove that for any integer $n > 0$ and any real number $r \neq 1$,

$$1 + r + r^2 + \ldots + r^n = \frac{1 - r^{n+1}}{1 - r}.$$

4. Complete the proof outlined in part (c) of the example above.

5. Let A be a $k \times k$ matrix. Prove that for all positive integers n and m, $A^{mn} = (A^m)^n$.

6. Prove the distributive property: $\sum_{i=1}^{k} a_i = \sum_{i=1}^{k} (aa_i)$, where $a_1, \ldots, a_n$ is any list of scalars.
   (See Example E.1(d).)

Appendix F

Additional Proofs

In this appendix we include several proofs that were omitted earlier. In general, these proofs are of greater complexity, greater length, and of less interest from a computational point of view. We include them for completeness.

In Section 1.5 we omitted the proof of the uniqueness of the reduced echelon form. Recall the definition: An $m \times n$ matrix $A = [a_{ij}]$ is said to be in **reduced echelon form** if and only if it satisfies the following three conditions:

1. For some integer r, the first r rows contain nonzero entries and the remaining $n - r$ rows contain only zeros.

2. For $i = 1, \ldots, r$, the first nonzero entry in row i is a 1 in column j_i and it is the only nonzero entry in column j_i.

3. $j_1 < j_2 < \ldots < j_r$.

The proof of the uniqueness of the reduced echelon form is simplified with the use of facts about rank and the row space of a matrix that were introduced in Sections 2.5 and 2.8. We will assume that we have these facts available, but it is important to note that the part of the theorem that we are about to prove was not used in establishing any of these later results.

The proof will be by induction on the rank r of a matrix and will make heavy use of the row space of the matrix; that is, the span of the row vectors of the matrix. Recall that (by 2.5.2) row equivalent matrices have the same row space and so the same rank. Recall also that for any $m \times n$ matrix M, M_i denotes the i-th row vector of M. The row space of the matrix M is then given by $\boldsymbol{R}\,(M) = \mathrm{span}\{M_1, \ldots, M_m\}$.

Theorem F.1 (Theorem 1.5.1). *Any $m \times n$ matrix A is row equivalent to one and only one $m \times n$ matrix B that is in reduced echelon form. B is called the* **reduced echelon form** *of A.*   *(Every matrix has a unique reduced echelon form.)*

Proof. (Uniqueness) Assume that the matrix A is similar to two matrices in reduced echelon form. Then these two matrices will themselves be row equivalent. Let B and C be $m \times n$

matrices and assume that B is row equivalent to C. We must prove that $B = C$. Note that B and C have the same rank, r, and $\mathbf{R}(B) = \mathbf{R}(C)$. We can state the result we need in a slightly different form: If B and C are matrices in reduced echelon form and $\mathbf{R}(B) = \mathbf{R}(C)$, then $B = C$. It is this result that we will prove by induction on the dimension of the row space.

If $r = 1$, then $\text{span}\{B_1\} = \text{span}\{C_1\}$, so $B_1 = aC_1$ for some scalar a. Since the first nonzero entry in each of the vectors is a 1, it must be that $a = 1$. It follows that $B_1 = C_1$ and so, $B = C$. This proves the theorem for matrices of rank 1.

Assume now that the theorem is true for matrices of rank $r - 1$ and let B and C be row equivalent matrices in reduced echelon form that have rank r. Now $\text{span}\{B_1, \ldots, B_r\} = \text{span}\{C_1, \ldots, Cr\}$. Let the constants associated with B be $r, j_1, \ldots, j_r$ and let those for C be $k_1, \ldots, k_r$. Assume that $k_r = j_r$. Since $B_r \in \text{span}\{C_1, \ldots, C_r\}$, $B_r = b_1 C_1 + \ldots + b_r C_r$, for some scalars $b_1, \ldots, b_r$. Then since $k_1, \ldots, k_{r-1} < j_r$, we see that $b_1, \ldots, b_{r-1}$ must all equal zero (these are the entries in columns $k_1, \ldots, k_{r-1}$ of the linear combination $b_1 C_1 + \ldots + b_r C_r$ and the vector B_r has zeros there). So, $B_r = b_r C_r$ and since both B_r and C_r have 1 as their first nonzero entries, we have that $b_r = 1$, and so $B_r = C_r$.

We have shown that the last nonzero rows in B and C are equal. It is not hard to see (see Exercise 1) that $\text{span}\{B_1, \ldots, B_{r-1}\} = \text{span}\{C_1, \ldots, C_{r-1}\}$ and applying the induction assumption, we see that

$$\begin{bmatrix} B_1 \\ \vdots \\ B_{r-1} \end{bmatrix} = \begin{bmatrix} C_1 \\ \vdots \\ C_{r-1} \end{bmatrix}.$$

We have shown that $B = C$, and so by induction, the statement holds for matrices of any rank. $\qquad\square$

Two cumbersome proofs on determinants were omitted. We include them here.

Theorem F.2 (Theorem 4.1.1). *Let $A = [a_{ij}]$ be an $n \times n$ matrix. Then:*

$$|A| = \sum_{j=1}^{n} a_{ij} A_{ij} \quad \text{(expansion along row i), and}$$

$$|A| = \sum_{i=1}^{n} a_{ij} A_{ij} \quad \text{(expansion down column j).}$$

(Expansion along any row or column gives the determinant.)

Proof. We know that this theorem is true for 1×1 and 2×2 matrices by Exercise 11. We will prove the theorem by mathematical induction and assume that the theorem is true for $(n-1) \times (n-1)$ matrices and prove it for $n \times n$. Let $A = [a_{ij}]$ be an $n \times n$ matrix and recall that M_{ij} is the determinant of the $(n-1) \times (n-1)$ matrix obtained by removing row i and column j of A. M_{ij} is called the **minor** of a_{ij}. By $(M_{ij})_{kh}$ we will mean the determinant of the $(n-2) \times (n-2)$ matrix obtained by removing rows i and k and columns j and h

of A. Note that these row and column numbers refer to the matrix A. Recall also that the **cofactors** of A are given by $A_{ij} = (-1)^{i+j}|M_{ij}|$.

We will first show that the expansion of $|A|$ along row 1 equals that down column 1. First, we will write the expansion using row 1 and then expand the minors using column 1, and then we will expand down column 1 with the minors expanded along row 1. Consider:

(*) $\displaystyle\sum_{j=1}^{n} a_{1j}A_{1j}$: expand $A_{1j} = (-1)^{1+j}|M_{1j}|$ down column 1

(**) $\displaystyle\sum_{i=1}^{n} a_{i1}A_{i1}$: expand $A_{i1} = (-1)^{1+i}|M_{i1}|$ across row 1

Using the induction assumption, the minors, since they are $(n-1)\times(n-1)$ determinants, can be expanded along any row or column and the results will be the same. Note that the term $a_{11}A_{11}$ is common to both of the above expansions. We will match the other terms in these expansions to show that they are the same. Consider:

(*) $a_{1j}(-1)^{1+j}a_{i1}(-1)^{1+i-1}(M_{1j})_{i1}$

(**) $a_{i1}(-1)^{1+i}a_{1j}(-1)^{1+j-1}(M_{i1})_{1j}$

The terms $i-1$ and $j-1$ come from the fact that the minor has one less row and one less column than A has. The expansions of each of the terms $a_{1j}A_{1j}$ and $a_{i1}A_{i1}$ involve a single term with the entries a_{1j}, and a_{i1}. Comparing these terms in the expansions (*) and (**) we see that they are equal. It follows that expansion along row 1 gives the same value as expansion down column 1.

We will next show that expansion along row 1 gives the same result as expansion along another row, say i. The method is basically the same, but slightly more complicated. As above we will consider two different expansions and compare terms.

(*) expand $A_{1j} = (-1)^{1+j}M_{1j}$ across row i

(**) expand $A_{ik} = (-1)^{i+k}M_{ik}$ across row 1

As before, the minors are $(n-1)\times(n-1)$ determinants and so they can be expanded along any row or column and the results will be the same. We will match terms, remembering that the column number in the minors might not agree with the column number in the matrix A since a column has been removed from A.

(*) $a_{1j}(-1)^{1+j}a_{ik}(-1)^{i+h}(M_{1j})_{ik}$, where $h = \begin{cases} k & \text{if } k < j \\ k-1 & \text{if } k > j \end{cases}$

(**) $a_{ik}(-1)^{i+k}a_{1j}(-1)^{1+h}(M_{ik})_{1j}$, where $h = \begin{cases} j & \text{if } k < j \\ j-1 & \text{if } k > j \end{cases}$

Let us compute the exponent of the term -1 above: if $k < j$ we get $i + k + 1 + j$ in each case, and if $k > j$, we get $i + k + j$ in each case. We see that the terms agree and so we see that expansion along row 1 produces the same result as expansion along row i.

The proof above could easily be modified to prove that expansion down any column gives the same result as expansion down column 1. Thus, by induction, the theorem is proved.

$\square$

Theorem F.3 (Theorem 4.1.2). *Let A be an $n \times n$ matrix and let B be the matrix obtained from A by interchanging two of the rows (or columns) of A. Then $|B| = -|A|$.*

(Switching rows or columns changes the sign of the determinant.)

Proof. We will prove the part of the theorem involving the switching of rows. Using the theorem above, a similar proof can be constructed for columns switches. We will prove the theorem by induction on n, the order of the matrix. For $n = 1$, there is not much to prove, and for $n = 2$, the result is proved in Exercise 11. Let us assume that the theorem is true for all $(n - 1) \times (n - 1)$ matrices and let $A = [a_{ij}]$ be an $n \times n$ matrix, where $n = 3$. Let B be the result of switching rows i and j in A, that is $B = R_{ij}A$. Let k denote a row of A different from i or j, and expand the determinant of B along row k. We get

$$|B| = \sum_{h=1}^{n} a_{kh} B_{kh},$$

since row k in B is the same as row k in A. Now note that the determinant in the cofactor B_{kh} is the same as that in A_{kh} except that two rows have been switched. This matrix is $(n - 1) \times (n - 1)$ and so by the induction assumption, $B_{kh} = -A_{kh}$. It follows that

$$|R_{ij}A| = |B| = \sum_{h=1}^{n} a_{kh} B_{kh} = \sum_{h=1}^{n} a_{kh}(-A_{kh}) = -\sum_{h=1}^{n} a_{kh} A_{kh} = -|A|,$$

and so by induction, $|R_{ij}A| = -|A|$ for any $n \times n$ matrix A. $\square$

The proof of the existence and uniqueness of the Jordan Canonical Form was omitted earlier. Here we will outline the proof of the existence. Recall the theorem:

Theorem F.4 (Jordan Canonical Form). *Let A be an $n \times n$ matrix and assume that $p_A(\lambda)$ factors completely. Then A is similar to a matrix J of the form*

$$J = \begin{bmatrix} J_1 & \cdots & 0 \\ \vdots & \ddots & \vdots \\ 0 & \cdots & J_k \end{bmatrix},$$

where $J_1, \ldots, J_k$ are Jordan blocks. The matrix J is unique except for the order of the blocks $J_1, \ldots, J_k$, which can occur in any order.

(If the characteristic polynomial factors completely, the matrix is similar to a matrix in Jordan form.)

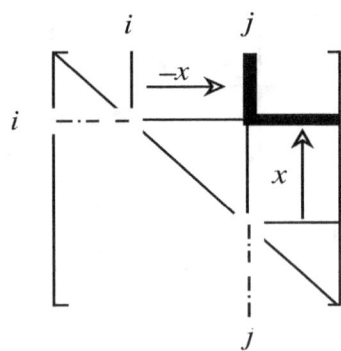

Figure F.1: Effect of a similarity transformation

Proof. Assume that A is as in the statement of the above theorem. We know that A is similar (in fact orthogonally similar) to an upper triangular matrix $U = [u_{ij}]$ by Theorem 5.5.2, Schur's Theorem. The diagonal of U contains the eigenvalues of A; recall that these eigenvalues may occur in any desired order. Suppose that the distinct eigenvalues are $\lambda_1, \ldots, \lambda_m$, and let us assume that these eigenvalues come in order λ_1's first, λ_2's next, etc. Then $U = [U_{ij}]$ is of the form

$$U = \begin{bmatrix} U_{11} & U_{12} & \ldots & U_{1m} \\ 0 & U_{22} & \ldots & U_{2m} \\ \vdots & \vdots & \ddots & \vdots \\ 0 & 0 & \ldots & U_{mn} \end{bmatrix}.$$

Let us try to make a zero above the diagonal. Let row i and column j be some position above the diagonal ($j > i$). We wish to perform the column operation $C_{ij}[a]$ in a similarity transformation. Note that $C_{ij}[a]^{-1} = C_{ij}[-a] = R_{ji}[-a]$. Thus, we can perform the similarity transformation $R_{ji}[-a]UC_{ij}[a]$. The entry in row i and column j becomes $u_{ij} - au_{jj} + au_{ii}$. We would like this expression to be zero. We can solve to find such an a exactly when $u_{jj} - u_{ii} \neq 0$. Remember that u_{jj} and u_{ii} are eigenvalues and so this difference will be nonzero when these entries lie in different blocks on the diagonal. We see that if the row lies in one block and the column in a different block, we can make a zero in that row and column.

Now consider the effect of the above similarity transformation. We see that only the entries above and to the left of the entry in row i and column j are affected. The picture below shows the situation.

What can be accomplished with such similarity transformations? If we start at the first of the "notches" between the diagonal blocks, we can make zeros in these regions. See the picture below. If we begin with region 1, we can make zeros along the bottom of the region, then along the next higher row, etc. Eventually, region 1 can be filled with zeros. Then proceed to do the same with region $2, 3, \ldots$.

The result of this sequence of similarity transformations is that U has been transformed into a matrix of the form

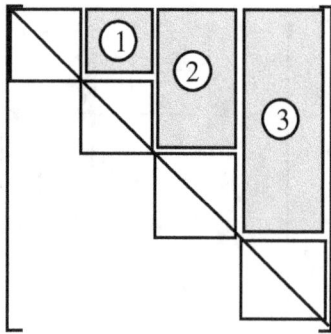

Figure F.2: Sequence of similarity transformations

$$\begin{bmatrix} U_{11} & \cdots & 0 \\ \vdots & \ddots & \vdots \\ 0 & \cdots & U_{mm} \end{bmatrix},$$

where each of the blocks U_{ii} on the diagonal contains a single eigenvalue. If we could find matrices S_i such that $S_i^{-1} U_{ii} S_i$ was in Jordan Form, then the block diagonal matrix $\mathrm{diag}(S_1, \ldots, S_m)$ would transform U into Jordan Form. So, to finish the proof we need only show that an upper triangular matrix with a single eigenvalue occurring on the diagonal can be reduced to Jordan Form. We will omit this final, difficult step.

$\square$

Appendix F Exercises

1. Prove the following: If B and C are $m \times n$ matrices in reduced echelon form with $\mathrm{span}\{B_1, \ldots, B_r\} = \mathrm{span}\{C_1, \ldots, C_r\}$ and $B_r = C_r$ then

$$\mathrm{span}\{B_1, \ldots, B_{r-1}\} = \mathrm{span}\{C_1, \ldots, C_{r-1}\}.$$

2. Show by example that the statement in Exercise 1 is false if the assumption that the matrices are in reduced echelon form is replaced by the assumption that the nonzero rows are linearly independent.

   In Exercises 3 - 5, reduce the given matrix to block diagonal form as described above.

3. $\begin{bmatrix} 1 & 3 & 2 \\ 0 & 1 & 2 \\ 0 & 0 & 2 \end{bmatrix}$

4. $\begin{bmatrix} 2 & 1 & 3 \\ 0 & 1 & 2 \\ 0 & 0 & 1 \end{bmatrix}$

5. $\begin{bmatrix} 2 & 1 & 3 & 1 \\ 0 & 2 & 2 & 1 \\ 0 & 0 & 1 & 3 \\ 0 & 0 & 0 & 1 \end{bmatrix}$

Hints, Comments, and Solutions for Selected Exercises

Chapter 1

Section 1.1

1. (a) $x = 2.5, y = 0.5$

   (b) $x = 0, y = 1.5$

2. (a) $x = 4.57142857, y = -1.71428571$

   (b) $x = 2, y = -1$

3. $x = 6, y = 2, z = -4$

4. There is no solution.

5. $w = 2.5, x = 1.5, y = 2.5, z = -3.5$

6. $I_1 = 2.6666667, I_2 = 0.6666667, I_3 = 2.0$

7. $I_1 = 8/11, I_2 = 3/11, I_3 = 5/11$

8. $x = -0.57142857, z = 6.42857143, y = 0.71428571, z = -0.28571429$

9. There is no solution when $a = 1/2$, otherwise there is a unique solution.

10. There is a unique solution when $a \neq 1$, no solution when $a = 1$.

11. $15, 5, 12$

12. $10, 6$

Section 1.2

In Exercises 1.-4., the augmented matrices are given. To obtain the coefficient matrix, remove the last column.

1. $\begin{bmatrix} 2 & 3 & 1 \\ 2 & -4 & 8 \end{bmatrix}$
2. $\begin{bmatrix} 1 & 1 & -1 & 3 \\ 2 & 1 & -3 & 2 \end{bmatrix}$
3. $\begin{bmatrix} -2 & 3 & 0 & 2 \\ 1 & 1 & 1 & 3 \\ -1 & 2 & 6 & 4 \end{bmatrix}$
4. $\begin{bmatrix} 2 & -3 & 0 & 2 \\ 1 & -1 & -1 & 0 \end{bmatrix}$

5. $\begin{bmatrix} 2 & 3 \\ 2 & -4 \end{bmatrix} \begin{bmatrix} x \\ y \end{bmatrix} = \begin{bmatrix} 1 \\ 8 \end{bmatrix}$

6. $\begin{bmatrix} 1 & 1 & -1 \\ 2 & 1 & -3 \end{bmatrix} \begin{bmatrix} x \\ y \\ z \end{bmatrix} = \begin{bmatrix} 3 \\ 2 \end{bmatrix}$

7. $\begin{bmatrix} -2 & 3 & 0 \\ 1 & 1 & 1 \\ -1 & 2 & 6 \end{bmatrix} \begin{bmatrix} x \\ y \\ z \end{bmatrix} = \begin{bmatrix} 2 \\ 3 \\ 4 \end{bmatrix}$

8. $\begin{bmatrix} 2 & -3 & 0 \\ 1 & -1 & -1 \end{bmatrix} \begin{bmatrix} x \\ y \\ z \end{bmatrix} = \begin{bmatrix} 2 \\ 0 \end{bmatrix}$

9. $\begin{bmatrix} -1 & 3 \\ 0 & 4 \end{bmatrix}$

10. $\begin{bmatrix} 1 & 4 & -1 \\ 8 & 6 & -1 \end{bmatrix}$

11. $\begin{bmatrix} -2 & 6 \\ 2 & -4 \end{bmatrix}$

12. $\begin{bmatrix} 6 & 5 & -2 \\ 11 & 1 & 1 \end{bmatrix}$

13. $\begin{bmatrix} 1 & 14 \end{bmatrix}$

14. $\begin{bmatrix} 2 & 3 & -5 \\ 1 & 3 & -7 \end{bmatrix}$

15. $\begin{bmatrix} 13 & -9 \\ -16 & 15 \end{bmatrix}$

16. $\begin{bmatrix} -6 & -11 & 21 \\ -1 & -5 & 13 \end{bmatrix}$

17. AB, BC, CB, CA

18. $B + AB$

19. $\begin{bmatrix} 1 & 2 \\ -1 & 1 \end{bmatrix}$

5. $\begin{bmatrix} 2 & 1 \\ -1 & 3 \\ 4 & -2 \end{bmatrix}$

5. $m = n$

5. $n = m', m = n'$

Section 1.3

1. $\begin{aligned} 2A + 3A &= (2+3)A && \text{by the distributive property} \\ &= 5A && \text{by addition of real numbers.} \end{aligned}$

4. Solve for C:

$$\begin{aligned} (-1)A + (A + C) &= (-1)A + B && \text{Theorem 1.3.1 e)} \\ ((-1)A + A) + C &= (-1)A + B && \text{Theorem 1.3.2 b)} \\ 0 + C &= (-1)A + B && \text{Theorem 1.3.6 d)} \\ C &= (-1)A + B && \text{Theorem 1.3.6 a)} \\ C &= \begin{bmatrix} 3 & -1 \\ -1 & -2 \end{bmatrix} && \text{Computation} \end{aligned}$$

5. (a) Clearly, $a_{ij} = a_{ij}$ for all i, j, and so $A = A$.

(b) If $A = B$ then $a_{ij} = b_{ij}$ for all i, j, so $b_{ij} = a_{ij}$ for all i, j; it follows that $B = A$.

(c) Let $A = [a_{ij}], B = [b_{ij}], C = [c_{ij}]$, and $D = [d_{ij}]$. then $A = B$ and $C = D$ imply that $a_{ij} = b_{ij}$ and $c_{ij} = d_{ij}$ for all i, j. Adding we get that $a_{ij} + c_{ij} = b_{ij} + d_{ij}$ for all i, j, and so $[a_{ij} + c_{ij}] = [b_{ij} + d_{ij}]$. Hence $A + C = B + D$.

6. (a) Let $A = [a_{ij}]$ and $B = [b_{ij}]$. Then $A + B = [a_{ij}] + [b_{ij}] = [a_{ij} + b_{ij}] = [b_{ij} + a_{ij}] = [b_{ij}] + [a_{ij}] = B + A$, Where the second and fourth equality follow from the defi-

nition of matrix addition, and the third equality follows from the commutativity of addition of scalars and the definition of equality of matrices.

7. Choose any A, B with $AB \neq BA$, say $A = \begin{bmatrix} 0 & 1 \\ 0 & 0 \end{bmatrix}$ and $B = \begin{bmatrix} 1 & 1 \\ 0 & 0 \end{bmatrix}$.

8. (a) Let $A = [a_{ij}]$ and let a and b be scalars. Then $a(bA) = a[ba_{ij}] = [a(ba_{ij})] = [(ab)a_{ij}] = (ab)[a_{ij}] = (ab)A$, where the third equality follows from the associativity of multiplication of scalars and the first, second, and fourth equalities follow from the definition of scalar multiplication.

9. (a) -1 (b) 6   (c) 5

10. (b) Let $A = [a_{ij}], B = [b_{ij}]$, and let a be a scalar. Assume that AB is the $m \times r$ matrix $[c_{ij}]$, where $c_{ij} = \sum_{k=1}^{n} a_{ik}b_{kj}$. Notice that $ac_{ij} = \sum_{k=1}^{n} a_{ik}b_{kj} = \sum_{k=1}^{n} a(a_{ik}b_{kj}) = \sum_{k=1}^{n}(aa_{ik}b_{kj}) = \sum_{k=1}^{n} a_{ik}(ab_{kj})$. The quantity ac_{ij} is the entry in row i and column j of the matrix $a(AB) = a[c_{ij}]$. The sum is the entry in row i and column j of the matrix $(aA)B$, and is the entry in row i and column j of the matrix $A(aB)$. Since these three expressions are equal, we see that $a(AB) = (aA)B = A(aB)$.

11. (a) The zero matrix   (b) the zero scalar   c) the zero matrix   d) Statement is ambiguous

12. Let $A = [a_{ij}]$ be an $m \times n$ matrix, and let $[z_{ij}]$ represent the $m \times n$ zero matrix. Then

    (a) $0 + A = [z_{ij}] + [a_{ij}] = [z_{ij} + a_{ij}] = [0 + a_{ij}] = [a_{ij}] = A$. Likewise $A + 0 = A$.

    (b) $0A = 0[a_{ij}] = [0a_{ij}] = [z_{ij}] = 0$.

    (c) $A + (-1)A = [a_{ij}] + (-1)[a_{ij}] = [a_{ij}] + [(-1)a_{ij}] = [a_{ij}] + [-a_{ij}] = [a_{ij} + (-a_{ij})] = [z_{ij}] = 0$. Likewise, $(-1)A + A = 0$.

13. Step 1) Theorem 1.3.1 (d), Step 2) Theorem 1.3.2 (b), Step 3) Theorem 1.3.6 (c) and the definition of $-A$, Step 4) Theorem 1.3.6 (a).

14. (b) Let $A = [a_{ij}]$ be $m \times n$ and let $I_n = [\delta_{ij}]$ be the $n \times n$ identity matrix. Then $AI_n = [a_{ij}][d_{ij}] = [c_{ij}]$, where $c_{ij} = \sum_{k=1}^{n} a_{ik}\delta_{kj} = a_{ij}$ since $\delta_{kj} = 0$ unless $k = j$, in which case $\delta_{jj} = 1$.

15. Assume that $\begin{bmatrix} 1 & -2 \\ -2 & 4 \end{bmatrix}\begin{bmatrix} a & b \\ c & d \end{bmatrix} = \begin{bmatrix} 1 & 0 \\ 0 & 1 \end{bmatrix}$. Then $a - 2c = 1$ and $-2a + 4c = 0$. Divide the second equation by -2 and get $a - 2c = 0$, a contradiction.

16. Assume that $\begin{bmatrix} 1 & 2 & -1 \\ 3 & 1 & 1 \\ 3 & -1 & 2 \end{bmatrix}\begin{bmatrix} a & b & c \\ d & e & f \\ g & h & i \end{bmatrix} = \begin{bmatrix} 1 & 0 & 0 \\ 0 & 1 & 0 \\ 0 & 0 & 1 \end{bmatrix}$. Show that the resulting system of equations is inconsistent.

17. The inverse is $\begin{bmatrix} 1/3 & 1/3 \\ -2/3 & 1/3 \end{bmatrix}$.

18. This exercise can be a challenge. Assume that A and B are some general 2×2 matrices, say $A = \begin{bmatrix} a & b \\ c & d \end{bmatrix}$ and $B = \begin{bmatrix} u & v \\ x & y \end{bmatrix}$. Using the assumption that $AB = I$, try to prove that $BA = I$. We will see later, in Chapter 4, that the result of this exercise is true for any $n \times n$ matrices. Consider trying to prove it!

Section 1.4

1. Equivalent, $x = y = 1$ is the unique solution of each system.

2. Not equivalent, $x = 1, y = 2$ is not a solution of the first system.

3. Not equivalent, $x = 1, y = z = 0$ is a solution of the first system, but not the second.

4. Equivalent, $x = 3y = 2, z = 1$ is the unique solution each system.

5. (b) Consider a linear equation $a_1x_1 + a_2x_2 + \ldots + a_nx_n = h$, and suppose that $(c_1, \ldots, c_n)$ is a solution. Then $a_1c1 + a_2c2 + \ldots + a_nc_n = h$, and so multiplying by c we see that $ca_1c1 + ca_2c2 + \ldots + ca_nc_n = ch$. Thus, $(c_1, \ldots, c_n)$ is a solution of the equation $ca_1x_1 + ca_2x_2 + \ldots + ca_nx_n = ch$. To see that solutions of $ca_1x_1 + ca_2x_2 + \ldots + ca_nx_n = ch$ are solutions of $a_1x_1 + a_2x_2 + \ldots + a_nx_n = h$, notice that dividing both sides of $ca_1x_1 + ca_2x_2 + \ldots + ca_nx_n = ch$ by c (it is here that we need the fact that $c \neq 0$) produces $a_1c_1 + a_2c_2 + \ldots + a_nc_n = h$.

6. 1) Notice that in the product $R_{ik}A$, multiplying A by row i of R_{ik} produces the same result as multiplying A by row k of I. That is, row i of $R_{ik}A$ is the same as row k of IA, but of course, this is row k of A. Similarly, row k of $R_{ik}A$ is row i of A. Thus in $R_{ik}A$ rows i and k of A are switched. Since the other rows of R_{ik} are equal to the corresponding rows in I, the remainder of A is unchanged.

   2) Since the rows of $R_i[a]$ are identical to the rows of the identity matrix except for row i, $R_i[a]A$ is the same as A except for row i. In row i there is an a on the diagonal in the matrix $R_i[a]$ and 0's elsewhere. Thus, in $R_i[a]A$ row i is row i of A multiplied by a.

7. $\begin{bmatrix} 0 & 1 & 0 \\ 1 & 0 & 0 \\ 0 & -1 & 3 \end{bmatrix}$ 9. $\begin{bmatrix} 1 & 0 & 0 \\ 0 & 1 & 0 \\ 0 & 0 & 1 \end{bmatrix}$ 11. $\begin{bmatrix} 1 & -2 & 0 \\ 0 & 1 & 0 \\ 4 & 0 & 1 \end{bmatrix}$

8. $\begin{bmatrix} 0 & 0 & 2 \\ 0 & 1 & 0 \\ 1 & -2 & 0 \end{bmatrix}$ 10. $\begin{bmatrix} 1 & 0 & 0 \\ 0 & 1 & 0 \\ 0 & 0 & 1 \end{bmatrix}$ 12. $\begin{bmatrix} 0 & 1 & 0 \\ -2 & 2 & 2 \\ 8 & 2 & -7 \end{bmatrix}$

13. (a) Not elementary. (b) $R_{12}(1)$. (c) R_{12}. (d) Not elementary.

14. The product is I. Thus the inverse of A is A.

15. Take $B = R_i(1/c)$.

16. Take $B = R_{ik}(-c)$.

Section 1.5

1. $\begin{bmatrix} 1 & 0 \\ 0 & 1 \end{bmatrix}$

2. $\begin{bmatrix} 1 & -1.5 \\ 0 & 0 \end{bmatrix}$

3. $\begin{bmatrix} 1 & 0 & 0 & -4.5 \\ 0 & 1 & 0 & 4 \end{bmatrix}$

4. $\begin{bmatrix} 0 & 1 & 1 & 0 & 0 \\ 0 & 0 & 0 & 1 & 0 \\ 0 & 0 & 0 & 0 & 1 \end{bmatrix}$

5. $\begin{bmatrix} 1 & 0 & 0 \\ 0 & 1 & 0 \\ 0 & 0 & 1 \end{bmatrix}$

6. $\begin{bmatrix} 1 & 0 & 6/10 & 2/10 \\ 0 & 1 & -8/10 & 14/10 \\ 0 & 0 & 0 & 0 \end{bmatrix}$

7. $\begin{bmatrix} 1 & 0 & 3 \\ 0 & 1 & -1 \end{bmatrix}$

8. $\begin{bmatrix} 1 & 0 & 7/3 & 17/3 \\ 0 & 1 & -4/3 & -5/3 \end{bmatrix}$

9. $\begin{bmatrix} 1 & 0 & 0 & 13/3 \\ 0 & 1 & 0 & -7/3 \\ 0 & 0 & 1 & 0 \end{bmatrix}$

10. $\begin{bmatrix} 1 & 2 & 0 & 1 & 3 \\ 0 & 0 & 1 & 3 & 2 \\ 0 & 0 & 0 & 0 & 0 \end{bmatrix}$

11. $\begin{bmatrix} 1 & 0 & 1 & 0 & 3 \\ 0 & 1 & 1 & 0 & 1 \\ 0 & 0 & 0 & 1 & 2 \\ 0 & 0 & 0 & 0 & 0 \end{bmatrix}$

12. $\begin{bmatrix} 1 & 0 & 5/3 & -1/3 \\ 0 & 1 & -2/3 & 7/3 \\ 0 & 0 & 0 & 0 \end{bmatrix}$

13. $\begin{bmatrix} 1 & 0 & 6/10 & 2/10 \\ 0 & 1 & -8/10 & 14/10 \\ 0 & 0 & 0 & 1 \end{bmatrix}$

14. $j_r \leq n$

15. $n - r$

Section 1.6

1. $\begin{bmatrix} 1 & 0 & 2 & -1 \\ 0 & 1 & -1 & 2 \end{bmatrix}, X_p = (-1, 2, 0)^t, X_h = (-2x_3, x_3, x_3)^t$

2. $\begin{bmatrix} 1 & 0 & -1 & 3 \\ 0 & 1 & 1 & 6 \end{bmatrix}, X_p = (3, 6, 0)^t, X_h = (x_3, -x_3, x_3)^t$

3. $\begin{bmatrix} 1 & 0 & 1 & 1 \\ 0 & 1 & 2 & 3 \end{bmatrix}, X_p = (1, 3, 0)^t, X_h = (-x_3, -2x_3, x_3)^t$

4. $\begin{bmatrix} 1 & 1 & 0 & 1 & 2 \\ 0 & 0 & 1 & -3 & -1 \end{bmatrix}, X_p = (2, 0, -1, 0)^t, X_h = (-x_2 - x_4, x_2, 3x_4, x_4)^t$

5. $\begin{bmatrix} 1 & 0 & 0 & 1 \\ 0 & 1 & 0 & -2 \\ 0 & 0 & 1 & -2 \end{bmatrix}, X_p = (1, -2, -2)^t, X_h = (0, 0, 0)^t$

6. $\begin{bmatrix} 1 & 0 & 0 & -3/2 & 1 \\ 0 & 1 & 0 & -4 & -1 \\ 0 & 0 & 1 & -9/2 & -4 \end{bmatrix}, X_p = (1, -1, -4, 0)^t, X_h = (1.5x_4, 4x_4, 4.5x_4, x_4)^t$

7. $X_h = (-2x_3, x_3, x_3)^t = x_3(-2, 1, 1)^t$

8. $X_h = (x_3, -x_3, x_3)^t = x_3(1, -1, 1)^t$

9. $X_h = (-x_3, -2x_3, x_3)^t = x_3(-1, -2, 1)^t$

10. $X_h = (-x_2 - x_4, x_2, 3x_4, x_4)^t = x_2(-1, 1, 0, 0)^t + x_4(-1, 0, 3, 1)^t$

11. $X_h = (0, 0, 0)^t$

12. $X_h = (1.5x_4, 4x_4, 4.5x_4, x_4)^t = x_4(1.5, 4, 4.5, 1)^t$

13. This is one of those results that is easier to see than to explain. Let the constants associated with B be r and $j_1, \ldots, j_r$. Note that for each of the matrices A and $[A|0]$, the three conditions in the definition of echelon form are true: 1) The nonzero rows in A and $[A|0]$ are at the top and the number of nonzero rows may be either r or $r - 1$. 2) the first nonzero entry in each of the first rows of A and $[A|0]$ is a 1 (for row i it occurs in column j_i) and it is the only nonzero entry in that column. 3) $j_1 \le j_2 \le \ldots$.

15. (a) All vectors are solutions - 3-dimensional space.

    (b) Two intersecting planes - solution set is a line through the origin.

    (c) No nonzero solutions - solution set consists only of the origin.

Section 1.7

1. $U = \begin{bmatrix} 2 & 1 \\ 0 & 1/2 \end{bmatrix}$, $L = \begin{bmatrix} 1 & 0 \\ 3/2 & 1 \end{bmatrix}$

2. $U = \begin{bmatrix} -1 & 2 \\ 0 & 5 \end{bmatrix}$, $L = \begin{bmatrix} 1 & 0 \\ 1 & 1 \end{bmatrix}$

3. $U = \begin{bmatrix} 2 & 1 & 3 \\ 0 & 0 & -1 \\ 0 & 0 & 3 \end{bmatrix}$, $L = \begin{bmatrix} 1 & 0 & 0 \\ -2 & 1 & 0 \\ -1 & 0 & 1 \end{bmatrix}$

4. $U = \begin{bmatrix} 1 & 0 & 2 \\ 0 & -1 & 1 \\ 0 & 0 & 1 \end{bmatrix}$, $L = \begin{bmatrix} 1 & 0 & 0 \\ 2 & 1 & 0 \\ 1 & -2 & 1 \end{bmatrix}$

5. $U = \begin{bmatrix} 2 & 1 & 3 & 4 \\ 0 & 4 & 0 & -3 \\ 0 & 0 & -5 & -10 \end{bmatrix}$,

$L = \begin{bmatrix} 1 & 0 & 0 \\ -1 & 1 & 0 \\ -3 & 0 & 1 \end{bmatrix}$

6. $U = \begin{bmatrix} 1 & 0 & 2 & 2 \\ 0 & 5 & 7 & 8 \end{bmatrix}$, $L = \begin{bmatrix} 1 & 0 \\ -3 & 1 \end{bmatrix}$

7. $U = \begin{bmatrix} 2 & -1 & 0 & 2 \\ 0 & 2 & -1 & 3 \\ 0 & 0 & 1 & 3 \\ 0 & 0 & 0 & 4 \end{bmatrix}$, $L = \begin{bmatrix} 1 & 0 & 0 & 0 \\ 1 & 1 & 0 & 0 \\ 1 & -2 & 1 & 0 \\ 2 & 1 & 3 & 1 \end{bmatrix}$, $A = \begin{bmatrix} 2 & -1 & 0 & 2 \\ 2 & 1 & -1 & 5 \\ 2 & -5 & 3 & -1 \\ 4 & 0 & 2 & 20 \end{bmatrix}$

8. $x = 3.25, y = -1.5$

9. $x = -0.5, y = -0.5$

10. $x = 1, y = 1, z = 0$

11. Hint: Let $A = [a_{ij}]$ be an $m \times n$ matrix and let $B = [b_{ij}]$ be an $n \times r$ matrix. Let $C = [c_{ij}]$ be the product of A and B. Then $c_{ij} = \sum_{k=1}^{n} a_{ik}b_{kj}$. Suppose that A and B are upper triangular. Then $a_{ij} = b_{ij} = 0$ when $i > j$. In the sum above, for any given k, either $i > k$ or $k > j$ so that either $a_{ik} = 0$ or $b_{kj} = 0$. If $i = j$, the sum contains only one nonzero term $a_{ii}b_{ii}$.

12. Hint: If A is lower triangular with nonzero entries on the diagonal, then row operations of the form $R_i(a_{ii}^{-1})$ will put ones on the diagonal, and row operations of the form $R_{ik}(-a_{ik})$, for $i < k$, will make zeros below the diagonal. The result will be the identity matrix and the product of these row elementary matrices will be lower triangular and will have the inverses, a_{ii}^{-1}, on the diagonal.

Section 1.8

1. $I_1 = 55/9$ amps, $I_2 = 10/9$ amps, $I_3 = 5$ amps.

2. $I_1 = 3$ amps, $I_2 = 0.5$ amps, $I_3 = 2.5$ amps.

3. $I_1 = 0.06300053$ amps, $I_2 = 0.03096636$ amps, $I_3 = 0.03203417$ amps, $I_4 = 0.00032034$ amps, $I_5 = 0.03128671$ amps, $I_6 = 0.03171383$ amps

4. $I_1 = 0.00360656$ amps, $I_2 = 0.00163934$ amps, $I_3 = 0.00196721$ amps, $I_4 = -0.00065574$ amps, $I_5 = 0.00098361$ amps, $I_6 = 0.00262295$ amps

5. $A = 45$ volts, $B = 30$ volts.

6. $I_1 = 0.36363636$ amps, $I_2 = 0.18181818$ amps, $I_3 = 0.18181818$ amps

7. $I_1 = 0.84507042$ amps, $I_2 = 0.56338028$ amps, $I_3 = 0.28169014$ amps

8. $I_1 = 0.46296296$ amps, $I_2 = 0.09259259$ amps, $I_3 = 0.37037037$ amps

Section 1.9

1. a), b), e) are stochastic; c), d) are not stochastic

2. There is not a unique answer. Fill in the blanks with numbers between 0 and 1 in any manner such that the sum of the entries in each column is 1.

3. $X_4 = (0.5750, 0.2428, 0.1822)^t$, $X_5 = (0.5875, 0.2336, 0.1789)^t$, $X_6 = (0.5938, 0.2292, 0.1770)^t$, $X_{20} = (0.6000, 0.2250, 0.1750)^t$

4. $X_1 = (0.4000, 0.6000)^t$, $X_2 = (0.4400, 0.5600)^t$, $X_3 = (0.4640, 0.5360)^t$, $X_4 = (0.4784, 0.5216)^t$, $X_5 = (0.4870, 0.5130)^t$, $X_6 = (0.4922, 0.5078)^t$, $X_7 = (0.4953, 0.5047)^t$, $X_8 = (0.4972, 0.5028)^t$

5. $X_2 = (0.2400, 0.4000, 0.3600)^t, X_3 = (0.2960, 0.3360, 0.3680)^t,$
   $X_4 = (0.3184, 0.3200, 0.3616)^t, X_5 = (0.3274, 0.3194, 0.3533)^t,$
   $X_6 = (0.3309, 0.3226, 0.3465)^t, X_7 = (0.3324, 0.3259, 0.3417)^t$

6. $P = X_1 = (0.9000, 0.1000)^t, X_2 = (0.7800, 0.2200)^t, X_3 = (0.7560, 0.2440)^t, X_4 = (0.7512, 0.2488)^t, X_5 = (0.7502, 0.2498)^t, X_6 = (0.7500, 0.2500)^t.$ Observations are made at intervals of one generation.

7. Use mathematical induction. Note that $P^n X = P(P^{n-1} X)$.

8. p_{22} must be 1. The entries in column 1 may be any numbers between 0 and 1 whose sum is 1.

9. $\begin{bmatrix} 0.9 & 1 \\ 0.1 & 0 \end{bmatrix}$

10. The matrix is $\begin{bmatrix} 0.8 & 0.2 \\ 0.2 & 0.8 \end{bmatrix}$

11. Sure, consider $\begin{bmatrix} 0.5 & 0.5 \\ 0.5 & 0.5 \end{bmatrix}$.

Chapter 2

Section 2.1

1. $(-2, 5)$

2. $(-2, 1)$

3. $(10, -18)$

4. $(-2, 1)$

5. $\theta = \arctan(2/3), r = \sqrt{13}$

6. $\theta = \arctan(-3), r = \sqrt{10}$

7. $\theta = \arctan(3), r = \sqrt{10}$

8. $\theta = \arctan(-1), r = \sqrt{8}$

9. $x = \sqrt{3}, y = 1$

10. $x = y = 3/\sqrt{2}$

11. $x = -5\sqrt{3}/2, y = -5/2$

12. $x = 4/\sqrt{2}, y = -4/\sqrt{2}$

13. Argue geometrically. Draw figures like Figures 2.2 and 2.3, and use similar and congruent triangles to prove the result. You will need to assume a given coordinate system.

14. $\theta = 176.989°$, time $= 1.922$ hours

15. $4.7636°$

16. Write the forces as ordered pairs: $F_1 = (5\sqrt{3}, 5) \approx (8.66, 5), F_2 = (0, 20), F_3 = (-15/\sqrt{2}, -15/\sqrt{2}) \approx (-10.61, -10.61)$. Then add coordinates: $F_1 + F_2 + F_3 \approx (-1.94, 14.39)$ (Note: $\approx$ is used to mean approximately equal.)

17. Use the fact that addition of pairs is given by $(a, b) + (c, d) = (a + c, b + d)$ and the fact that addition of real numbers is commutative. We can then see that $(a, b) + (c, d) = (a + c, b + d) = (c + a, d + b) = (c, d) + (a, b)$.

Section 2.2

1. Let (x, y) and (x', y') be in $\mathbf{V}$, and let r be a real number. Then $x = 2y$ and $x' = 2y'$. Now add: $(x, y) + (x', y') = (x + x', y + y')$, and this vector is in $\mathbf{V}$ since adding corresponding sides of the equalities $x = 2y$ and $x' = 2y'$ produces $x + x' = 2y + 2y' = 2(y + y')$. A similar argument shows closure under scalar multiplication.

2. The argument here is similar to that in in Exercise 1. Take two vectors in $\mathbf{V}$ and a real number and show that the sum and scalar product satisfy the two equations which determine membership in $\mathbf{V}$.

3. Observe that $(1, 0)$ and $(0, 1)$ are in $\mathbf{V}$, but the sum $(1, 0) + (0, 1) = (1, 1)$ is not. Also $2(1, 0) = (2, 0)$ is not in $\mathbf{V}$.

4. Find examples as in Exercise 3.

5. Argue geometrically that this set must contain the lines through the vectors $(1, 1)$ and $(1, 2)$ and that any vector may be expressed as a sum of two vectors, one lying on each of these nonparallel lines.

6. Let $\mathbf{V}$ be the line containing the vector $(-2, 1)$, that is, $\mathbf{V} = \{(x, y)|y = -2x\}$

7. Note that, by definition, $(X - Y) + (-(X - Y)) = 0$. Also, using the definition of difference, $(X - Y) + (-X + Y) = (X + (-Y)) + (-X + Y) = (X + (-X)) + (-Y + Y) = 0 + 0 = 0$. Thus, we have $(X - Y) + (-(X - Y)) = (X - Y) + (-X + Y)$, and we can cancel the term $(X - Y)$ to obtain $-(X - Y) = -X + Y$.

8. b) Assume that $a \neq 0$ and that $aX + Y = Z$. Add $-Y$ to both sides to obtain $aX = Z - Y$. Now multiply both sides by $(1/a)$ to show that $X = (1/a)(Z - Y)$.

9. To show closure under addition, let $r = s = 1$ and apply the condition to show that $X + Y = rX + sY \in \mathbf{V}$. For closure under scalar multiplication, let $Y = 0$. Then $rX + sY = rX + 0 = rX \in \mathbf{V}$. Hence $\mathbf{V}$ is closed under addition and scalar multiplication.

10. $X = (-7/2, -1/2)$

11. $X = (-13/3, 8)$

12. Here's one example: $\{(x, y)|x > 1 \text{ and } y > 1\}$

13. In $\mathbb{R}^2$, two nonparallel lines through the origin will have this property, for example: $\{(x, y)|x = 0 \text{ or } y = 0\}$. Try something similar for $\mathbb{R}^3$.

Section 2.3

    1.

$$x^2 + x + 1 \qquad\qquad \textbf{(b)}\ a^2 + 3a - 1 \qquad\qquad \text{(c)}\ 5$$

2. (a) 2 $\qquad\qquad$ (b) $\dfrac{3}{\sqrt{2}}$ $\qquad\qquad$ (c) $\dfrac{1 - \sqrt{3}}{2}$

3. (c), (d)

4. Note that z is a constant function and so it is continuous, that is $z \in C[a,b]$. We must show that $z + f = f$ or that $z(x) + f(x) = f(x)$ for all x in $[a,b]$. This is easy; $z(x) + f(x) = 0 + f(x) = f(x)$ for any x in $[a,b]$, and so $z + f = f$.

5. As above, we note that $-f \in C[a,b]$ and that $f(x) + (-f)(x) = f(x) + (-f(x)) = 0 = z(x)$. Hence A4) holds.

6. Let $f \in C[a,b]$ and let r and s be scalars. Then for any x in $[a,b]$, $((r+s)f)(x) = (r+s)f(x) = rf(x) + sf(x) = (rf + sf)(x)$. Hence $(r+s)f = rf + sf$ and so M1) holds.

7. $5 + 12x + 9x^2$

8. This is a direct application of the definition of equality. Notice that the definition says that two polynomials are equal if corresponding coefficients are equal and those coefficients that "don't correspond" are zero. Because of this result, whenever one chooses two arbitrary polynomials, it may be assumed that the highest degree term is the same in each case. Of course, it may not be assumed that the coefficient is nonzero.

9. If $p(x) = a_0 + a_1 x + \ldots + a_n x^n$ then $z(x) + p(x) = (a_0 + 0) + a_1 x + \ldots + a_n x^n = a_0 + a_1 x + \ldots + a_n x^n = p(x)$.

10. Using the hint, note that $f(x) + (-f)(x) = (a_0 + a_1 x + \ldots + a_n x^n) + ((-a_0) + (-a_1)x + \ldots + (-a_n)x^n) = (a_0 + (-a_0)) + (a_1 + (-a_1))x + \ldots + (a_n + (-a_n))x^n) = 0 + 0x + \ldots + 0x_n = z(x)$. This shows that A4) holds.

11. This is laborious, but not too difficult. Apply the definitions of the operations: Let $f(x) = a_0 + a_1 x + \ldots + a_n x_n$ and $g(x) = b_0 + b_1 x + \ldots + b_n x_n$ be two polynomials and let r be a scalar. Then

$$\begin{aligned}
r(f(x) + g(x)) &= r((a_0 + b_0) + (a_1 + b_1)x + \ldots + (a_n + b_n)x^n) \\
&= r(a_0 + b_0) + r(a_1 + b_1)x + \ldots + r(a_n + b_n)x^n \\
&= (ra_0 + rb_0) + (ra_1 + rb_1)x + \ldots + (ra_n + rb_n)x^n \\
&= (ra_0 + ra_1 x + \ldots + ra_n x_n) + (rb_0 + rb_1 x + \ldots + rb_n x^n) \\
&= rf(x) + rg(x).
\end{aligned}$$

12. (d) This uses a familiar trick. Notice that $0X = (0+0)X = 0X + 0X$, and so we see that $0X + 0 = 0X + 0X$. Applying cancellation (part (a)), we see that $0X = 0$.

    (e) By definition, $X + (-X) = 0$. Also $X + (-1)X = 1X + (-1)X = (1 + (-1))X = 0X = 0$. Thus, we see that $X + (-X) = X + (-1)X$, and by cancellation, $-X = (-1)X$.

(f) Assume that $rX = 0$. If $r = 0$, the conclusion is true. If $r \neq 0$, then r^{-1} exists and we can multiply, obtaining: $0 = r^{-1}0 = r^{-1}(rX) = (r^{-1}r)X = 1X = X$.

13. Since $\mathbb{R} \subseteq \mathbb{C}$, scalar multiplication is clearly defined and it is not hard to see that the properties of a vector space relating to scalars hold true when one considers elements only from $\mathbb{R}$. It follows that a vector space over $\mathbb{C}$ is naturally also a vector space over $\mathbb{R}$.

14. The integers do not form a field, and so there are no vector spaces "over $\mathbb{Z}$."

15. $f(0) = 0 + 0 \cdot 0 = 0$, and $f(1) = 1 + 1 \cdot 1 = 1 + 1 = 0$, so $f(x) = g(x)$ for all x in Z_2.

16. Over an arbitrary field, which definition of equality implies the other - equality as polynomials implies equality as functions or conversely?

Section 2.4

1. We must show that $\mathbf{V}$ is closed under addition and scalar multiplication and that $\mathbf{V}$ is nonempty. $\mathbf{V}$ is nonempty since $(2, 1)$ is in $\mathbf{V}$. Let (a, b) and (c, d) be two vectors in $\mathbf{V}$ and let r be a scalar. Then $a = 2b$ and $c = 2d$, and so $(a, b) + (c, d) = (a + c, b + d)$ is in $\mathbf{V}$ since $a + c = 2b + 2d = 2(b + d)$. Likewise, $r(a, b) = (ra, rb)$ is in $\mathbf{V}$ since $ra = r(2b) = 2(rb)$. Thus, $\mathbf{V}$ is closed under addition and scalar multiplication, so $\mathbf{V}$ is a subspace of $\mathbb{R}^2$.

2. The proof here is almost identical to the one above. The fact that no condition is made regarding the coordinate z is not a problem.

3. We can show that this set $\mathbf{V}$ is not closed by giving an example. Notice that $(2, 1, 0)$ and $(2, 0, 2)$ are in $\mathbf{V}$, but $(2, 1, 0) + (2, 0, 2) = (4, 1, 2)$ is not in $\mathbf{V}$ since neither condition holds ($x \neq 2y$ and $x \neq z$).

4. Find an example as in Exercise 3. You will be able to show that this set is not closed under scalar multiplication.

5. Since $(0, 0)$ is in $\mathbf{V}$, $(0, 0, 0)$ is in $\mathbf{W}$, so $\mathbf{W}$ is not the empty set. Let (x, y, z) and (u, v, w) be two vectors in $\mathbf{W}$, and let r be a scalar. Then $(x, y, z) + (u, v, w) = (x+u, y+v, z+w) \in \mathbf{W}$ since $z+w = 0$ and (x, y) and (u, v) in $\mathbf{V}$ imply $(x+u, y+v)$ is in $\mathbf{V}$. Likewise (x, y) in $\mathbf{V}$ implies $r(x, y) = (rx, ry)$ in $\mathbf{V}$, and so $r(x, y, z) = (rx, ry, 0)$ is in $\mathbf{W}$.

6. Yes, $\mathbf{V}$ can be shown to be nonempty and closed.

7. No, since $0 \in \mathbf{U}$ and $0 \in \mathbf{W}$ the intersection of these sets cannot be empty.

8. $\mathbf{U} \cap \mathbf{W}$ is a subspace of $\mathbf{V}$. By Exercise 7, this set is nonempty, and it can be shown to be closed under addition and scalar multiplication. $\mathbf{U} \cup \mathbf{W}$ is not, in general, a subspace of $\mathbf{V}$. To prove this, give a counterexample. Find two subspaces of $\mathbb{R}^2$ whose union is not a subspace. What about two nonparallel lines through the origin?

9. The set is nonempty since $0 = 0X_1 + 0X_2$ is in the set. To show closure, let $aX_1 + bX_2$ and $cX_1 + dX_2$ be two members of span$\{X_1, X_2\}$. Then $(aX_1 + bX_2) + (cX_1 + dX_2) = (a+c)X_1 + (b+d)X_2$ is in span$\{X_1, X_2\}$, and likewise $r(aX_1 + bX_2) = r(aX_1) + (rb)X_2$ is in span$\{X_1, X_2\}$. Hence span$\{X_1, X_2\}$ is closed and nonempty and therefore a subspace of $\mathbf{V}$.

10. It is clear that the zero function is in $\mathbf{W}$, and so $\mathbf{W} \neq \emptyset$. Assume $y_1, y_2 \in \mathbf{W}$, and let r be a scalar. Then $y_1' - y_1 = 0$, and $y_2' - y_2 = 0$. Now $(y_1 + y_2)' - (y_1 + y_2) = y_1' - y_1 + y_2' - y_2 = 0 + 0 = 0$, and so $y_1 + y_2 \in \mathbf{W}$. Also, $(ry_1)' - ry_1 = r(y_1' - y_1) = r \cdot 0 = 0$, and so $ry_1 \in \mathbf{W}$. Hence, $\mathbf{W}$ is a subspace of $C^1[a, b]$.

    e^x is a nonzero function in $\mathbf{W}$. From differential equations we know that the solution set of this differential equation is $\mathbf{W} = \{re^x | r \text{ is real}\}$.

11. (a) Show closure by using properties of the derivative as in Exercise 10, above.

    (b) The functions $\sin x$ and $\cos x$ will work.

    (c) The theory from differential equations will show that every solution is of the form $c_1 \sin x + c_2 \cos x$.

12. This line is either the y-axis, $\{(x, y) | x = 0\}$, or a set of the form $\{(x, y) | y = mx\}$. Show that these sets are closed under addition and scalar multiplication.

13. See Figures 2.2-2.2 in Section 2.2. Argue informally as follows: Suppose that $\mathbf{U}$ is a subspace of R^2. Certainly, if $\mathbf{U} = \{0\}$ then $\mathbf{U}$ is a subspace. Suppose that $\mathbf{U} \neq \{0\}$ and say X is a nonzero vector in $\mathbf{U}$. Since $\mathbf{U}$ is closed under scalar multiplication, $\{rX | r \text{ a scalar}\} \subseteq \mathbf{U}$. If $\{rX | r \text{ a scalar}\} = \mathbf{U}$ then $\mathbf{U}$ is a subspace as in Exercise 11. If $\mathbf{U} \neq \{rX | r \text{ a scalar}\}$, then there is a vector Y which is in $\mathbf{U}$, but not in $\{rX | r \text{ a scalar}\}$. Now we claim, as in Figure 2.2 of Section 2.2, that $\mathbf{U}$ must be $\mathbb{R}^2$.

14. Use Exercise 13. If $\mathbf{U} = \{0\}$, take $X = Y = 0$. If $\mathbf{U}$ is a line through the origin, take X and Y to be any nonzero vectors lying along that line. If $\mathbf{U} = \mathbb{R}^2$, let $X = (1, 0)$ and $Y = (0, 1)$.

15. The operations in $\mathbf{U}$ and $\mathbf{W}$ are the restrictions of the operations in $\mathbf{V}$, so the operations in $\mathbf{U}$ are the same as those in $\mathbf{W}$. Since $\mathbf{U}$ is a subspace of $\mathbf{V}$, $\mathbf{U}$ is closed under addition and scalar multiplication. Thus, $\mathbf{U}$ is a subspace of $\mathbf{W}$.

16. $\mathbf{W}$ is nonempty since $\mathbf{U}$ is nonempty. Let $Y_1, Y_2 \in \mathbf{W}$. Then $Y_1 = PX_1$ and $Y_2 = PX_2$ for some $X_1, X_2 \in \mathbf{U}$. Now $\mathbf{U}$ is a subspace, so $X_1 + X_2 \in \mathbf{U}$, and so $Y_1 + Y_2 = PX_1 + PX_2 = P(X_1 + X_2) \in \mathbf{W}$. Likewise, from $rX_1 \in \mathbf{U}$ it follows that $rY_1 = rPX_1 = P(rX_1) \in \mathbf{W}$. Hence $\mathbf{W}$ is a subspace of $\mathbf{V}$.

Section 2.5

1. We must prove that the two sets, span$\{(1, -1), (2, 2)\}$ and $\mathbb{R}^2$, are equal. It is clear that span$\{(1, -1), (2, 2)\} \in \mathbb{R}^2$. We must show the set containment in the reverse

order. Let $(x, y) \in R^2$. Then we see that (after a lot of work on scratch paper) $(x, y) = \dfrac{x - y}{2}(1, -1) + \dfrac{x + y}{4}$. Hence, $(x, y) \in \text{span}\{(1, -1), (2, 2)\}$.

2. Try $X = (1, 0, 0)$. Show that $X \neq a(1, -1, 2) + b(2, 2, 1)$ for any a, b.

3. $(1, -2, 1) \notin \text{span}\{(1, 3, 1), (1, 0, -1)\}$. Show that $(1, -2, 1) = a(1, 3, 1) + b(1, 0, -1)$ has no solution.

4. Note that $(2, -4) = 2(1, -2)$ and $(-1, 2) = -1(1, -2)$. From this it follows that $(1, -2) \in \text{span}\{(2, -4), (-1, 2)\}$ and $(2, -4), (-1, 2) \in \text{span}\{(1, -2)\}$. Applying Theorem 2.5.1, we see that $\text{span}\{(2, -4), (-1, 2)\} = \text{span}\{(1, -2)\}$.

5. $(5, 1, 6) \in \text{span}\{(1, 2, 3), (-1, 1, 0)\}$ since $(5, 1, 6) = 2(1, 2, 3) + (-3)(-1, 1, 0)$.

6. $\boldsymbol{R}(A) = \text{span}\{(1, -1), (2, 0)\}$. It is not hard to show that this set spans $\mathbb{R}^2$.

7. By Theorem 2.5.2, we see that $\boldsymbol{R}(A) = \boldsymbol{R}(B)$.

8. (c) Let $X \in \text{span}(S)$. By part (b), we know that $\text{span}(S) \in \text{span}(S \cup \{X\})$. We must show the reverse containment. Assume that $X = a_1 X_1 + \ldots + a_n X_n$ for some $X_1, \ldots, X_n \in S$, and let $Y \in \text{span}(S \cup \{X\})$. Assume that for some $Y_1, \ldots, Y_n \in S, Y = b_1 Y_1 + \ldots + b_m Y_m + aX$. Then $Y = b_1 Y_1 + \ldots + b_m Y_m + aX = b_1 Y_1 + \ldots + b_m Y_m + a(a_1 X_1 + \ldots + a_n X_n) \in \text{span}(S)$. Hence, $\text{span}(S) = \text{span}(S \cup \{X\})$. If $\text{span}(S) = \text{span}(S \cup \{X\})$, then we see that $X \in \text{span}(S \cup \{X\}) = \text{span}(S)$, so $X \in \text{span}(S)$.

   (d) If $\boldsymbol{W}$ is a subspace of $\boldsymbol{V}$, then $\boldsymbol{W}$ is closed under addition and scalar multiplication. If $S \subseteq \boldsymbol{W}$, then all linear combinations of elements of S must lie in $\boldsymbol{W}$. Hence, $\text{span}(S) \subseteq \boldsymbol{W}$.

9. The row vectors of $R_{ik}[a]A$ are $A_1, \ldots, A_i, \ldots, A_k + aA_i, \ldots, A_n$. All of these vectors lie in $\text{span}\{A_1, \ldots, A_n\} = \boldsymbol{R}(A)$. Thus, by Theorem 2.5.1, $\text{span}\{A_1, \ldots, A_i, \ldots, A_k + aA_i, \ldots, A_n\} = \boldsymbol{R}(R_{ik}[a]A) \subseteq \boldsymbol{R}(A)$.

10. The statement "$\text{span}(S)$ is the smallest subspace of $\boldsymbol{V}$ containing S" means that:

   (a) $\text{span}(S)$ is a subspace of $\boldsymbol{V}$ containing S, and

   (b) If $\boldsymbol{W}$ is any subspace of $\boldsymbol{V}$ containing S, then $\text{span}(S)$ is contained in $\boldsymbol{W}$. We will see that $\text{span}(S)$ satisfies (a) and (b) by Theorem 2.5.1, parts a) and d).

11. Geometrically, this is clear, $\text{span}\{X\}$ is the line through the origin determined by X. We can approach the problem algebraically as follows: Let $X = (x, y)$. If $\mathbb{R}^2 = \text{span}\{X\}$, then $(1, 0) = a(x, y)$ and $(0, 1) = b(x, y)$ for some a and b. From the first we see that $a \neq 0$ and $y = 0$ and from the second we see that $b \neq 0$ and $x = 0$. This means that $x = y = 0$ and this is clearly a contradiction.

13. Let A be an $m \times n$ matrix and let B be the echelon form of A. Let r be the number of nonzero rows in B. Then $\boldsymbol{R}(A) = \boldsymbol{R}(B) = \text{span}\{B_1, \ldots, B_r\}$ and r is the least number with this property.

14. Assume that $\boldsymbol{R}(A) = \boldsymbol{R}(B)$ where both A and B are 2×2 and in echelon form. We must consider cases: If A has two nonzero rows, then $A = I$, the identity matrix. In this case, we can see that B cannot have a just one nonzero row so $B = I$, and $A = B$. If A has only one nonzero row, then $A_1 = rB_1$ for some scalar r, and since the first nonzero entry in each is a 1, we see that $r = 1$. This implies that $A = B$.

15. A proof using mathematical induction is given in Appendix F in the completion of the proof of Theorem 1.5.1.

Section 2.6

1. $(1,2) = 4(1,1) + (-1)(3,2)$. Theorem 2.6.5 applies since $\mathbb{R}^2 = \text{span}\{(1,0), (0,1)\}$.

2. Assume that $a(1,0,0) + b(1,1,0) + c(1,0,1) = (0,0,0)$. Then $a + b + c = 0, b = 0$, and $c = 0$. Thus, $a = b = c = 0$ and so the vectors are linearly independent.

3. It is clear that $\text{span}\{E_1, \ldots, E_n\} \subseteq R^n$. To prove the reverse containment, we note that $(a_1, \ldots, a_n) = a_1 E_1 + \ldots + a_n E_n$, proving that an arbitrary vector in R^n lies in $\text{span}\{E_1, \ldots, E_n\}$.

4. Assume that $a(1,2,-1,0) + b(1,0,-1,0) + c(0,1,0,1) = (0,0,0,0)$. Then $a + b = 0, 2a + c = 0, -a - b = 0$, and $c = 0$. We solve and see that $a = b = c = 0$. Hence the vectors are linearly independent.

5. Linearly dependent: $-2(1,1,2) + 1(2,-1,1) + 1(0,3,3) = (0,0,0)$.

6. Use Theorem 2.6.2. Form the matrix $\begin{bmatrix} 1 & 2 & -1 & 2 \\ 3 & 1 & 1 & 1 \\ -4 & 2 & -4 & 2 \end{bmatrix}$ and reduce to echelon form.

We get $\begin{bmatrix} 1 & 0 & 6/10 & 0 \\ 0 & 1 & -8/10 & 1 \\ 0 & 0 & 0 & 0 \end{bmatrix}$. We see that the rows of A are linearly dependent since the echelon form contains a row of zeros.

7. By Theorem 2.6.5, $X_1, \ldots, X_4$ must be linearly dependent since we know that $\mathbb{R}^3$ is spanned by three vectors. Thus, there is a linear combination $a_1 X_1 + \ldots + a_4 X_4 = 0$ with a nonzero coefficient. Now write $X = a_1 X_1 + a_2 X_2 = -a_3 X_3 - a_4 X_4$. Since one of the coefficients is nonzero, $X \neq 0$. Clearly, $X \in \text{span}\{X_1, X_2\} \cap \text{span}\{X_3, X_4\}$.

8. $1Z = 0$, but $1 \neq 0$, so $\{Z\}$ is a linearly dependent set.

10. (a) Assume that S' is linearly independent. If we have a linear combination of elements of S that is zero, we have a linearly combination of elements of S' that is zero. Since S' is linearly independent, all of the coefficients must be zero. Hence, S is linearly independent.

(b) If S is linearly dependent, then there exists a linear combination of elements of S (with at least one nonzero coefficient) that is equal to zero. Since $S \subseteq S'$, this linear combination is also a linear combination of elements of S'. Hence, S' is linearly dependent.

Note that these statements are contrapositives of one another. Since contrapositives are logically equivalent, we didn't have to prove both statements.

11. Assume that $X_{n+1} \notin \operatorname{span}\{X_1, \ldots, X_n\}$, and assume that we have a linear combination of the $n + 1$ vectors that is zero, say $a_1 X_1 + \ldots + a_n X_n + a_{n+1} X_{n+1} = 0$. If $a_{n+1} \neq 0$, we can solve for X_{n+1} in terms of $X_1, \ldots, X_n$ which will show that $X_{n+1} \in \operatorname{span}\{X_1, \ldots, X_n\}$. Thus, $a_{n+1} = 0$, and so $a_1 X_1 + \ldots + a_n X_n = 0$. But $X_1, \ldots, X_n$ are linearly independent, so $a_1 = \ldots = a_n = 0$, and it follows that all $n + 1$ vectors are linearly independent.

12. If $X_1 = X_2$, then $1X_1 + (-1)X_2 + 0X_3 + \ldots + 0X_n = 0$. This shows that $X_1, \ldots, X_n$ are linearly dependent.

13. No, they would be linearly independent by Theorem 2.6.7, but $\mathbb{R}^2$ can contain at most two independent vectors by Theorem 2.6.5.

14. Let X_1 and X_2 be linearly independent vectors in $\mathbb{R}^2$. If $X \notin \operatorname{span}\{X_1, X_2\}$, then X_1, X_2, X are linearly independent by Theorem 2.6.7. Since this is not possible by Theorem 2.6.5, $X \in \operatorname{span}\{X_1, X_2\}$.

15. If X_1 and X_2 are linearly dependent, then $X_2 \in \operatorname{span}\{X_1\}$. It follows that $R^2 = \operatorname{span}\{X_1\}$ and this implies that any two vectors in $\mathbb{R}^2$ are linearly dependent, which is a contradiction. Hence, X_1 and X_2 are linearly independent.

16. Assume that $X_1, \ldots, X_n$ are linearly independent. If $a_1 P X_1 + \ldots + a_n P X_n = 0$, then $P(a_1 X_1 + \ldots + a_n X_n) = 0$. Since P is nonsingular, we can multiply both sides by P^{-1}, obtaining $a_1 X_1 + \ldots + a_n X_n = 0$. But $X_1, \ldots, X_n$ are linearly independent so we must have $a_1 = \ldots = a_n = 0$. It follows that $P X_1, \ldots, P X_n$ are linearly independent.

Now assume that $P X_1, \ldots, P X_n$ are linearly independent. If $a_1 X_1 + \ldots + a_n X_n = 0$, then multiplying both sides by P, we see that $P(a_1 X_1 + \ldots + a_n X_n) = a_1 P X_1 + \ldots + a_n P X_n = 0$. Since $P X_1, \ldots, P X_n$ are linearly independent, we must have $a_1 = \ldots = a_n = 0$. Thus, $X_1, \ldots, X_n$ are linearly independent. (Note: We have proved a more general result in the second part of this proof. What is the statement of this result?)

Section 2.7

1. These vectors are linearly independent since $a(1, 0, 0) + b(1, 1, 0) + c(1, 1, 1) = (0, 0, 0)$ implies that $c = 0, b + c = 0$, and $a + b + c = 0$, and so $a = b = c = 0$. To see that these vectors span $\mathbb{R}^3$, note that if $X \in R^3$ then $X \notin \operatorname{span}\{(1, 0, 0), (1, 1, 0), (1, 1, 1)\}$ implies that all four vectors are linearly independent by Theorem 2.6.7. Hence, $(1, 0, 0), (1, 1, 0), (1, 1, 1)$ form a basis for $\mathbb{R}^3$.

2. $(1, -1, 2)(0, 1, 0), (0, 0, 1)$ is one possibility.

3. Since $(2,1,1) = (1,0,1) + (1,1,0)$, the vectors are not linearly independent and cannot form a basis.

4. $(1,3), (1,0)$ works.

5. Since $Y \neq 0, \{Y\}$ is a linearly independent set. Since $X \neq kY, X \notin \text{span}\{Y\}$. By Theorem 2.6.7, $\{X,Y\}$ is a linearly independent set. $\{X,Y\}$ must span $\mathbb{R}^2$, for otherwise we could find a vector $Z \notin \text{span}\{X,Y\}$ and this would produce a set of three independent vectors in $\mathbb{R}^2$. Hence $\{X,Y\}$ is a basis for $\mathbb{R}^2$.

6. We must prove that $\text{span}\{X_1,\ldots,X_n\} = \mathbf{V}$. Clearly, $\text{span}\{X_1,\ldots,X_n\} \subseteq \mathbf{V}$. If $X \in \mathbf{V}$, then $X \in \text{span}\{X_1,\ldots,X_n\}$, for otherwise, $X_1,\ldots,X_n,X$ are linearly independent and this contradicts the assumption $\dim \mathbf{V} = n$. Thus, $\mathbf{V} \subseteq \text{span}\{X_1,\ldots,X_n\}$ and so $X_1,\ldots,X_n$ forms a basis for $\mathbf{V}$.

7. Because of Exercise 6, we need only prove that the vectors are linearly independent. Assume that $b_1 E_1 + \ldots + b_{k-1}E_{k-1} + b_k X + b_{k+1}E_{k+1} + \ldots + b_n E_n = 0$. Using the representations for X and the E_i's, we see that $(b_1 + b_k a_1, \ldots, b_{k-1} + b_k a_{k-1}, b_k a_k, b_{k+1} + b_k a_{k+1}, \ldots, b_n + b_k a_n) = (0,\ldots,0)$. Now $b_k a_k = 0$ and $a_k \neq 0$ imply that $b_k = 0$, and given this we see that $b_i = 0$ for all $i \neq k$ as well. It follows that the vectors are linearly independent.

8. $(0,0,1)$ and $(1,-1,0)$ are in $\mathbf{W}$ and they are linearly independent. Since $\mathbf{W} \neq R^3$, these vectors must form a basis for $\mathbf{W}$. $\dim \mathbf{W} = 2$.

9. $(1,1,0,0), (0,0,2,1)$ form a basis for $\mathbf{U}$ since for any vector in

$$\mathbf{U}, (x,y,z,t) = (x,x,2t,t) = x(1,1,0,0) + t(0,0,2,1).$$

Clearly, $\dim \mathbf{U} = 2$. The vectors $(1,0,0,0)$ and $(0,0,0,1)$ will extend this basis for $\mathbf{U}$ to a basis for $\mathbb{R}^4$.

10. Assume that $X_1,\ldots,X_n$ are linearly dependent. Then by Theorem 2.6.1, one of the vectors is in the span of the others, say $X_1 \in \text{span}\{X_2,\ldots,X_n\}$. But then $\dim \mathbf{V} = n-1$, which is a contradiction. Thus, $X_1,\ldots,X_n$ are linearly independent.

11. Let $X_1,\ldots,X_n$ be a basis for $\mathbf{V}$ with $X_1,\ldots,X_k$ a basis for $\mathbf{U}$. This is possible by Theorem 2.7.6. Show that $X_1 + X_n,\ldots,X_k + X_n, X_{k+1},\ldots X_n$ is a basis with the desired property. (There are many other possibilities, as well.)

12. The polynomials $1,x,x^2,\ldots,x^n,\ldots$ form a basis for $\mathbb{R}[x]$.

13. Assume that $a\sin x + b\cos x = 0$. Choosing $x = 0$, we see that $b = 0$ and, similarly, $x = \pi/2$ shows that $a = 0$. Hence these functions are linearly independent and so their span must have dimension 2.

14. By Exercise 2.4.16, $\mathbf{V}$ is a subspace. If $X_1,\ldots,X_n$ is a basis for $\mathbf{U}$, then by Exercise 2.6.16, $PX_1,\ldots,PX_n$ are linearly independent. To see that these vectors span $\mathbf{V}$, note that $P(a_1 X_1 + \ldots + a_n X_n) = a_1 PX_1 + \ldots + a_n PX_n$.

15. ($\implies$) Assume that $X_1, \ldots, X_n$ forms a basis for $\mathbf{V}$. Then since a basis is a spanning set, every vector $X \in \mathbf{V}$ is a linear combination of the vectors $X_1, \ldots, X_n$. Assume that X is expressible in two different ways, say $X = a_1 X_1 + \ldots + a_n X_n = b_1 X_1 + \ldots + b_n X_n$. Then subtracting, we see that $0 = (a_1 - b_1) X_1 + \ldots + (a_n - b_n) X_n$ and so $a_1 - b_1 = \ldots = (a_n - b_n) = 0$ since $X_1, \ldots, X_n$ are linearly independent. Thus, $a_1 = b_1, \ldots, a_n = b_n$ and the coefficients are unique.

($\impliedby$) Assume that every vector $X \in \mathbf{V}$ can be expressed as a linear combination of $X_1, \ldots, X_n$ in one and only one way. Clearly, then $X_1, \ldots, X_n$ span $\mathbf{V}$. To show that the vectors are linearly independent, assume that $a_1 X_1 + \ldots + a_n X_n = 0$. Note that $0 = 0X_1 + \ldots + 0X_n$. Since the expression of the zero vector as a linear combination of the vectors $X_1, \ldots, X_n$ is unique we must have $a_1 = 0, \ldots, a_n = 0$. Hence $X_1, \ldots, X_n$ are linearly independent.

16. By Theorem 2.6.4, the nonzero rows of $B, B_1, \ldots, B_r$, are linearly independent. Since A is row equivalent to B, $\mathbf{R}(A) = \mathbf{R}(B)$. Since $\mathbf{R}(B)$ is spanned by $B_1, \ldots, B_r$, it is clear that $B_1, \ldots, B_r$ form a basis for $\mathbf{V}$.

17. Assume that B is $m \times n$ and let $j_1, \ldots, j_r$ be the constants associated with B. Notice that columns $j_1, \ldots, j_r$ are linearly independent; in fact, these columns are the transposes of the first r standard basis basis vectors in $\mathbb{R}^m$. Since the rows of B below row r are all zeros, we know that $\mathbf{C}(B)$ has dimension less than or equal to r. Since we have r independent vectors in $\mathbf{C}(B)$ we know that these vectors form a basis.

Section 2.8

1. Echelon form is $\begin{bmatrix} 1 & 0 & 2 \\ 0 & 1 & -1 \\ 0 & 0 & 0 \end{bmatrix}$, so rank = 2.

2. Echelon form is $\begin{bmatrix} 1 & 0 & -2 & -3 \\ 0 & 1 & 1 & 1 \\ 0 & 0 & 0 & 0 \end{bmatrix}$, so rank = 2.

3. Echelon form is $\begin{bmatrix} 1 & 0 \\ 0 & 1 \\ 0 & 0 \end{bmatrix}$, so rank = 2.

4. Echelon form is $\begin{bmatrix} 1 & 0 & -2/3 & 1 & 0 \\ 0 & 1 & -5/6 & 1 & 0 \\ 0 & 0 & 0 & 0 & 1 \end{bmatrix}$, so rank of coefficient matrix is 2, but augmented matrix has rank 3. Thus, system is inconsistent.

5. Echelon form is $\begin{bmatrix} 1 & 0 & 0 & 0 & -0.88 \\ 0 & 1 & -1 & 0 & 2.38 \\ 0 & 0 & 0 & 1 & 0.25 \end{bmatrix}$, augmented and coefficient matrices have same rank (3), therefore system is consistent.

6. Echelon form is $\begin{bmatrix} 1 & 0 & 0 & 0 & 2.50 \\ 0 & 1 & 0 & 0 & -0.50 \\ 0 & 0 & 1 & 0 & -0.50 \\ 0 & 0 & 0 & 1 & -0.50 \end{bmatrix}$. Consistent as above.

7. Form the augmented matrix $\begin{bmatrix} 1 & -3 & 2 & 4 \\ 2 & 1 & -1 & 1 \\ 3 & -2 & 1 & \alpha \end{bmatrix}$ and reduce to echelon form.

8. Let $H = \begin{bmatrix} a \\ b \\ c \end{bmatrix}$. The system will have a solution when $c - b - a = 0$.

9. Echelon form is $\begin{bmatrix} 1 & 0 & 0 & 1 \\ 0 & 1 & 0 & 1 \\ 0 & 0 & 1 & -3 \end{bmatrix}$, so this matrix has rank 3, while the matrix in Exercise 2 has rank 2. The matrices cannot be equivalent since equivalent matrices have the same rank.

10. No. For example, $\begin{bmatrix} 1 & 1 \\ 0 & 0 \end{bmatrix}$ and $\begin{bmatrix} 1 & 0 \\ 0 & 0 \end{bmatrix}$ are two matrices in echelon form with rank 1, but we can see that they are not equivalent since they are not equal (using the uniqueness of the echelon form).

11. Since row operations do not change the row space, we may assume that both A and B are in echelon form. If A and B are in echelon form and $\mathbf{R}(A) = \mathbf{R}(B)$, it can be shown that $A = B$. See the proof of Theorem 1.5.1 in Appendix F.

12. No. Consider the echelon form of the matrix. If the rank is r, then there are columns $j_1, \ldots, j_r$ containing the first nonzero entries in rows $1, \ldots, r$. Since this matrix has only two columns, $r = 2$.

13. $0 = \text{rank}(A) = \min\{n, m\}$.

14. $AX = H$ has a unique solution when no variables can be arbitrarily chosen. The number of arbitrarily chosen variables is $n - r$. Thus, there is a unique solution when $n - r = 0$.

15. If A is $m \times n$ and r is the rank of is the rank of A then we must have $r = n$. This implies that the nonzero rows of A look like the identity matrix, and so B is of the form

$$\begin{bmatrix} 1 & 0 & \ldots & 0 \\ \vdots & \vdots & \ddots & \vdots \\ 0 & 0 & \ldots & 1 \\ 0 & 0 & \ldots & 0 \\ \vdots & \vdots & \ddots & \vdots \\ 0 & 0 & \ldots & 0 \end{bmatrix}$$

Section 2.9

1. $n - r = 4 - 2 = 2$

2. 0

3. $x = 2y, y$ arbitrary. Solution space has dimension 1. $\{(2,1)\}$ is a basis.

4. $x = y = 0$. The solution space is $\{(0,0)\}$ and has dimension 0. The empty set $(\emptyset)$ is, by convention, the basis.

5. The dimension is 1 and $\{(-3, 2, 1)\}$ is a basis.

6. The dimension is 3. The following vectors will form a basis:

$$(1, 1, 0, 0, 0, 0), (-3, 0, 2, -3, 1, 0), (0, 0, -1, 2, 0, 1).$$

7. General solution: $x_2 = -2x_3 - 2x_5, x_4 = -x_5$ with x_1, x_3, x_5 arbitrary. Basis: $(1, 0, 0, 0, 0), (0, -2, 1, 0, 0), (0, -2, 0, -1, 1)$.

8. $x_1 = -(3/2)x_4, x_2 = (1/2)x_4, x_3 = -3x_4$, basis: $(-3/2, 1/2, -3, 1)$.

9. Yes, $r = 3$, and so $n - r = 4 - 3 = 1$. Thus, the dimension of the solution space is greater than or equal to one. This means that nonzero solutions may be found.

10. $r = 4$, and so $n - r = 7 - 4 = 3$, so the solution space must have dimension 3, or more.

11. This is clear since the solution must have dimension two $(= 6 - 4)$ or more.

12. The dimension lies between 0 and 4.

13. They are the same.

14. Assume that $x(a, e, b) + y(c, f, d) = 0$. Then $(xa + yc, xe + yf, xb + yd) = (0, 0, 0)$, but then $(xa + yc, xb + yd) = x(a, b) + y(c, d) = (0, 0)$. Since (a, b) and (c, d) are linearly independent vectors, $x = y = 0$. It follows that the vectors are linearly independent.

15. (a, b, e) and (f, c, d) needn't be linearly independent: Consider $(a, b) = (1, 0)$ and $(c, d) = (0, 1)$. Let $e = f = 1$. Then (a, b, e) and (f, c, d) are both equal to $(1, 0, 1)$, so they are not linearly independent. The vectors (a, b, e) and (c, d, f) will be linearly independent as in Exercise 14 above.

Chapter 3

Section 3.1

1. $T(1, 2) = (-1, 7), T(1, 0) = (1, 1), T(0, 2) = (-2, 6), T(1, 0) + T(0, 2) = (-1, 7)$

2. $T(2, 3) = (8, 1), T(2, 0) = (2, 4), T(0, 3) = (6, -3), T(2, 0) + T(0, 3) = (8, 1)$

3. $T(2,1,1)=(3,0), T(2,3,1)=(5,4)$

4. $T(3,2,1)=(4,1), T(1,3,1)=(2,2)$

5. The vector $(0,0,1)$ works - so does $(0,0,2)$.

6. It's not possible. If $T(x,y)=(0,0)$ then $x-y=0$ and $x+3y=0$. The only possible solution is $x=y=0$.

7. $T(1,2,3)=(-1,-4,-1)$

8. Yes, $(1,0)$ and $(0,1)$ form a basis for $\mathbb{R}^2$, so the method of Example 3.1.1 (j) may be applied.

9. No. If T is a linear transformation satisfying the first two conditions given, then $T(1,1)=T(1,0)+T(0,1)=(2,3)+(-1,2)=(1,5)\neq(2,0)$, so the third condition cannot be satisfied.

10. Let X be any vector in $\mathbf{U}$. Then for some scalars $a_1,\ldots,a_n$, $X=\sum_{i=1}^{n}a_iX_i$. Now

$$T(X)=T\left(\sum_{i=1}^{n}a_iX_i\right)=\sum_{i=1}^{n}a_iT(X_i)=\sum_{i=1}^{n}a_iS(X_i)=S\left(\sum_{i=1}^{n}a_iX_i\right)=S(X).$$

Hence $S=T$.

11. Here are the tricks: $T(0)=T(0+0)=T(0)+T(0)$, but $T(0)=T(0)+0$, so $T(0)+0=T(0)+T(0)$. Now cancel $T(0)$ and get $T(0)=0$. Cancellation is permitted since the equality is an equality of vectors. For the second, note that $40=T(0)=T(X+(-X))=T(X)+T(-X)4$. Add $-T(X)$ to both sides to obtain $-T(X)=T(-X)$.

12. Since $I_{\mathbf{U}}(X)=X$ for all X in $\mathbf{U}$, $TI_{\mathbf{U}}(X)=T(I_{\mathbf{U}}(X))=T(X)$. For the second, note that $T(X)$ is a vector in $\mathbf{V}$, so $I_{\mathbf{V}}(T(X))=T(X)$.

13. We will prove the statement $T\left(\sum_{i=1}^{n}a_iX_i\right)=\sum_{i=1}^{n}a_iT(X_i)$ by induction on n. For $n=1$, The statement says that $T(a_1X_1)=a_1T(X_1)$, which is true since T is a linear transformation.

Assume that the statement is true for some number n. Using this assumption we must prove that the statement is true for $n+1$. Note that $\sum_{i=1}^{n}a_iX_i=\sum_{i=1}^{n-1}a_iX_i+a_nX_n$, and so

$$T\left(\sum_{i=1}^{n} a_i X_i\right) = T\left(\sum_{i=1}^{n-1} a_i X_i + a_n X_n\right)$$

$$= T\left(\sum_{i=1}^{n-1} a_i X_i\right) + T(a_n X_n)$$

$$= \sum_{i=1}^{n-1} a_i T(X_i) + a_n T(X_n)$$

$$= \sum_{i=1}^{n} a_i T(X_i).$$

Thus, using the truth of the statement for $n - 1$, we have proved the statement for n, and so by mathematical induction, the statement is true for all positive integers n.

14. Let $Z = X - Y$. Then by Exercise 11 above, $T(Z) = T(X - Y) = T(X) - T(Y) = 0$ since $T(X) = T(Y)$.

15. If (a, b) is any vector in $\mathbb{R}^2$, then $T(a, b) = aT(1, 0) + bT(0, 1)$. Thus, for any vector X in R^2, $T(X) \in \text{span}\{T(1, 0), T(0, 1)\}$. By the results of Chapter 2, $\text{span}\{T(1, 0), T(0, 1)\} \neq R^3$, so there is a vector Y in $\mathbb{R}^3$ with $T(X) \neq Y$ for all X in $\mathbb{R}^2$.

Section 3.2

1. Set $T(x, y) = (2, -1)$ and attempt to solve. You'll find $T(0, 1) = (2, -1)$, and so $(2, -1) \in \text{Im}(T)$.

2. As above, $T(1, 0) = (1, 1, 2)$.

3. Set $T(x, y, z) = (0, 0)$ and attempt to solve. There are many solutions, but $(1, 1, -1)$ is one.

4. As in 3, $(-2, 1, -4/3)$.

5. $\text{null}(T) = \{0\}, \text{Im}(T) = \mathbb{R}^2, \text{rank}(T) = 2, \text{nullity}(T) = 0$.

6. $\text{null}(T) = \{0\}, Im(T) = \text{span}\{(1, 1, 2), (2, 0, -1)\}, \text{rank}(T) = 2, \text{nullity}(T) = 0$.

7. If $T(x, y, z) = (0, 0, 0)$ then $2x + y = 0, z = 0, 2y = 0$. It follows that $x = y = z = 0$ and so $\text{nullity}(T) = 0$. This implies that $\text{rank}(T) = 3$ and it follows that T is nonsingular. To find T^{-1}, set $(u, v, w) = T(x, y, z) = (2x + y, z, 2y)$ and solve for x, y, z in terms of u, v, w. We get $x = u/2 - w/4, y = w/2, z = v$. Now $T^{-1}(u, v, w) = (u/2 - w/4, w/2, v)$.

8. Let Y_1 and Y_2 be in $\text{Im}(T)$, and let r be a scalar. Then $Y_1 = T(X_1)$ and $Y_2 = T(X_2)$ for some $X_1, X_2 \in \mathbf{U}$. Now $Y_1 + Y_2 = T(X_1) + T(X_2) = T(X_1 + X_2) \in \text{Im}(T)$ since $X_1 + X_2 \in \mathbf{U}$. Also, $rY_1 = rT(X_1) = T(rX_1) \in \text{Im}(T)$ since $rX_1 \in \mathbf{U}$. It is clear that $\text{Im}(T) \neq \emptyset$ since $T(0) = 0 \in \text{Im}(T)$. It follows that $\text{Im}(T)$ is a subspace of $\mathbf{V}$. To show that $\text{null}(T)$ is a subspace, show that it is also closed and non empty.

9. Assume that $dim(\mathbf{U}) = n$. Since null$(T) = \{0\}$, nullity$(T) = 0$. From Theorem 3.2.2, rank$(T) = n$, so Im$(T) = \mathbf{U}$. Hence T is nonsingular.

10. Assume that $Y_1 = T(X_1)$ and $Y_2 = T(X_2)$ so that $S(Y_1) = X_1$ and $S(Y_2) = X_2$. Then $Y_1 + Y_2 = T(X_1) + T(X_2) = T(X_1 + X_2)$ and so $S(Y_1 + Y_1) = X_1 + X_2 = S(Y_1) + S(Y_1)$. Likewise we can show that $S(rY) = rS(Y)$ for a scalar r. If $Y = T(X)$ (so that $S(Y) = X$), then $S(T(X)) = S(Y) = X$, and $T(S(Y)) = T(X) = Y$. It follows that $TS = I_{\mathbf{V}}$ and $ST = I_{\mathbf{U}}$.

11. Since Im(T) is a subspace of $\mathbb{R}^2$, rank$(T) = 2$. By Theorem 3.2.2, nullity$(T) = 3 - rank(T) = 1$. It follows that null$(T) \neq \{0\}$ and that T is singular.

14. By Theorem 3.2.3, a transformation T is nonsingular if and only if there is a transformation S with the property that both composite transformations TS and ST are identity transformations. Now consider the transformation S. We see that there is a transformation, T, with the property that TS and ST are identity transformations. It follows that S also satisfies this condition and so S is nonsingular. Now S is denoted by T^{-1} and from the later comment, S^{-1} is T (here we invoke Theorem 3.2.4: T behaves like the inverse of S, so it must *be* the inverse of S). Thus, $S^{-1} = (T^{-1})^{-1} = T$.

15. First note that for X in $\mathbf{U}$, $(T^{-1}S^{-1})(ST)(X) = T^{-1}(S^{-1}(S(T(X)))) = T^{-1}(I_{\mathbf{V}}(T(X))) = T^{-1}(T(X)) = I_{\mathbf{U}}(X) = X$. Likewise, $(ST)(T^{-1}S^{-1})(X) = X$. Applying Theorems 3.2.3 and 3.2.4, we see that $(ST) - 1 = T^{-1}S^{-1}$.

Section 3.3

1. $\begin{bmatrix} 1 & 1 \\ 1 & -1 \end{bmatrix}$

2. $\begin{bmatrix} 2 & -1 \\ 1 & 3 \end{bmatrix}$

3. $\begin{bmatrix} 1 & 1 & 1 \\ 1 & -1 & 0 \\ 0 & 0 & 2 \end{bmatrix}$

4. $\begin{bmatrix} 1 & 1 & 0 \\ 1 & 1 & -2 \end{bmatrix}$

5. $[T : \mathbf{B}, \mathbf{E}] = \begin{bmatrix} 1 & 2 \\ 2 & 1 \end{bmatrix}, [T : \mathbf{E}, \mathbf{B}] = -\dfrac{1}{3}\begin{bmatrix} 1 & -2 \\ -2 & 1 \end{bmatrix}, [T : \mathbf{B}, \mathbf{C}] = \begin{bmatrix} -1/2 & 1/2 \\ 3/2 & 3/2 \end{bmatrix}$

6. $[T : \mathbf{E}, \mathbf{E}] = \begin{bmatrix} 1 & -1 \\ 1 & 2 \end{bmatrix}, [T : \mathbf{E}, \mathbf{E}_1] = \begin{bmatrix} 1 & 2 \\ 1 & -1 \end{bmatrix}, [T : \mathbf{E}, \mathbf{E}_2] = \begin{bmatrix} 1 & -1 \\ 1/2 & 1 \end{bmatrix},$

$[T : \mathbf{E}, \mathbf{E}_3] = \begin{bmatrix} 0 & -3 \\ 1 & 2 \end{bmatrix}$

7. $[T : \mathbf{E}, \mathbf{E}] = \begin{bmatrix} 2 & 1 \\ 0 & 3 \end{bmatrix}, [S : \mathbf{E}, \mathbf{E}] = \begin{bmatrix} 1 & 1 \\ 1 & -1 \end{bmatrix}, [TS : \mathbf{E}, \mathbf{E}] = \begin{bmatrix} 3 & 1 \\ 3 & -3 \end{bmatrix}$

$= [T : \mathbf{E}, \mathbf{E}][S : \mathbf{E}, \mathbf{E}]$

8. $T(1,0) = (1,2), T(0,1) = (-1,3), T(-2,3) = (-5,5)$

9. $\mathbf{B} = \{(-1,0), (-1,1)\}$ (don't forget that the order is important)

10. Let $\mathbf{B} = \{(a,b), (c,d)\}$. Then $[I : \mathbf{B}, \mathbf{E}] = \begin{bmatrix} a & c \\ b & d \end{bmatrix}$.

11. $[T : \mathbf{B}, \mathbf{E}] = [I : \mathbf{B}, \mathbf{E}][T : \mathbf{B}, \mathbf{B}] = \begin{bmatrix} 1 & -1 \\ 1 & 2 \end{bmatrix} [T : \mathbf{B}, \mathbf{B}]$.

12. $[I_\mathbf{V} : \mathbf{B}, \mathbf{B}] = I_n$, where $n = \dim \mathbf{V}$.

13. $[T : \mathbf{B}, \mathbf{C}]^{-1} = [T^{-1} : \mathbf{C}, \mathbf{B}]$

14. Note that $[I : \mathbf{B}, \mathbf{C}][I : \mathbf{C}, \mathbf{B}] = [I : \mathbf{C}, \mathbf{C}] = I_n = [I : \mathbf{B}, \mathbf{B}] = [I : \mathbf{C}, \mathbf{B}][I : \mathbf{B}, \mathbf{C}]$. It follows that $[I : \mathbf{B}, \mathbf{C}]^{-1} = [I : \mathbf{C}, \mathbf{B}]$.

Section 3.4

1. Product is: $\begin{bmatrix} 0 & -1 & 2 & 1 \\ 1 & 0 & 1 & -5 \\ 0 & 1 & 1 & 2 \end{bmatrix}$

2. Product is: $\begin{bmatrix} 5 & 1 & 11 & -7 \\ 1 & 3 & 1 & -5 \\ 2 & 12 & 4 & -4 \end{bmatrix}$

3. Product is: $\begin{bmatrix} 2 & 1 \\ 2 & -2 \end{bmatrix}$

4. Product is: $\begin{bmatrix} 2 & 12 \\ 4 & -4 \end{bmatrix}$

5. Product is: $\begin{bmatrix} 0 & -1 \\ 0 & 1 \end{bmatrix}$

6. Product is: $\begin{bmatrix} 3 & -1 \\ 5 & 5 \end{bmatrix}$

7. Let $A = [a_{ij}]$, $B = [b_{ij}]$, and $AB = [c_{ij}]$, where $a_{ii} = a_i$, $b_{ii} = b_i$, $a_{ij} = b_{ij} = 0$ for $i \neq j$ and $c_{ij} = \sum_{k=1}^{n} a_{ik}b_{kj}$. Then, since $a_{ij} = b_{ij} = 0$ for $i \neq j$, $c_{ij} = 0$ for $i \neq j$, and $c_{ii} = a_i b_i$.

8. Inverse is: $\begin{bmatrix} -1 & 2 \\ 1 & -1 \end{bmatrix}$

9. Inverse is: $\begin{bmatrix} 1 & 1/2 \\ 0 & -1/2 \end{bmatrix}$

10. Echelon form is: $\begin{bmatrix} 1 & -2 \\ 0 & 0 \end{bmatrix}$

11. Nonsingular, inverse is: $\begin{bmatrix} -1 & 1 \\ 1 & 0 \end{bmatrix}$

12. Inverse is: $[I : \mathbf{B}, \mathbf{A}] = \begin{bmatrix} -1 & 0 \\ 2 & 2 \end{bmatrix}$

13. This exercise appeared in Section 1.3. If you didn't get it there, here's another chance. Can the result be generalized to $n \times n$ matrices?

Section 3.5

Remember our convention wherein $(x_1, \ldots, x_n)^t = \begin{bmatrix} x_1 \\ \vdots \\ x_n \end{bmatrix}$.

1. $(3/2, 1/2)^t$

2. $(1, -2)^t$

3. $(-1, -1, 2)^t$

4. $(-1, 4, -1)^t$

5. $(-3, 2)^t$

6. We see that $X = (x, y) = x(1, 0) + y(0, 1)$, so $X_\mathbf{E} = (x, y)^t$.

7. $[I : \mathbf{B}, \mathbf{D}] = \begin{bmatrix} 1 & 0 \\ -1 & 1 \end{bmatrix}$, $[T : \mathbf{B}, \mathbf{D}] = \begin{bmatrix} 1 & -1 \\ 1 & 1 \end{bmatrix}$, $[I : \mathbf{D}, \mathbf{B}] = \begin{bmatrix} 1 & 0 \\ 1 & 1 \end{bmatrix}$, $[T : \mathbf{D}, \mathbf{D}] = \begin{bmatrix} 0 & -1 \\ 2 & 1 \end{bmatrix}$.

8. $P = Q = [I : \mathbf{B}, \mathbf{C}] = \begin{bmatrix} 0 & 1 \\ 1 & -2 \end{bmatrix}$

9. $[T : \mathbf{C}, \mathbf{B}] = \begin{bmatrix} 0 & -1 & 2 \\ 1 & 1 & -1 \\ 0 & 0 & 2 \end{bmatrix}$, $P = Q = [I : \mathbf{B}, \mathbf{C}] = \begin{bmatrix} -2 & 1 & 1 \\ 2 & 0 & -1 \\ 1 & 0 & 0 \end{bmatrix}$

10. $[T : \mathbf{C}, \mathbf{B}] = \begin{bmatrix} -3 & 1 & 4 \\ 2 & -1 & 0 \\ 1 & 0 & -1 \end{bmatrix}$, $P = Q = [I : \mathbf{B}, \mathbf{C}] = \begin{bmatrix} 1 & 1 & 2 \\ 0 & 1 & -2 \\ 0 & 0 & 1 \end{bmatrix}$

11. $A = \begin{bmatrix} 1 & 1 & -2 \\ -2 & 0 & 2 \\ 0 & 0 & 1 \end{bmatrix}$, $X_{\mathbf{B}} = \begin{bmatrix} y \\ x \\ z \end{bmatrix}$, $AX_{\mathbf{B}} = \begin{bmatrix} y + x - 2z \\ -2y + 2z \\ z \end{bmatrix} = (x - y, z, 2y)_{\mathbf{C}} = T(X)_{\mathbf{C}}$

since $(y + x - 2z)(1, 0, 0) + (-2y + 2z)(1, 0, -1) + z(0, 1, 2) = (x - y, z, 2y) = T(X)$.

12. Assume that A is nonsingular. Let $P = A^{-1}$ and let $Q = I_n$. Then $PAQ = A^{-1}AI_n = I_n$, and thus A is equivalent to I_n.

13. Assume that A is similar to I_n, say $S^{-1}AS = I_n$. Then $AS = SI_n$, and so $A = SI_nS^{-1} = I_n$. Thus, $A = I_n$.

Section 3.6

1. Product is: $\begin{bmatrix} -2 & 5 \\ 1 & -3 \end{bmatrix}$ 3. Product is: $\begin{bmatrix} 3 & -8 \\ 1 & -2 \end{bmatrix}$

2. Product is: $\begin{bmatrix} -3 & 0 & 0 \\ 2 & 0 & 1 \\ -2 & 1 & 0 \end{bmatrix}$ 4. Product is: $\begin{bmatrix} 0 & -1 & 2 \\ 0 & 0 & 1 \\ 1 & -3 & 0 \end{bmatrix}$

5. a) C_{12} b) $C_2(1/2)$ c) $C_{12}(-1)$ d) $C_{21}(2)$

6. a) R_{12} b) $R_2(-1)$ c) $R_{21}(2)$ d) $R_{12}(-2)$

7. a) R_{13} b) $R_1(-3)$ c) $R_{32}(-2)$ d) $R_{21}(3)$

8. a) R_{12} b) $R_2(2)$ c) $R_{21}(1)$ d) $R_{12}(-2)$

9. a) C_{12} b) $C_2(-1)$ c) $C_{12}(-2)$ d) $C_{21}(2)$

10. a) C_{13} b) $C_1(-1/3)$ c) $C_{23}(2)$ d) $C_{12}(-3)$

11. The $n \times n$ identity matrix I_n has rank n, and performing row operations on a matrix does not change the rank of a matrix. Left multiplication of a matrix by an elementary matrix perform the corresponding row operation on the matrix, thus $P = PI_n = E_1 \ldots E_m I_n$ is a matrix of rank n.

12. *Proof.* Assume that B has the given conditions and let $r, j_1, \ldots, j_r$ be the constants associated with B. Now we are given that $r = n$ and $1 = j_1 < \ldots < j_n = n$. This can only happen when $j_1 = 1, j_2 = 2, \ldots, j_n = n$, and this implies that $B = I_n$. $\square$

13. Inverse is: $\begin{bmatrix} 3/5 & -2/5 \\ 1/5 & 1/5 \end{bmatrix}$

14. Matrix equals: $R_{12}(-1)R_2(5)R_{21}(2)$

15. Matrix equals: $R_{31}R_1(2)R_{12}(1)R_2(1/2)R_{21}(-1/2)R_3(2)R_{31}(1)R_{32}(-1)$

Inverse is: $\begin{bmatrix} -1/2 & 1 & 0 \\ 1/2 & 2 & -1 \\ 1/2 & 0 & 0 \end{bmatrix}$

16. By Theorem 3.4.1, the inverse of a nonsingular matrix is unique. Since $AA^{-1} = A^{-1}A = I_n$, it must be the case that A^{-1} is nonsingular and that its inverse is A. That is, $(A^{-1})^{-1} = A$.

Chapter 4

Section 4.1

1. 7

3. -52

5. 50

7. -36

2. 7

4. 86

6. -10

8. 10

9. determinant is 22, cofactor is 0

10. $|A| = 5, A_{11} = 3, A_{21} = -2, A_{12} = 1, A_{22} = 1$.

11. Assume that you are given a general 2×2 matrix, say $\begin{bmatrix} a & b \\ c & d \end{bmatrix}$. To prove Theorem 4.1.1, compute the expansions of the determinant of the matrix expanding along row 2 and columns 1 and 2. Compare with the expansion along row 1. For Theorem 4.1.2, switch rows, compute the determinant and compare the result with the determinant of the unchanged matrix.

12. $|A| = -3, |B| = -4, |AB| = 12, |BA| = 12$.

13. Assume that the determinant of a matrix equals the determinant of its transpose for all $(n-1) \times (n-1)$ matrices, and let $A = [a_{ij}]$ be $n \times n$. Let M_{ik} be the minor of the entry in row 1 and column k of A, and let N_{k1} be the minor of the entry in row k and column 1 of A^t. We see that the matrices in these minors are transposes of each other. By the induction assumption, their determinants are equal; that is, $M_{ik} = N_{k1}$. Expanding $|A|$ along row 1, we get $|A| = \sum_{k=1}^{n} a_{1k}(-1)^{1+k}M_{1k}$, and expanding $|A^t|$ down column 1, we get $|A^t| = \sum_{k=1}^{n} a_{1k}(-1)^{1+k}N_{k1}$. Since $M_{ik} = N_{k1}$ for all k, we see that $|A| = |A^t|$. By induction, the theorem must be true for all positive integers n.

14. Observe that expanding $|D|$ along row 1 yields $|D| = a_1|\text{diag}(a_2, \ldots, a_n)|$. This allows induction on the size of the matrix.

16. Let $A = [a_{ij}]$ and $B = [b_{ij}]$ and assume that $A^t = [c_{ij}]$ and $B^t = [d_{ij}]$, where $c_{ij} = a_{ji}$ and $d_{ij} = b_{ji}$. Now compute: The ij entry in $B^t A^t$ is $\sum_{k=1}^{n} d_{ik}c_{kj} = \sum_{k=1}^{n} b_{ki}a_{jk}$. The ij entry in $(AB)^t$ is the entry in row j and column i in AB. This is $\sum_{k=1}^{n} a_{jk}b_{ki} = \sum_{k=1}^{n} b_{ki}a_{jk}$. Since corresponding entries are equal, the matrices $(AB)^t$ and $B^t A^t$ are equal.

Section 4.2

1. $\text{Adj}(A) = \begin{bmatrix} 3 & -1 \\ 1 & 2 \end{bmatrix}$, $A\text{Adj}(A) = \begin{bmatrix} 7 & 0 \\ 0 & 7 \end{bmatrix}$, $A^{-1} = \dfrac{1}{7}\begin{bmatrix} 3 & -1 \\ 1 & 2 \end{bmatrix}$

2. $\text{Adj}(A) = \begin{bmatrix} 1 & -2 \\ 2 & 3 \end{bmatrix}$, $A\text{Adj}(A) = \begin{bmatrix} 7 & 0 \\ 0 & 7 \end{bmatrix}$, $A^{-1} = \dfrac{1}{7}\begin{bmatrix} 1 & -2 \\ 2 & 3 \end{bmatrix}$

3. $\text{Adj}(A) = \begin{bmatrix} 3 & 0 & -2 \\ -7 & -5 & 3 \\ -1 & 0 & -1 \end{bmatrix}$, $A^{-1} = \begin{bmatrix} -3/5 & 0 & 2/5 \\ 7/5 & 1 & -3/5 \\ 1/5 & 0 & 1/5 \end{bmatrix}$

4. $\text{Adj}(A) = \begin{bmatrix} 3 & 0 & 2 \\ -7 & 5 & -3 \\ -1 & 0 & 1 \end{bmatrix}$, $A\text{Adj}(A) = \begin{bmatrix} 5 & 0 & 0 \\ 0 & 5 & 0 \\ 0 & 0 & 5 \end{bmatrix}$

5. $\text{Adj}(A) = \begin{bmatrix} 3 & -2 & 2 \\ 0 & 5 & 0 \\ -1 & -1 & 1 \end{bmatrix}$, $A\text{Adj}(A) = \begin{bmatrix} 5 & 0 & 0 \\ 0 & 5 & 0 \\ 0 & 0 & 5 \end{bmatrix}$

6. $|AB| = 2 \cdot 3 = 6$.

8. Notice that if one row and one column are canceled from P, the resulting matrix has the same property; that is, each row and each column contains exactly one 1, and all other entries are zero. This allows us to prove this result using induction.

9. We know that AB is nonsingular and $(AB)^{-1} = B^{-1}A^{-1}$ (see Exercise 14 of Section 3.4). Using Theorem 4.2.6, $(AB)^{-1} = \dfrac{1}{|AB|}\text{Adj}(AB)$, and $B^{-1}A^{-1} = \dfrac{1}{|B|}\text{Adj}(B)\dfrac{1}{|A|}\text{Adj}(A)$. Since $|AB| = |A||B|$, we can cancel and obtain $\text{Adj}(AB) = \text{Adj}(B)\text{Adj}(A)$.

10. Assume that A is a singular $n \times n$ matrix and let PA be the echelon form of A with P nonsingular ($P = E_n \ldots E_1$ is the product of elementary matrices corresponding to the row operations performed in the reduction of A to echelon form - see Section 3.6.) Since PA is an $n \times n$ matrix in echelon form, it must be the case that either $PA = I$ or PA has a row of zeros. We cannot have $PA = I$, for then $A = E_1^{-1} \ldots E_n^{-1}I$ would be a product or nonsingular matrices and hence nonsingular. We see that PA has a row of zeros. If AB were nonsingular, then there would be a matrix C with $ABC = I$. But then, multiplying by P, we see that $PABC = PI$. Now PA has a row of zeros, so $PABC$ has a row of zeros. But then PI has a row of zeros, and this is impossible since $PI = P$ is nonsingular. Hence, AB must be singular.

11. Using Theorem 4.2.4, $|AB| = |A||B| = |B||A| = |BA|$, where the second equality follows from the commutativity of scalar multiplication (remember that $|A|$ and $|B|$ are scalars).

12. If $\text{rank}(B) = n$, then $B = I$, and so $|B| = 1$. If $\text{rank}(B) < n$, then B has a row of zeros and so the determinant will be 0. Thus, $|B| = 0$.

13. Let B be the echelon form of the $n \times n$ matrix A, say with $B = E_1 \ldots E_n A$, where $E_1, \ldots, E_n$ are elementary matrices. Then $|B| = |E_1| \ldots |E_n||A|$, so $|A| = |E_n|^{-1} \ldots |E_1|^{-1}|B|$. To compute $|B|$, just multiply the diagonal elements. Actually, since B is $n \times n$, we see that B is either the identity matrix ($|B| = 1$) or B has a row of zeros ($-B| = 0$) and therefore the entry in the lower right hand corner of B will be the determinant of B. Evaluating $|E_n|^{-1} \ldots |E_1|^{-1}$ is also straightforward. One can compute the product of these elementary matrices as follows. Define a variable D and start with $D = 1$. Modify D at each step as follows: if a row operation of the form R_{ik} is performed, change the sign of D. If a row operation of the form $R_i[a]$ is performed, multiply D by $1/a$. If a row operation of the form $R_{ik}[a]$ is performed, leave D unchanged. When the row operations $E_1, \ldots, E_n$ have been performed, the value of D will be $|E_n|^{-1} \ldots |E_1|^{-1}$.

Section 4.3

In Exercises 1 - 4, we give the quotient of the two determinants.

1. $x_1 = -1/1, x_2 = 0/1$

3. $x_1 = 1/11, x_2 = -7/11$

2. $x_1 = 5/18, x_2 = -1/18, x_3 = 11/18$

4. $x_1 = 1/2, x_2 = 0, x_3 = 1/2$

In Exercises 5 – 8, we give the inverse to the coefficient matrix. The solutions are given in Exercises 1 – 4, above.

5. Inverse $= \begin{bmatrix} 3 & -1 \\ -2 & 1 \end{bmatrix}$

7. Inverse $= \dfrac{1}{-11} \begin{bmatrix} 2 & -5 \\ -3 & 2 \end{bmatrix}$

6. Inverse $= \dfrac{1}{18} \begin{bmatrix} 1 & -5 & 7 \\ 7 & 1 & -5 \\ -5 & 7 & 1 \end{bmatrix}$

8. Inverse $= \dfrac{1}{-8} \begin{bmatrix} -8 & -2 & 9 \\ 8 & 4 & -14 \\ 0 & -2 & 1 \end{bmatrix}$

9. Echelon form $= \begin{bmatrix} 1 & 0 & 2/3 & 1 \\ 0 & 1 & -5/3 & 0 \end{bmatrix}$.

    (a) $z = 1, x = 1/3, y = 5/3$

    (b) $y = 2, x = 1/5, z = 6/5$

    (c) $x = 0, y = 5/2, z = 3/2$

12. $AX = 0$ has a nonzero solution only when A is singular; that is, when $|A| = 0$. But if $|A| = 0$, then $|AB| = |A||B| = 0$, and so $ABX = 0$ also has a nonzero solution.

Section 4.4

1. Apply the law of cosines to the triangle formed by the vectors A, B, and the line joining the endpoints of the vectors. This line can be represented by the vector $A - B$. We get:

$$||A - B||^2 = ||A||^2 + ||B||^2 - 2||A||\,||B||\cos\theta.$$

Express each of the magnitudes in terms of the coordinates of the vectors. The squared terms will cancel, leaving the cross products with a coefficient of -2. The result will follow.

2. This just involves computation. For example in part (a),

$$A \times B = (A_y B_z - A_z B_y)\mathbf{i} + (A_z B_x - A_x B_z)\mathbf{j} + (A_x B_y - A_y B_x)\mathbf{k}$$

and

$$B \times A = (B_y A_z - B_z A_y)\mathbf{i} + (B_z A_x - B_x A_z)\mathbf{j} + (B_x A_y - B_y A_x)\mathbf{k},$$

so $A \times B = -(B \times A)$.

3. $x = t + 1, y = 2t + 1, z = -t - 1$

4. To find the normal, take the cross product of the differences of the vectors: $N = (2, 1, 2)$. Since the plane passes through $(0, 0, 0)$, an equation is: $2x + y + 2z = 0$.

5. Proceeding as in 4, we get $-(x - 1) - (y - 2) - (z + 1) = 0$.

6. Find a vector (a, b, c) that is perpendicular to a normal to the plane, say $(2, 3, -1)$. That is, find a solution to $2a + 3b - c = 0$. We take the solution $(0, 1, 3)$. An equation of the plane is: $(y + 2) + 3(z - 3) = 0$.

7. We need to "shift" so that the origin is at the point $(-1, 3, 2)$, so subtract this vector from the other two. We get $\Omega = \dfrac{6}{\sqrt{6}}(2, -1, -1)$ and $R = (1, -3, -2)$. It follows that $V = \Omega \times R = \sqrt{6}(-1, 3, -5)$.

8. We need only compute the component of the force in the direction of motion: $F = (0, 0, 4)$ and $D = (1, 1, 1)$, so we divide D by its magnitude and take the dot product. The result is $\dfrac{4}{\sqrt{3}}$.

9. Just compute using the components of the vectors. For example:

(b) $(rA) \cdot B = (rA_x)B_x + (rA_y)B_y + (rA_z)B_z = r(A_x B_x + A_y B_y + A_z B_z) = r(A \cdot B)$.

10. Assume that a linear combination is zero, say $aU + bV + cW = 0$. To show that $a = 0$, form the dot product $U \cdot (aU + bV + cW) = aU \cdot U + bU \cdot V + cU \cdot V = U \cdot 0 = 0$. Since $U \neq 0, U \cdot U \neq 0$, and so $a = 0$. Likewise $b = c = 0$ and it follows that U, V, and W are linearly independent.

11. By Exercise 10, we know that U, V, and W form a basis for $\mathbb{R}^3$, and so A is a linear combination of these vectors, say $A = aU + bV + cW$. To find a, take the dot product with U: $A \cdot U = U \cdot (aU + bV + cW) = aU \cdot U + bU \cdot V + cU \cdot V = aU \cdot U = a \cdot 1 = a$. Therefore, $a = A \cdot U$. The other coefficients can be found in a similar manner.

12. Using Exercise 9 parts (b) and (c), we see that $U \cdot U = \dfrac{A}{||A||} \cdot \dfrac{A}{||A||} = \dfrac{1}{||A||^2}(A \cdot A) = 1$.

13. As above, just compute using the components of the vectors. For example:

    (a)

$$A \cdot (B \times C) = (A_x\mathbf{i} + A_y\mathbf{j} + A_z\mathbf{k}) \cdot ((B_yC_z - B_zC_y)\mathbf{i} + (B_zC_x - B_xC_z)\mathbf{j}$$
$$+ (B_xC_y - B_yC_x)\mathbf{k})$$
$$= A_x(B_yC_z - B_zC_y) + A_y(B_zC_x - B_xC_z) + A_z(B_xC_y - B_yC_x)$$
$$\vdots$$
$$= (A_yB_z - A_zB_y)C_x + (A_zB_x - A_xB_z)C_y + (A_xB_y - A_yB_x)C_z$$
$$= (A \times B) \cdot C.$$

Section 4.5

1. (a) Cramer's Rule: 7.1514×10^{35} seconds, Gauss Reduction: 83.7 seconds.

   (b) Cramer's Rule: 1.1919×10^{29} seconds, Gauss Reduction: 0.01395 seconds.

   (c) Cramer's Rule: 2.3838×10^{25} seconds, Gauss Reduction: 0.00000279 seconds.

2. Cramer's Rule: $11! \cdot 9 = 359,251,200$ multiplications, 199,584 seconds. Gauss Reduction: 1100 multiplications, 0.00061111 seconds.

3. A crude estimate is given by: $m(m-1)n$ additions $+ m(mn)$ multiplications $= mn(2m-1)$ operations. This estimate assumes that the entire length of each row must be processed at each step. Actually, in row i only the last $n-i$ entries must be computed. For a 50×50 system, the time required would be $50 \cdot 50 \cdot 99/2,000,000 = 0.12375$ seconds. Try to find a better estimate!

4. (a) Here we must perform calculate $n^2(n-1)(n-1)$. The required multiplications would be $n^2(n-1)!$.

   (b) Here we must reduce an $n \times 2n$ matrix to echelon form. From Exercise 3, this would require $2n^3$ multiplications.

5. If the matrix is $n \times n$, the reduction to echelon form requires $2n^3 - n^2$ operations, as in Exercise 3. Keeping track of the determinant requires at most 2 operations for each pivot, and so a total of $2n$ operations. Thus, $2(n^3 - n^2 + n)$ are needed.

6. A is arbitrary: n^3 multiplications. A is diagonal: n^2 multiplications. A is upper-triangular: $n(n + (n-1) + (n-2) + \ldots + 1) = n\dfrac{n(n-1)}{2} = \dfrac{n^3 - n^2}{2}$ multiplications.

Chapter 5

Section 5.1

1. $X \cdot Y = 4, X \cdot Z = 8$

2. $Y \cdot Y = 38, Z \cdot X = 8$

3. $||X|| = \sqrt{11}, ||Y|| = \sqrt{38}, ||Z|| = \sqrt{14}$.

4. $||U|| = 1$

5. $X_1 \cdot X_2 = -1, ||X_2|| = \sqrt{6}$

6. $Z_1 = \dfrac{1}{\sqrt{2}}(1, -1, 0), Z_2 = \dfrac{1}{\sqrt{18}}(1, 1, -4)$

7. $X_1 \cdot X_2 = (1 - i)i + (-i)(1 - i) + 0 = 0, X_2 \cdot X_3 = -i(1 - i) + (1 + i), X_1 \cdot X_3 = (1 - i)(1 - i) + (-i)1 + 1(3i) = 0$

8. $Z_1 = \dfrac{1}{\sqrt{2}}(1, 0, 1, 0), Z_2 = (0, 1, 0, 0), Z_3 = \dfrac{1}{\sqrt{2}}(1, 0, -1, 0)$ is the result of the Gram-Schmidt Process, but $(1, 0, 0, 0), (0, 1, 0, 0), (0, 0, 1, 0)$ is a more straightforward choice.

9. $\dfrac{1}{\sqrt{5}}(1, 2), \dfrac{1}{\sqrt{5}}(2, -1)$

10. $\dfrac{1}{\sqrt{13}}(-2, 3), \dfrac{1}{\sqrt{13}}(3, 2)$

11. $\dfrac{1}{\sqrt{13}}(0, -3, 2), \dfrac{1}{\sqrt{13}}(0, 2, 3), (1, 0, 0)$

12. $\dfrac{1}{3}(-1, 2, -2), \dfrac{1}{\sqrt{5}}(2, 1, 0), \dfrac{1}{\sqrt{45}}(-2, 4, 5)$

13. (c) Let $z = a + bi$ and $w = c + di$ be complex numbers. Then

$$\overline{zw} = (ac - bd) - (ad + bc)i = (a - bi)(c - di) = \overline{z}\,\overline{w}.$$

14. c) Let $X = (x_1, \ldots, x_n), Y = (y_1, \ldots, y_n)$ and let c be a scalar. Then $X \cdot cY = (x_1, \ldots, x_n) \cdot (cy_1, \ldots, cy_n) = x_1 cy_1 + \ldots + x_n cy_n = c(x_1 y_1 + \ldots + x_n y_n) = c(X \cdot Y)$.

15. See the comments for Exercise 4.4.1.

16. If $z = a + bi$ is a nonzero complex number, then $z\overline{z} = a^2 + b^2 > 0$ since either a or b is nonzero. Since $X \cdot X$ is a sum of terms each of which is either zero or positive, and since at least one is positive, $X \cdot X > 0$.

18. Use Exercise 17: $||U|| = ||(1/||X||)X|| = (1/||X||)||X|| = 1$.

19. Use Exercise 17 again: $\dfrac{1}{||cX||}cX = \dfrac{1}{|c|||X||}cX = \dfrac{1}{||X||}X$ since $c \in \mathbb{R}^+$.

20. Let $A = [a_{ij}]$ and $B = [b_{ij}]$. Assume that $AB = [c_{ij}]$, where $c_{ij} = \displaystyle\sum_{k=1}^{n} a_{ik}b_{kj}$. Then using Theorem 5.1.1, $\overline{c}_{ij} = \displaystyle\sum_{k=1}^{n} \overline{a}_{ij}\overline{b}_{ij}$, but this is the i, jth entry in the product $\overline{A}\,\overline{B}$. Hence, $\overline{AB} = \overline{A}\,\overline{B}$.

21 These results follow from straightforward calculations.

22 Refer to Section 5.1.

23 If Z is parallel to Y, then $Z = cY$ for some scalar c. Thus $X \cdot Z = X \cdot (cY) = c(X \cdot Y) = c(0) = 0$.

24 Suppose that for some $v_1, \ldots, v_n \in S$ and $a_1, \ldots, a_n \in F$, $a_1 v_1 + \ldots + a_n v_n = 0$. Then $(a_1 v_1 + \ldots + a_n v_n) \cdot v_j = 0 \cdot v_j = 0$ for each j, and thus $a_j ||v_j||^2 = 0$. Since $0 \notin S$, $||v_j|| \neq 0$, so $a_j = 0$. Since j was arbitrary, each $a_j = 0$, so S is linearly independent.

25 Use the trigonometric identities $\cos A \cos B = \dfrac{1}{2}(\cos(A-B)+\cos(A+B))$ and $\sin A \sin B = \dfrac{1}{2}(\cos(A - B) - \cos(A + B))$.

26 Use the trigonometric identity $\sin A \cos B = \dfrac{1}{2}(\sin(A - B) + \sin(A + B))$.

27 $<f, g> = \displaystyle\int_{-\pi}^{-\pi} x \cos x \, dx = 0$, so x and $\cos(x)$ are orthogonal in this inner product space.

28 $<f, g> = \displaystyle\int_{-\pi}^{-\pi} x \sin x \, dx = 2\pi$, $<f, f> = \displaystyle\int_{-\pi}^{\pi} x^2 \, dx = \dfrac{2}{3}\pi^3$, and $<g, g> = \displaystyle\int_{-\pi}^{\pi} \sin^2(x) \, dx = \pi$, so $\cos(\theta) = \dfrac{2\pi}{\sqrt{(2\pi^3/3)(\pi)}} = \dfrac{2}{\pi^3}$. Therefore $\theta \approx 1.4738$ radians or $84.4477°$.

29 $<f, g> = \displaystyle\int_{0}^{1} (4x^2 + 3)(7x - 1) \, dx = 79/6$, $<f, f> = \displaystyle\int_{0}^{1} (4x^2 + 3)^2 \, dx = 101/5$, and $<g, g> = \displaystyle\int_{0}^{1} (7x - 1)^2 \, dx = 31/3$. Therefore, $\cos \theta = \dfrac{79/6}{\sqrt{(101/5)(31/3)}} \approx 0.9113$, so $\theta \approx 0.4242$ radians or $24.3049°$.

30 $\cos \theta = \dfrac{<X, X>}{\sqrt{<X, X><X, X>}} = \dfrac{<X, X>}{<X, X>} = 1$. Thus $\theta = 0$.

31 Since $<0, X> = 0$ for all $X \in V$, $0 \in W^\perp$, so $W^\perp \neq \emptyset$. Suppose that $X, Y \in W^\perp$ and $r \in F$. If $Z \in W$, then $<X + Y, Z> = <X, Z> + <Y, Z> = 0 + 0 = 0$, and $<rX, Z> = \overline{r}<X, Z> = r0 = 0$. Therefore, $W^\perp$ is a subspace of V.

32   (a) For $v, w \in V$,

$$T(v + w) = (v + w) - ((v + w) \cdot w_1)w_1 - \ldots - ((v + w) \cdot w_k)w_k$$
$$= (v + w) - (v \cdot w_1 + w \cdot w_1)w_1 - \ldots - (v \cdot w_k + w \cdot w_k)w_k$$
$$= (v - (v \cdot w_1)w_1 - \ldots - (v \cdot w_k)w_k) + (w - (w \cdot w_1)w_1 - \ldots - (w \cdot w_k)w_k)$$
$$= T(v) + T(w).$$

   (b) Let $v \in V$. Then $< T(v), w_j >=< v - (v \cdot w_1)w_1 - \ldots - (v \cdot w_k)w_k, w_j >= 0$, so $T(v) \in W^\perp$. On the other hand, if $w \in W^\perp$, $T(w) = w - (w \cdot w_1)w_1 - \ldots - (w \cdot w_k)w_k = w$, so $w \in \mathrm{Im}(T)$. Therefore, $\mathrm{Im}(T) = W^\perp$.

   (c) If $w \in W$, then $T(w) = w - (w \cdot w_1)w_1 - \ldots - (w \cdot w_k)w_k$. But if $w = a_1 w_1 + \ldots + a_k w_k$, then $w \cdot w_j = a_j$, so $T(w) = 0$. Therefore, $W \subseteq \ker(T)$. On the other hand, if $T(v) = 0$, then we must have $v = (v \cdot w_1)w_1 + \ldots + (v \cdot w_k)w_k \in W$. Therefore, $\ker(T) = W$.

   (d) $\dim V = \mathrm{rank}(T) + \mathrm{nullity}(T) = \dim\mathrm{Im}(T) + \dim \ker(T) = \dim W^\perp + \dim W$.

   (e) If $w \in W, v \in W^\perp$, then $< v, w >= 0$ by definition. Thus $< w, v >= 0$ as well, so $w \in (W^\perp)^\perp$ and $W \subseteq (W^\perp)^\perp$. We know that $\dim W^\perp + \dim W = n$ and $\dim W^\perp \dim(W^\perp)^\perp = n$ for the same reason, so $\dim(W^\perp)^\perp = \dim W$. But W is a subspace of $(W^\perp)^\perp$ of the same dimension as $(W^\perp)^\perp$, so we must have $W = (W^\perp)^\perp$.

33 Write $A = \begin{bmatrix} A_1 \\ \vdots \\ A_n \end{bmatrix} = \begin{bmatrix} A_1^T & \cdots & A_n^T \end{bmatrix}$. Notice that if $w \in F^n$, then $Aw = \begin{bmatrix} A_1 w \\ : A_n w \end{bmatrix} = \begin{bmatrix} A_1^T \cdot w \\ \vdots \\ A_n^T \cdot w \end{bmatrix}$. If $w \in N(A)$, then this is the zero vector, so each $A_j^T \cdot w = 0$. Thus w is orthogonal to each A_j^T, which means that if we treat $A_1, \ldots, A_n$ as vectors in F^n without worrying about whether they are row vectors or column vectors, we have w orthogonal to the span of $A_1, \ldots, A_n$, which is $R(A)$. Likewise, if $w \in R(A)^\perp$, then $A_j^T \cdot w = 0$ for each j, so $Aw = 0$ and $w \in N(A)$.

Section 5.2

In Exercises 1 - 8, remember that while eigenvalues are unique, corresponding eigenvectors are not uniquely determined - in fact, any nonzero linear combination of eigenvectors for a given eigenvalue is an eigenvector.

1. $\lambda_1 = 2, \lambda_2 = -2, X_1 = \begin{bmatrix} 1 \\ 0 \end{bmatrix}, X_2 = \begin{bmatrix} 1 \\ 4 \end{bmatrix}$.

2. $\lambda_1 = -1$ (multiplicity 2), $X_1 = \begin{bmatrix} 1 \\ 0 \end{bmatrix}$.

3. $\lambda_1 = -1, \lambda_2 = 2, X_1 = \begin{bmatrix} -1 \\ 5 \end{bmatrix}, X_2 = \begin{bmatrix} 1 \\ -2 \end{bmatrix}.$

4. $\lambda_1 = 1$ (multiplicity 2), $X_1 = \begin{bmatrix} 1 \\ 1 \end{bmatrix}.$

5. $\lambda_1 = 1$ (multiplicity 2), $\lambda_2 = 2, X_1 = \begin{bmatrix} 1 \\ 0 \\ 0 \end{bmatrix}, X_2 = \begin{bmatrix} 5 \\ 3 \\ 1 \end{bmatrix}.$

6. $\lambda_1 = -2, \lambda_2 = -1, \lambda_3 = 3, X_1 = \begin{bmatrix} 1 \\ 0 \\ 0 \end{bmatrix}, X_2 = \begin{bmatrix} 3 \\ 1 \\ 0 \end{bmatrix}, X_3 = \begin{bmatrix} 2 \\ 2 \\ 4 \end{bmatrix}.$

7. $\lambda_1 = 1$ (multiplicity 2), $\lambda_2 = 2, X_1 = \begin{bmatrix} 1 \\ -1 \\ 0 \end{bmatrix}, X_1' = \begin{bmatrix} 0 \\ 2 \\ -1 \end{bmatrix}, X_2 = \begin{bmatrix} -2 \\ -1 \\ 1 \end{bmatrix}.$

8. $\lambda_1 = -1, \lambda_2 = 2, \lambda_3 = 1, X_1 = \begin{bmatrix} 1 \\ 1 \\ 1 \end{bmatrix}, X_2 = \begin{bmatrix} 2 \\ 0 \\ 1 \end{bmatrix}, X_3 = \begin{bmatrix} 1 \\ 0 \\ 1 \end{bmatrix}.$

9. The first has characteristic polynomial $\lambda^2 - 6\lambda + 1$, the second $\lambda^2 - \lambda + 2$. By Theorem 5.2.3 they cannot be similar.

11. $p_A(A) = \begin{bmatrix} 0 & 0 \\ 0 & 0 \end{bmatrix}$

12. Multiply by A:

$$
\begin{aligned}
AX &= A(a_1 X_1 + \ldots + a_k X_k) \\
&= Aa_1 X_1 + \ldots + Aa_k X_k \\
&= a_1 AX_1 + \ldots + a_k AX_k \\
&= a_1 \lambda X_1 + \ldots + a_k \lambda X_k \\
&= \lambda(a_1 X_1 + \ldots + a_k X_k) \\
&= \lambda X.
\end{aligned}
$$

Hence, $a_1 X_1 + \ldots + a_k X_k$ is an eigenvector of A associated with λ.

13. $\begin{bmatrix} -1 & 0 \\ 0 & 3 \end{bmatrix}$ 15. $\begin{bmatrix} -1 & 1 \\ -1 & -3 \end{bmatrix}$ 17. $\begin{bmatrix} 1 & 1 \\ -1 & 1 \end{bmatrix}$

14. $\begin{bmatrix} 1 & 0 \\ 0 & 1 \end{bmatrix}$ 16. $\begin{bmatrix} 1 & 2 \\ -1/2 & -1 \end{bmatrix}$

Section 5.3

1. Yes, two distinct eigenvalues, and so, two independent eigenvectors.

2. Yes, two distinct eigenvalues, and so, two independent eigenvectors.

3. No, not similar to a diagonal matrix: 1 is a repeated eigenvalue, but $A - 1I$ has rank 2 and so there is only one independent eigenvector.

4. Yes, similar to a diagonal matrix since the eigenvalues are distinct.

5. Yes, the matrix has a repeated eigenvalue, 1, but $A - 1I$ has rank 1 and so there are two independent eigenvectors.

6. No. The eigenvalue 1 has multiplicity 3, but $A - 1I$ has rank 1 and so there are only two independent eigenvectors.

7. $S = \begin{bmatrix} 1 & 1 \\ 0 & -1 \end{bmatrix}$ 9. $S = \begin{bmatrix} 3 & 0 \\ -2 & 1 \end{bmatrix}$ 11. $S = \begin{bmatrix} -1 & 0 & 0 \\ 2 & 1 & 2 \\ 0 & 0 & 1 \end{bmatrix}$

8. $S = \begin{bmatrix} 1 & 1 \\ -1 & 1 \end{bmatrix}$ 10. $S = \begin{bmatrix} 1 & 1 & 1 \\ 0 & 2 & 2 \\ 0 & 0 & -2 \end{bmatrix}$ 12. $S = \begin{bmatrix} 0 & 1 & 1 \\ 1 & 0 & 0 \\ 0 & -1 & 1 \end{bmatrix}$

13. 1 is the only eigenvalue, and $A - 1I$ has rank 1, so there is only 1 independent eigenvector.

14. The eigenvalue 1 has multiplicity 2, but $A - 1I$ has rank 2, so as above there is only 1 independent eigenvector.

15. The eigenvector 1 has multiplicity 3, but $A - 1I$ has rank 1, so there are only 2 independent eigenvectors.

16. Yes. Since $A - 1I$ has rank 1, there are two independent eigenvectors. The remaining eigenvector (note that it must be real) has an eigenvector, and so, A has three independent eigenvectors. It follows that A is similar to a diagonal matrix.

19. (a) A is similar to A, since $A = I^{-1}AI$, where I is the $n \times n$ identity matrix.

     (b) If A is similar to B, then $B = S^{-1}AS$ for some nonsingular matrix S. Multiplying on the left by S and on the right by S^{-1}, we obtain $A = SBS^{-1} = (S^{-1})^{-1}BS^{-1}$. Here we have used the fact that $(S^{-1})^{-1} = S$. Hence, B is similar to A.

     (c) If A is similar to B and B is similar to C, then $B = S^{-1}AS$ and $C = T^{-1}BT$ for some nonsingular matrices S and T. Then $C = T^{-1}BT = T^{-1}S^{-1}AST = (ST)^{-1}A(ST)$, and so A is similar to C.

Section 5.4

1 $X = (4/7, 3/7)^T$

2 $X = (0.3028, 0.3267, 0.3705)^T$

3 $A = \begin{bmatrix} 0 & 1/2 & 1/3 & 1/2 \\ 1/2 & 0 & 1/3 & 0 \\ 0 & 0 & 0 & 1/2 \\ 1/2 & 1/2 & 1/3 & 0 \end{bmatrix}$, $PR = (6/19, 4/19, 3/19, 6/19)^T$

4 $A = \begin{bmatrix} 0 & 1/3 & 1/3 & 0 & 0 \\ 1/3 & 0 & 1/3 & 1/2 & 0 \\ 0 & 0 & 0 & 1/2 & 0 \\ 1/3 & 1/3 & 0 & 0 & 1 \\ 1/3 & 1/3 & 1/3 & 0 & 0 \end{bmatrix}$, $X = 1/160(21, 39, 24, 48, 28)^T$

Section 5.5

In Exercises 1 - 12, note that the orthogonal matrix P and the upper triangular matrix $P^t A P$ are not unique - we give here one solution.

1. $P = \begin{bmatrix} 0 & 1 \\ 1 & 0 \end{bmatrix}$

2. $P = \dfrac{1}{\sqrt{2}} \begin{bmatrix} 1 & 1 \\ -1 & 1 \end{bmatrix}$

3. $P = \begin{bmatrix} 0 & 1 \\ 1 & 0 \end{bmatrix}$

4. $P = \dfrac{1}{\sqrt{2}} \begin{bmatrix} 1 & -1 \\ 1 & 1 \end{bmatrix}$

5. $P = \begin{bmatrix} 1 & 0 & 0 \\ 0 & 0 & 1 \\ 0 & 1 & 0 \end{bmatrix}$

6. $P = \begin{bmatrix} 0 & 0 & 1 \\ 1 & 0 & 0 \\ 0 & 1 & 0 \end{bmatrix}$

7. $P = \begin{bmatrix} 0 & 1 & 0 \\ 0 & 0 & 1 \\ 1 & 0 & 0 \end{bmatrix}$

8. $P = \begin{bmatrix} 0 & 1 & 0 \\ 0 & 0 & 1 \\ 1 & 0 & 0 \end{bmatrix}$

9. $P = \dfrac{1}{\sqrt{2}} \begin{bmatrix} 1 & 1 \\ 1 & -1 \end{bmatrix}$

10. $P = \dfrac{1}{\sqrt{2}} \begin{bmatrix} 1 & 1 \\ -1 & 1 \end{bmatrix}$

11. $P = \dfrac{1}{\sqrt{2}} \begin{bmatrix} 0 & \sqrt{2} & 0 \\ 1 & 0 & 1 \\ -1 & 0 & 1 \end{bmatrix}$

12. $P = \dfrac{1}{\sqrt{2}} \begin{bmatrix} 0 & 1 & 1 \\ \sqrt{2} & 0 & 0 \\ 0 & 1 & -1 \end{bmatrix}$

For each of the matrices in Exercises 13 - 16, determine by inspection whether the matrix is similar to a diagonal matrix, orthogonally similar to a diagonal matrix, orthogonally similar to an upper triangular matrix, or unitarily similar to an upper triangular matrix.

13. This matrix has two distinct eigenvalues and so it is similar to a diagonal matrix. It is not symmetric, so it is not orthogonally similar to a diagonal matrix, but it is orthogonally similar to an upper triangular matrix.

14. This matrix has three distinct eigenvalues and so it is similar to a diagonal matrix. It is not symmetric, so it is not orthogonally similar to a diagonal matrix, but it is orthogonally similar to an upper triangular matrix.

15. This matrix is symmetric, and so, it is orthogonally similar to a diagonal matrix. Thus, it is also similar to a diagonal matrix and orthogonally similar to an upper triangular matrix.

16. This matrix is symmetric and so we make the same conclusion as in Exercise 15.

17. The eigenvalues are $-1, i$, and $-i$. Let $P_1 = \dfrac{1}{\sqrt{2}} \begin{bmatrix} 0 & 0 & \sqrt{2} \\ 1 & 1 & 0 \\ -1 & 1 & 0 \end{bmatrix}$,

$P_2 = \dfrac{1}{2} \begin{bmatrix} 2 & 0 & 0 \\ 0 & 1-i & -\sqrt{2} \\ 0 & \sqrt{2} & 1+i \end{bmatrix}$. If $U = P_1 P_2$, then $U^{-1}AU = \begin{bmatrix} -1 & 0 & 0 \\ 0 & i & \sqrt{2}+\sqrt{2}i \\ 0 & 0 & -i \end{bmatrix}$.

18. $\begin{bmatrix} 1 & i \\ -i & 1 \end{bmatrix}$

19. $\begin{bmatrix} 1 & i \\ i & 1 \end{bmatrix}$

20. (a) A formal proof requires induction; we'll give an informal proof. We know that for two $n \times n$ matrices A and B, $(AB)^t = B^t A^t$, and if A and B are nonsingular, $(AB)^{-1} = B^{-1}A^{-1}$. Using these properties, we see that if $P_1, \ldots, P_n$ are orthogonal matrices, then

$$(P_1 \ldots P_n)^{-1} = P_n^{-1} \ldots P_1^{-1} = P_n^t \ldots P_1^t = (P_1 \ldots P_n)^{-1}.$$

It follows that $P_1 \ldots P_n$ is an orthogonal matrix.

(b) The proof is similar to that in (a).

21. $U = \dfrac{1}{\sqrt{2}} \begin{bmatrix} i & i \\ -1 & 1 \end{bmatrix}$.

Section 5.6

1. $A^{-1} = \dfrac{1}{2}(A^2 - 2A - 3I), A^4 = 7A^2 + 8A + 4I.$

2. $A^{-1} = \dfrac{1}{2}(A^2 + 2A - I), A^4 = 5A^2 - 4I.$

3. $A^{-1} = -\dfrac{1}{5}(A - 4I), A^6 = 44A - 205I.$

4. $A^{-1} = \dfrac{1}{2}(-A^2 + 2A + I) = \dfrac{1}{2} \begin{bmatrix} 2 & 0 & -2 \\ 0 & -2 & 3 \\ 0 & 0 & 1 \end{bmatrix}$

5. $p_A(\lambda) = -\lambda(\lambda - 2)(\lambda - 1)$, so $m_A(\lambda) = -p_A(\lambda)$ since all linear factors must appear in $m_A(\lambda)$.

6. $p_A(\lambda) = -(\lambda - 1)^3$, by trial we see that $m_A(\lambda) = -p_A(\lambda)$.

8. $p_A(\lambda) = -(\lambda - 1)^3(\lambda - 2)^2, m_A(\lambda) = (\lambda - 1)(\lambda - 2)$.

9. $p_A(\lambda) = (\lambda - 1)^2 = m_A(\lambda)$

10. $p_A(\lambda) = -(\lambda - 1)^3 = -m_A(\lambda)$

11. $m_D(\lambda) = (\lambda - b_1)\ldots(\lambda - b_k)$

12. If $b_1, \ldots, b_k$ are the distinct entries on the diagonal of the diagonal matrix, then, using the fact that similar matrices have the same minimum polynomial, we see that $m_D(\lambda) = (\lambda - b_1)\ldots(\lambda - b_k)$.

Section 5.7

1. The angle of rotation satisfies $\cot 2\alpha = 3/4$, so $\alpha \approx 26.57°$. The substitution is: $x = \dfrac{1}{\sqrt{5}}(2x_1 - y_1), y = \dfrac{1}{\sqrt{5}}(x_1 + 2y_1)$. The new equation is $30x_1^2 + 5x_2^2 = 45$.

2. The translation is given by: $x_1 = x + 2, y_1 = y - 1$, the new equation is $2x_1^2 + y_2^2 = 21$.

3. The quadratic forms are (a) and (b).

4. $q = [xy]\begin{bmatrix} 2 & 1/2 \\ 1/2 & -1 \end{bmatrix}\begin{bmatrix} x \\ y \end{bmatrix}$

7. $q = [x_1 x_2 x_3]\begin{bmatrix} 3 & 0 & 0 \\ 0 & 1 & 2 \\ 0 & 2 & 1 \end{bmatrix}\begin{bmatrix} x_1 \\ x_2 \\ x_3 \end{bmatrix}$

5. $q = [xyz]\begin{bmatrix} 2 & 1/2 & 2 \\ 1/2 & -1 & 0 \\ 2 & 0 & 3 \end{bmatrix}\begin{bmatrix} x \\ y \\ z \end{bmatrix}$

8. $S = \dfrac{1}{\sqrt{2}}\begin{bmatrix} 0 & \sqrt{2} & 0 \\ 1 & 0 & 1 \\ -1 & 0 & 1 \end{bmatrix}$

6. $q = [x_1 x_2 x_3]\begin{bmatrix} 1 & 0 & 0 \\ 0 & 1 & 1 \\ 0 & 1 & 1 \end{bmatrix}\begin{bmatrix} x_1 \\ x_2 \\ x_3 \end{bmatrix}$

9. $S = \dfrac{1}{\sqrt{2}}\begin{bmatrix} \sqrt{2} & 0 & 0 \\ 0 & 1 & -1 \\ 0 & 1 & 1 \end{bmatrix}$

10. The eigenvalues are 1, 2, 0, so the quadratic form is positive semidefinite.

11. The rotation $x = \dfrac{1}{\sqrt{5}}(2x_1 + y_1), y = \dfrac{1}{\sqrt{5}}(-x_1 + 2y_1)$ transforms the equation into the new equation: $10x_1^2 - 15y_1^2 = 20$. The curve is a hyperbola.

12. There is a critical point at $(0, 0)$. The matrix of second partials is $A = \dfrac{1}{2}\begin{bmatrix} 4 & 1 \\ 1 & 2 \end{bmatrix}$, the eigenvalues are $\dfrac{6 \pm \sqrt{8}}{2}$, and since both are positive, there is a minimum at $(0, 0)$.

13. There is a critical point at $(0,0,0)$. The matrix A is $\dfrac{1}{2}\begin{bmatrix} 2 & 0 & 0 \\ 0 & -4 & -1 \\ 0 & -1 & 0 \end{bmatrix}$, the eigenvalues are 2 and $\dfrac{-4 \pm \sqrt{20}}{2}$, and since some are positive and some are negative, there is no maximum or minimum.

Section 5.8

1. $R(0) = (1,0,0), R(1) = (-1,0,1), V(t) = R'(t) = (-\pi \sin \pi t, \pi \cos \pi t, 1), A(t) = R''(t) = (-\pi^2 \cos \pi t, \pi^2 \sin \pi t, 0)$. The path is a spiral upward around the cylinder $x^2 + y^2 = 1$.

2. $R(0) = (1,0,0), R(1) = (-1,0,0), V(t) = R'(t) = (-\pi \sin \pi t, \pi \cos \pi t, 0), A(t) = R''(t) = (-\pi^2 \cos \pi t, \pi^2 \sin \pi t, 0)$. The path is circular in the $x, y-$plane, around the circle $x^2 + y^2 = 1$. $\Omega = (0,0,\pi), \Omega \times R = (-\pi \sin \pi t, \pi \cos \pi t, 0)$.

3. $P = mV = 3V = 3(-\pi \sin \pi t, \pi \cos \pi t, 0)$. $L = R \times P = (0,0,3\pi)$.

4. $I_T = \begin{bmatrix} 23 & 3 & 16 \\ 3 & 37 & -2 \\ 16 & -2 & 24 \end{bmatrix}$

5. $(-11/9, 5/9, 10/9)$

6. $I_T = \begin{bmatrix} 0 & 0 & 0 \\ 0 & 30 & 0 \\ 0 & 0 & 30 \end{bmatrix}$. Since I_T is diagonal, the principal axes are the standard basis vectors $(1,0,0), (0,1,0), (0,0,1)$.

7. $I_T = \begin{bmatrix} 6 & -3 & 0 \\ -3 & 6 & 0 \\ 0 & 0 & 12 \end{bmatrix}$. The eigenvalues are $12, 3,$ and 9. Choosing corresponding eigenvectors, we obtain the principal axes: $(0,0,1), (1/\sqrt{2})(1,1,0),$ and $(1/\sqrt{2})(1,-1,0)$.

Chapter 6

Section 6.1

1. Similar to a diagonal matrix since the eigenvalues, -1 and 2, are distinct.

2. Similar to a diagonal matrix since the matrix is symmetric.

3. In this matrix, 1 is an eigenvalue of multiplicity 2, but $A - 1I$ has rank 1 and so there is only 1 independent eigenvector. Thus, the matrix is not similar to a diagonal matrix.

4. The characteristic polynomial is $-(\lambda + 1)(\lambda^2 - 3\lambda + 1)$. The roots are real and distinct since $3^2 - 4 \cdot 1 \cdot 1 > 0$ and so the matrix is similar to a diagonal matrix.

5. This matrix is hermitian and so similar to a real diagonal matrix.

6. (b) and (c) are Jordan blocks, but (a) is not since there are two different eigenvalues, and (e) fails since the superdiagonal has a 0.

7. The characteristic polynomial is $(\lambda - 2)^2$, so the Jordan block must be $J = \begin{bmatrix} 2 & 1 \\ 0 & 2 \end{bmatrix}$.

8. The characteristic polynomial is $-(\lambda - 1)^3$, so the Jordan block is $J = \begin{bmatrix} 1 & 1 & 0 \\ 0 & 1 & 1 \\ 0 & 0 & 1 \end{bmatrix}$.

9. Take $S = \begin{bmatrix} 1 & -1 \\ -1 & 0 \end{bmatrix}$; then $S^{-1}AS = \begin{bmatrix} 2 & 1 \\ 0 & 2 \end{bmatrix}$.

10. Take $S = \begin{bmatrix} 0 & 0 & 1 \\ 1 & 0 & 0 \\ 0 & 1 & -1 \end{bmatrix}$; then $S^{-1}AS = \begin{bmatrix} 1 & 1 & 0 \\ 0 & 1 & 1 \\ 0 & 0 & 1 \end{bmatrix}$.

11. It remains to show that the reverse implication holds true. Assume that there are independent $m \times 1$ column vectors $X_1, \ldots, X_m$ satisfying

$$(A - aI)X_1 = 0$$
$$(A - aI)X_2 = X_1$$
$$(A - aI)X_3 = X_2$$
$$\vdots$$
$$(A - aI)X_m = X_{m-1}.$$

and let $S = [X_1 \ldots X_m]$. Then S is nonsingular and as in the discussion before the theorem, we can see that $AS = SJ$; thus $S^{-1}AS = J$.

12.

$$(A - 2I)X_1 = 0$$
$$(A - 2I)X_2 = X_1$$
$$(A - 2I)X_3 = X_2.$$

13.

$$(A - 2I)X_1 = 0$$
$$(A - 2I)X_2 = X_1$$
$$(A - 2I)X_3 = 0.$$

Section 6.2

In Exercises 1- 12, we list the Jordan blocks that lie on the diagonal of the Jordan canonical form of the matrix. Recall that these blocks may occur in any order.

1. $\begin{bmatrix} -2 & 1 \\ 0 & -2 \end{bmatrix}, [-2]$

2. $\begin{bmatrix} 2 & 1 & 0 \\ 0 & 2 & 1 \\ 0 & 0 & 2 \end{bmatrix}$

3. $\begin{bmatrix} 2 & 1 \\ 0 & 2 \end{bmatrix}, [2]$

4. $\begin{bmatrix} 2 & 1 \\ 0 & 2 \end{bmatrix}, [2], [3], [3]$

5. $\begin{bmatrix} 2 & 1 \\ 0 & 2 \end{bmatrix}, \begin{bmatrix} 2 & 1 \\ 0 & 2 \end{bmatrix}, \begin{bmatrix} 3 & 1 \\ 0 & 3 \end{bmatrix}$

6. $\begin{bmatrix} -2 & 1 \\ 0 & -2 \end{bmatrix}, \begin{bmatrix} 1 & 1 \\ 0 & 1 \end{bmatrix}, [1], [1]$

7. $[0], [2]$

8. $[0], [1]$

9. $[2], \begin{bmatrix} 1 & 1 \\ 0 & 1 \end{bmatrix}$

10. $\begin{bmatrix} 1 & 1 & 0 \\ 0 & 1 & 1 \\ 0 & 0 & 1 \end{bmatrix}$

11. $[-1], \begin{bmatrix} 3 & 1 \\ 0 & 3 \end{bmatrix}$

12. $\begin{bmatrix} 1 & 1 & 0 \\ 0 & 1 & 1 \\ 0 & 0 & 1 \end{bmatrix}$

13. $J = \begin{bmatrix} 2 & 0 & 0 \\ 0 & 1 & 1 \\ 0 & 0 & 1 \end{bmatrix}, S = \begin{bmatrix} 0 & 1 & 0 \\ 0 & 0 & 1 \\ 1 & -1 & 0 \end{bmatrix}$

14. $J = \begin{bmatrix} 1 & 1 & 0 \\ 0 & 1 & 1 \\ 0 & 0 & 1 \end{bmatrix}, S = \begin{bmatrix} 0 & 1 & -2 \\ 1 & 0 & 0 \\ 0 & 0 & 1 \end{bmatrix}$

15. $J = \begin{bmatrix} 2 & 1 & 0 \\ 0 & 2 & 0 \\ 0 & 0 & 2 \end{bmatrix}$

16. $J = \begin{bmatrix} 1 & 1 & 0 \\ 0 & 1 & 1 \\ 0 & 0 & 1 \end{bmatrix}$

17. The conditions imply that the Jordan form of A must be the identity matrix I. Thus, there is a nonsingular matrix S with $S^{-1}AS = I$. It follows that $A = I$.

Section 6.3

1. $y = Ce^{2t} - e^t, y = 2e^{2t} - e^t$

2. $y = \left(\dfrac{t}{3} - \dfrac{1}{9}\right) + Ce^{-3}t, y = \left(\dfrac{t}{3} - \dfrac{1}{9}\right) - \dfrac{8}{9}e^{-3t}$

3. $y = -1 + Ce^{t^2/2}, y = -1 + e^{t^2/2}$

4. $y = \dfrac{t^3}{2} + Ct, y = \dfrac{t^3}{2} - \dfrac{3}{2}t$

5. $X = \begin{bmatrix} x_1(t) \\ x_2(t) \end{bmatrix}, A = \begin{bmatrix} 2 & 1 \\ 0 & 3 \end{bmatrix}, F(t) = \begin{bmatrix} e^t \\ e^{2t} \end{bmatrix}$

6. $X = \begin{bmatrix} x_1(t) \\ x_2(t) \end{bmatrix}, A = \begin{bmatrix} 3 & 1 \\ 2 & -2 \end{bmatrix}, F(t) = \begin{bmatrix} \sin t \\ t \end{bmatrix}$

7. $x_1 = \dfrac{1}{4}e^t + (Ct + C')e^{-t} + t - 1, x_2 = e^t + Ce^{-t}$. For the solution of the initial condition,
   $C = -\dfrac{3}{2}, C' = \dfrac{7}{4}$.

8. $x_1 = (Ct + C')e^{2t} - \dfrac{1}{4}(2t^2 + t), x_2 = -\dfrac{1}{4}(2t + 1) + Ce^{2t}$. For the solution of the initial
   condition, $C = \dfrac{1}{4}, C' = -1$.

9. $x_1 = -\dfrac{1}{2} + e^{2t}\left(t + \dfrac{1}{2}\right), x_2 = e^{-2t}, x_3 = 0$.

10. $x_1 = -e^t + \dfrac{e^{-t}}{9} + e^{2t}(C't + C''') - \dfrac{1}{2}, x_2 = e(t + C - 1) + \dfrac{e^{-t}}{9} + e^{2t}(C't + C'') + \dfrac{1}{2}, x_3 = $
    $e(t + C) - \dfrac{2e^{-t}}{9} + e^{2t}(C'(t + 1) + C'') + \dfrac{1}{2}$

11. $x_1 = Ce^t - C'e^{-3t}, x_2 = 2Ce^t + 2C'e^{-3t}$

Chapter 7

Section 7.1

4. maximum $P = 9000$ at either point 3 or 4.

5. maximum $P = 6000$ at point 3.

6. maximum $P = 3225$ at point 2.

7. maximum $P = 4250$ at point 2.

8. maximum $P = 19,800$ at $x = 120, y = 300$.

9. maximum $P = 18,000$ at $x = 120, y = 300$, or at $x = 257.1428571, y = 128.5714287$

10. $A = \begin{bmatrix} 1 & -1 & 3 & 1 & 0 & 0 \\ 2 & 1 & 3 & 0 & 1 & 0 \\ 1 & 3 & 1 & 0 & 0 & 1 \end{bmatrix}, X = \begin{bmatrix} x_1 \\ \vdots \\ x_6 \end{bmatrix}, C = \begin{bmatrix} 2 & 4 & 1 & 0 & 0 & 0 \end{bmatrix}, B = \begin{bmatrix} 3 \\ 4 \\ 6 \end{bmatrix}$.

11. $A = \begin{bmatrix} 1 & 1 & 2 & 1 & 0 & 0 \\ 2 & -1 & 2 & 0 & 1 & 0 \\ 1 & 3 & 1 & 0 & 0 & 1 \end{bmatrix}, X = \begin{bmatrix} x_1 \\ \vdots \\ x_6 \end{bmatrix}, C = \begin{bmatrix} 2 & 1 & 2 & 0 & 0 & 0 \end{bmatrix}, B = \begin{bmatrix} 4 \\ 6 \\ 6 \end{bmatrix}$.

12. The variable z is essentially unbounded; that is, it may take on arbitrarily large values. Thus, P has no maximum value.

13. From the last two constraints we see that $2 \geq x, y, z \geq 0$, and so any point satisfying the last two constraints must satisfy the first. It follows that the maximum value of P is 6 when $y = 2$, and $x = y = 0$.

14. Substitute the value of a point on the line into the expression for the function z. We get $z = c_1 x_1 + \ldots + c_n x_n = (c_1 a_1 + \ldots + c_n a_n) + (1 - t)(c_1 b_1 + \ldots + c_n b_n)$. Now it is not hard to see that if $c_1 b_1 + \ldots + c_n b_n = 0$, the function is constant; if $c_1 b_1 + \ldots + c_n b_n > 0$, the function has its maximum when $t = 0$; and if $c_1 b_1 + \ldots + c_n b_n < 0$, the function has its maximum when $t = 1$.

15. $z = \dfrac{41x}{4} \approx 64.03124238$ at $x = \sqrt{\dfrac{1600}{41}} \approx 6.246950475, y = \sqrt{\dfrac{2500}{41}} \approx 7.808688095$

Section 7.2

1. The maximum value of P is 14, when $x_2 = 7, x_4 = 4$, and $x_1 = x_3 = 0$.

2. The maximum value of P is 27, when $x_1 = 9, x_4 = 2$, and $x_2 = x_3 = 0$.

3. The maximum value of P is 10,000, when $x_1 = 200, x_4 = 800, x_5 = 300$, and $x_2 = x_3 = 0$.

4. The maximum value of P is 16,800, when $x_2 = 240, x_4 = 120, x_5 = 660$, and $x_1 = x_3 = 0$.

5. A pivot on the coefficient of x in the first equation produces the system

$$x + (b/a)y = h/a$$
$$(d - cb/a)y = k - ch/a$$

Since $h, a > 0, h/a > 0$, and since $h/a < k/c, ch/a < k$ and so $k - ch/a > 0$.

Section 7.3

1. Maximum is 3 when $x_1 = 1, x_2 = 2, x_3 = 1, x_4 = x_5 = x_6 = 0$.

2. Maximum is 3 when $x_2 = 5/2, x_3 = 1/2, x_5 = 1/2, x_1 = x_4 = x_6 = 0$.

3. No maximum value - x_5 may be made as large as desired..

4. Maximum is 1 when $x_1 = 1, x_3 = 1, x_6 = 2, x_2 = x_4 = x_5 = 0$.

5. Maximum is 3 when $x_1 = 1, x_4 = 1, x_2 = x_3 = 0$.

6. Maximum is 2 when $x_2 = 1, x_3 = 1, x_5 = 1, x_1 = x_4 = 0$.

7. Maximum $z = 2100$ when $x_1 = 15, x_2 = 25, x_3 = 15, x_4 = x_5 = 0$.

8. Maximum $z = 1480$ when $x_1 = 28, x_2 = 6, x_3 = 36, x_4 = x_5 = 0$.

9. Maximum $z = 104$ when $x_1 = 16, x_2 = 8, x_5 = 2, x_3 = x_4 = x_6 = 0$.

10. Maximum $z = 220/3$ when $X = x_1 = 216, Y = x_2 = 252, x_4 = 144, x_3 = x_5 = 0$.

11. Maximum $P = \$21,240$ when $x_2 = 20/3, x_3 = 40/3, x_6 = 20/6, x_1 = x_4 = x_5 = 0$.

12. If we let x_1 be the number of microscopes and x_2 be the number of telescopes, then solving the problem in the usual fashion we get: $P = \$250$ when $x_2 = 12.5, x_4 = 35, x_1 = x_3 = x_5 = 0$. Of course making 12.5 telescopes is difficult. We could change the time period to two weeks, then the solution would be to manufacture 25 telescopes for a profit of $500.

Section 7.4

1. The row player gains 3. The row player loses 1.

2. If the column player chooses 1, the row player gains 2. If 2 is chosen by the column player, the row player loses 1; if column 3 is chosen, the row player gains 3.

3. The row player never loses, and so will be the winner.

4. The optimal strategy is $(1/3, 2/3)$ and the value of h is $-1/3$.

5. The optimal strategy is $(0, 5/8, 3/8)$ and the value of h is $1/2$.

6. -0.08 7. $1/2$ 8. $1/2$

9. Optimal strategy for the column player is $(1/3, 2/3)$ from Exercise 4. For the row player, the optimal strategy is $(5/8, 3/8)$. The expected value is $1/4$.

364

Index